Designer Surfaces

Designer Surfaces

Alexei A. Maradudin
University of California,
Irvine, USA

Eugenio R. Méndez
CICESE,
Ensenada, B.C., Mexico

and

Tamara A. Leskova
University of California,
Irvine, USA

ELSEVIER

Amsterdam • Boston • Heidelberg • London • New York • Oxford
Paris • San Diego • San Francisco • Singapore • Sydney • Tokyo

Elsevier
Radarweg 29, PO Box 211, 1000 AE Amsterdam, The Netherlands
Linacre House, Jordan Hill, Oxford OX2 8DP, UK

Library of Congress Cataloging-in-Publication Data
A catalog record for this book is available from the Library of Congress

British Library Cataloguing in Publication Data
A catalogue record for this book is available from the British Library

ISBN: 978-0-444-53048-6

For information on all Elsevier publications
visit our website at books.elsevier.com

Transferred to Digital Printing 2009

Dedicated to
Peggy, Citlali, and Aleksandr

CONTENTS

PREFACE

The fabrication of optical elements that can manipulate light in desirable ways has been the object of many investigations during the past four decades. Such elements can be utilized in imaging systems, for the correction of various optical aberrations, in projection systems, in the fabrication of vertical-cavity surface-emitting lasers, and in a variety of other applications where the transformation of light with one set of properties, e.g., intensity profile, spectrum, and coherence, into light with a different set of properties is required.

Many, if not most, of these optical elements have been based on phase-coherent surface-relief gratings. These diffractive structures have come to be known as diffractive optical elements. They consist of superpositions of periodic structures, and effect optical transformations by acting only on the phase of an incident wave-front. The phase change brought about by the element can be converted into a surface profile function. A great deal of the work on these elements in recent years has been devoted to the development of accurate and efficient computational methods for solving Maxwell's equations for the scattering of light from periodically corrugated surfaces, when, driven by the desire to produce ever smaller devices, the linear dimensions of the features on these surfaces, e.g., amplitudes and periods, become of the order of, or smaller than, the wavelength of the incident light. The results of such calculations are used to demonstrate that a particular diffractive optical element transforms light in the manner desired.

The design of a diffractive element that effects a desired optical transformation is commonly carried out using concepts of Fourier optics. This approach, which is based on scalar wave theory, breaks down when the linear dimensions of the topographic features of the element are of the order of the wavelength. In this regime a different approach to the design of diffractive optical elements is needed.

The problem of designing diffractive optical elements has given rise to a subfield of diffractive optics devoted to what is called the *optimal design problem*, an inverse scattering problem. However, in contrast to the classical inverse problem of determining an unknown surface profile function from a given diffractive field, the optimal design problem is that of determining a surface profile function that gives the "best" approximation to some desired diffractive field. Several approaches have been used in establishing existence theorems for the solution of this problem, and for obtaining solutions to it. The latter generally involve an iterative optimization (minimization) of some cost function that depends on the difference between the desired diffracted field and the diffracted field calculated for some choice of the surface profile function that can be varied systematically. The existence theorems and solution methods are well summarized in the review by Bao and Dobson [1]. However, at the present time they appear to be restricted to one-dimensional surface profile functions that are periodic, namely to classical gratings. (An exception is provided by the work of Prather [2].) The optimal design problem for biperiodic gratings remains a challenge to optical designers.

In contrast to the essentially deterministic approach to the design of light manipulating elements provided by diffractive optics, in this book we present an approach that is stochastic in nature, and is based not on superpositions of periodic or quasi-periodic surface reliefs but on the assumption that the surface of such an element is randomly rough. It is based on two ingredients. The first is the use of a representation of the field scattered from a rough surface provided by a single-scattering approximation such as the geometrical optics limit of the Kirchhoff approximation, phase perturbation theory, or the Fraunhofer approximation, to express the property of that field that we wish the surface to produce. The second is the representation of the random surface by a continuous sequence of straight-line segments with random slopes, in the case of a one-dimensional surface, or by a continuous array of triangular facets, whose slopes along two perpendicular directions within each facet are random variables, in the case of a two-dimensional surface. The design problem then simplifies to the determination of the probability density function of the slope of each segment in the one-dimensional case, or the joint probability function of the two slopes characterizing each triangular facet in the two-dimensional case. From a knowledge of these probability density functions as many realizations of the random surfaces can be generated as is desired.

Because of the approximations that go into the derivations of these probability density functions, and hence into the generation of the random surfaces, it is necessary to validate the approaches used by showing that the surfaces they produce indeed scatter light in the manner they are designed to do. We do this by generating an ensemble of realizations of the surface profile function. For each realization of the surface we then solve the direct problem of the scattering of light from it by a method that is more rigorous than the one used in obtaining that realization, e.g., by the use of the Kirchhoff approximation, but without passing to the geometrical optics limit of it, or by a rigorous computer simulation approach. This is because if we used the same approximation in calculating the scattering from a surface that was used in generating it we would be testing only the quality of our programming skills but not the quality of the surface.

We have chosen to work with random surfaces in designing optical elements that scatter or transmit light in a prescribed fashion because the use of such surfaces leads to precise algorithms for designing them, in the form of the representations of their surface profile functions and the probability density functions of the slopes entering these representations. These algorithms thus allow us to generate a realization of a random surface in its entirety at once, rather than iteratively by the use of an optimization procedure. Such algorithms are something that, for the most part, we have been unable to find in dealing with deterministic surfaces.

The advantages or disadvantages of the solution to the design problem developed in this book in comparison with that provided by diffractive optics will be discussed at a later point in this book. However, we note here that it can be used to design two-dimensional surfaces with specified scattering properties almost as easily as it can be used to design one-dimensional surfaces with such properties.

The fact that in applications of the surfaces designed by the methods described here only a single realization of a random surface drawn from an ensemble of an infinite number of possible realizations of that surface is fabricated is not an impediment to its use. In fact, we show that the ensemble averages used in the calculations validating these design procedures can be replaced by the use of a broadband source in illuminating a single realization of a random surface.

In presenting the methods for designing randomly rough surfaces that scatter or transmit light in specified ways, we have adopted a frankly didactic approach, and have tried to make this book as self-contained as possible. As a result, a reader will become acquainted with concepts, theoretical and computational methods, and experimental techniques that can be useful in dealing with rough surface scattering problems beyond those to which they are applied in this book. We have also not striven for mathematical rigor in our derivations, but hope nonetheless that they contain no errors.

After a brief introduction to designer surfaces in Chapter 1, we turn in Chapter 2 to a discussion of the design of one-dimensional randomly rough surfaces that scatter or transmit light in specified ways. We consider first the case of perfectly conducting surfaces that produce a scattered field with a specified angular dependence of its mean intensity, both because of the relative simplicity of the scattering theory for such surfaces, and because a perfect conductor is a good approximation to a metal in the infrared range of the optical spectrum, a range that is of interest in information technology. The design of penetrable surfaces—dielectric and metallic surfaces—with specified scattering properties is considered next. This discussion of penetrable surfaces leads to the study of random metallic surfaces that suppress the conversion of the surface electromagnetic waves they support into volume electromagnetic waves in the vacuum region, or that utilize these surface waves to produce a scattered field that displays the phenomenon of enhanced backscattering for only a single, specified, angle of incidence. We then turn our attention away from the design of surfaces that produce scattered fields with specified angular dependencies toward the design of surfaces that produce a scattered field at fixed angles of incidence and scattering with a specified dependence of its intensity on the wavelength of the incident light. This discussion prompts a discussion of opaque surfaces whose thermal emissivities at specified angles of emission have prescribed wavelength dependencies. The design of surfaces that, when illuminated by an incident beam with a suitable intensity profile, produce a scattered beam with a specified intensity distribution in the mean scattering surface and a specified spectral degree of coherence in that surface, is considered next. It is also shown that such surfaces can transform an incident field with a specified intensity profile into a beam with a different specified intensity profile.

The discussion up to this point has been devoted to surfaces with specified scattering properties. It is important for applications to be able to design surfaces that transmit light in a prescribed way. This we do for the case of transmission through a film with a planar illuminated surface and a randomly rough back surface. It is shown how the latter surface can be designed so as to produce a transmitted field with a prescribed angular dependence of its mean intensity.

In the design of all of the randomly rough surfaces considered up to this point, an average over the ensemble of realizations of the surface profile function is carried out in obtaining the probability density function of the slopes of the straight line segments from which the surface is generated. In validating the design of each type of surface considered, by solving the direct scattering problem for that surface for a large number of realizations of the surface, a second ensemble average is employed. This average eliminates the speckles that would be present if a single realization of the surface were illuminated by a monochromatic beam. An experimentalist, however, does not fabricate a large number of random surfaces, all with the same statistical properties, and hence cannot average the results measured for each of these realizations. We show how the ensemble average can be replaced in theoretical calculations of rough surface scattering, and in experimental studies of it, by the use of a broadband beam to illuminate a single realization of the random surface.

To validate the design approach developed in this chapter, the scattering of light from the surfaces generated by its use is solved by a computational approach that is more rigorous than the approach used in solving the design problem, to avoid the circularity of the demonstration that would occur if the same method was used to solve the design problem and to validate it. The more rigorous approaches to the solution of the scattering problem used in the validation process are described in some detail in each case. Finally, techniques for fabricating in the laboratory one-dimensional randomly rough surfaces of the type considered in this chapter are described. Where they exist, experimental results demonstrating that the surfaces designed in the ways described here indeed possess the scattering or transmission properties they were designed to have, are also presented. However, in the present case theory has outrun experiment, and the number of such experimental results is small.

The more difficult problem of designing two-dimensional randomly rough Dirichlet surfaces that scatter a scalar plane wave in specified ways is discussed in Chapter 3. Here a succession of three approaches is presented that are applicable to surfaces that on average have a rectangular, circular, or no particular symmetry. The order in which these approaches are presented reflects the order in which they were developed by us, as our understanding of the design problems involved grew. These approaches are illustrated by applying them to the design of surfaces that produce a scattered field with a specified angular dependence of its mean intensity, that produce a transmitted field with a specified spatial dependence of its mean intensity, and that produce a scattered field with specified coherence properties. Methods for solving the direct problem of scattering from a two-dimensional rough surface, which are needed for validating the approaches derived in this chapter, are presented. Techniques for fabricating such surfaces on photoresist are also described, and experimental results obtained with their use are presented.

In Chapter 4 we present conclusions drawn from the results we obtained in the preceding chapters, point to outstanding but as yet unsolved problems, and suggest directions in which future work on designer surfaces could go.

Many individuals have contributed significantly to the work whose results are presented in this book. We would like to acknowledge in particular Böern Baumeier, Héctor M. Escamilla, Efren García-Guerrero, Manuel Leyva-Lucero, Javier Muñoz-Lopez, Igor V. Novikov, Andrei V. Schegrov, Ingve Simonsen, and Wolfgang Zierau.

We are grateful to E. Efrén García-Guerrero for providing the experimental figures.

The research of T.A.L. and A.A.M. on the topics considered in this book was supported in large part by the US Army Research Office. We are grateful for that support, and to Dr. Mikael Ciftan of the ARO for his confidence in our ability to design randomly rough surfaces with desirable scattering and transmission properties. The work of E.R.M. on the subject matter of this book was partially supported by CONACYT.

Finally, we wish to express our appreciation and thanks to Ms. Jeannie M. Brown, for the excellent job of typing and re-typing the many drafts of a difficult manuscript.

Alexei A. Maradudin
Eugenio R. Méndez
Tamara A. Leskova

References

[1] G. Bao, D.C. Dobson, Variational methods for diffraction optics modeling, in: G. Bao, L. Cowsar, W. Masters (Eds.), Mathematical Modeling in Optical Science, SIAM, Philadelphia, 2001, pp. 37–69.

[2] D.W. Prather, Analysis and Synthesis of Finite Aperiodic Diffractive Optical Elements Using Rigorous Electromagnetic Models, Ph.D. Thesis, Department of Electrical Engineering, University of Maryland, College Park, 1997.

Chapter 1

INTRODUCTION

The appearance of objects illuminated by natural light depends on the kind of illumination employed. Diffuse illumination is the kind produced by an overcast sky, while direct illumination is the kind provided by a distant source, like the Sun on a clear day. The low contrast scenes obtained with diffuse illumination are considered more pleasant and easier to record than the high contrast ones obtained with direct illumination, which produces strongly defined shadows. Perhaps for this reason, illumination systems often combine light sources with diffusing screens (e.g., bulbs and lampshades).

Optical diffusers are optical elements that produce diffuse light by modifying the angular divergence of the incident light. They are used to smooth and homogenize the illumination provided by filament lamps,, LEDs, arc lamps, and other sources, and are useful in illumination systems, microscopes, projection systems, computer screens, bar code scanners, and radiometers, to mention but a few applications.

The conventional methods to produce diffuse light are based on volume and/or surface scattering. An opal glass diffuser, for instance, consists of a slab of glass that contains small particles or inhomogeneities. A ground glass diffuser, on the other hand, is a plane-parallel window of transparent glass with a polished front face and a ground rear face. Diffusers based on volume scattering produce nearly Lambertian scattering distributions. That is, the angular intensity distribution of the scattered light follows a cosine law as a function of the scattering angle. Diffusers of this kind are often used as diffuse reflectance standards [1.1]. On the other hand, diffusers based on surface scattering diffuse light less thoroughly, producing typically Gaussian-like angular intensity distributions. For the purposes of illuminating a scene evenly, opal glass diffusers are better, but the scattering losses are much higher than those associated with ground glass diffusers.

Given the ubiquity of diffuse light and its importance in everyday life, it is not surprising that there have been many attempts to produce diffusers that have a high transmission efficiency and, at the same time, provide uniform illumination over a given area (see, e.g., [1.2–1.7]). The design of optical elements to tailor the illumination conditions in different practical situations is a classical problem in optics. With the advances in diffractive optics and computer generated holography, cheap optical diffusers designed for a variety of tasks have become commercially available.

The diffractive optical diffusers are deterministic periodic structures that produce, essentially, an array of diffractive orders over the desired region in the far field [1.6]. In order to produce some overlap between the orders, the cell size must be greater than about 500 wavelengths. The design is usually based on scalar diffraction theory and paraxial optics [1.8]. The smallest features impressed in the plate must then be larger than the wavelength of the light employed and the angles of scattering must be small (so that the cosine of the angle of scattering can be approximated by unity), imposing some limitations on the achievable designs. The structure within the cell is not obtained analytically, and must be found through some kind of optimization algorithm.

Most of the designs employ binary phase distributions. Thus, the resulting diffusers work well only in the neighborhood of the design wavelength, and would not be useful in applications that use broadband illumination. The sensitivity to the wavelength can be decreased by using multilayer designs and subwavelength structures [1.9], but the presence of even a small fraction of the incident light in the undiffracted order can be unacceptable for some applications. So, although diffractive optical elements have been successful in solving some problems and have replaced opal and ground glass diffusers in many applications, there are situations for which they would not be appropriate.

In this book, we describe an alternative approach to the design of optical diffusers. It is based on refraction, rather than diffraction. This kind of diffuser is well suited for applications in which broadband illumination is required. Our involvement on this research topic started a few years ago, when we became interested in the possibility of designing random surfaces that would act as uniform diffusers. It was not clear at the time if this was possible, and if so, what kind of statistics the surfaces would have. It was clear that the surfaces would have to be sufficiently rough to extinguish the coherent or specular component, and it soon became evident that in this limit the statistics of the surface would have to be non-Gaussian. This conclusion can be reached by noting that the derivative of a Gaussian process is itself Gaussian and, as explained later in the book, in the geometrical optics approximation the mean intensity is a fairly direct map of the probability density function of slopes on the surface. It is however not necessary to invoke the geometrical optics approximation to reach this conclusion (see, e.g., the discussion in Section 2.1.2.1). Since not much is known about scattering by non-Gaussian surfaces this constituted an interesting challenge and research topic.

The kind of approach described here has been successful in tackling many problems, and complements diffractive optics. The method has been investigated in a series of papers, and has evolved throughout the last few years. In contrast to the case of diffractive optical elements, which are better known and for which there are a few reviews and books [1.8, 1.10], there is no comprehensive or unified account of the methods and the problems that have been considered with this approach. This book attempts to fill this void.

From the beginning, due perhaps to our background, our approach was to search for random surfaces. It is of course possible to approach the design in terms of deterministic procedures, as one does with diffractive optical elements. We shall see however, that the random approach provides simple and well-defined procedures for designing the diffusers, without having to resort to iterative techniques that might not converge for some designs.

Optical diffusers are related to coherent beam shapers [1.11,1.12], which are optical elements employed to modify the spatial profile of laser beams. A typical application for

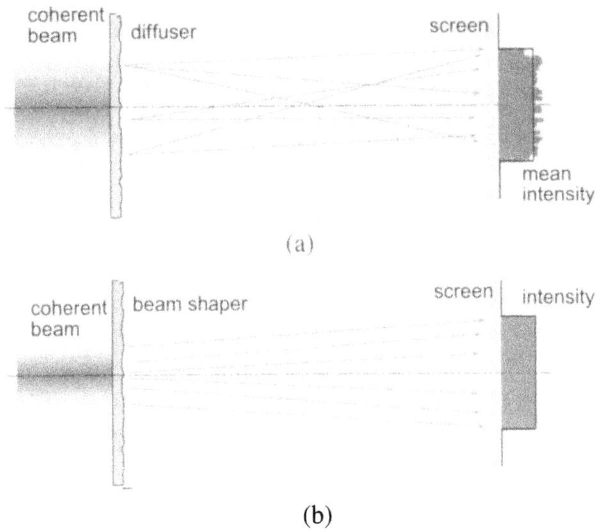

Figure 1.1. Illustration of the differences between diffusers and beam shapers.

a beam shaper consists in the conversion of a Gaussian laser beam into a flat-top beam without a significant modification of its propagation properties. Although this can be achieved by the use of an absorbing mask that compensates for the intensity variations in the beam, the interesting problem is the design of lossless optical elements that produce similar results. Since beam shapers can also be used to produce a prescribed intensity profile on a given plane and can work with spatially partially coherent beams [1.13], the differences between a diffuser and a coherent beam shaper are not well defined.

In general terms, it can be said that coherent beam shapers redistribute the intensity without scrambling the phase of the output beam, which then retains some kind of directionality. With diffusers, in contrast, each element scatters light (on average) within a well-defined cone, and the output complex amplitude results from the superposition of many such contributions. So, beam shapers are based on a 1 to 1 mapping between the input and output, whereas optical diffusers work on a 1 to N mapping. This difference is illustrated in Fig. 1.1. Beam shapers retain phase coherence and should not produce speckle when illuminated by coherent light. Diffusers, on the other hand, produce intensity fluctuations that depend on the coherence of the illumination. The kind of averaging provided by the coherence of the illumination and its connection with ensemble averages will be a subject of discussions in various sections of the book. When diffusers are used in beam shaping problems, it should be clear that this is done on the basis of a 1 to N mapping.

References

[1.1] F. Grum, G.W. Luckey, Optical sphere paint and a working standard of reflectance, Appl. Opt. 7 (1968) 2289–2294.

[1.2] C.N. Kurtz, H.O. Hoadley, J.J. DePalma, Design and synthesis of random phase diffusers, J. Opt. Soc. Am. 63 (1973) 1080–1092.
[1.3] W.J. Dallas, Deterministic diffusers for holography, Appl. Opt. 12 (1973) 1179–1187.
[1.4] R.W. Cohn, A.A. Vasiliev, W. Liu, D.L. Hill, Fully complex diffractive optics by means of patterned diffuser arrays: Encoding concept and implications for fabrication, J. Opt. Soc. Am. A 14 (1997) 1110–1123.
[1.5] D.J. Schertler, N. George, Uniform scattering patterns from grating-diffuser cascades for display applications, Appl. Opt. 38 (1999) 291–303.
[1.6] A.S. Fedor, Binary optic diffuser design, in: J.M. Karam, J.A. Yasaitis (Eds.), Micromachining and Microfabrication Process Technology VII, Proc. SPIE 4557 (2001) 378–385.
[1.7] Y.T. Lua, S. Chi, Fabrication of light-shaping diffusion screens, Opt. Commun. 214 (2002) 55–63.
[1.8] J.N. Mait, Understanding diffractive optic design in the scalar domain, J. Opt. Soc. Am. A 12 (1995) 2145–2158.
[1.9] H. Tsukamoto, M. Nishiyama, Diffractive white-light diffuser with subwavelength structure, Japan. J. Appl. Phys. 45 (2006) 6678–6682.
[1.10] B. Kress, P. Meyrueis, Digital Diffractive Optics: An Introduction to Planar Diffractive Optics and Related Technology, John Wiley & Sons, New York, 2001.
[1.11] J. Cordingley, Application of a binary diffractive optic for beam shaping in semiconductor processing by lasers, Appl. Opt. 32 (1993) 2538–2542.
[1.12] J.S. Liu, M.R. Taghizadeh, Iterative algorithm for the design of diffractive phase elements for laser beam shaping, Opt. Lett. 27 (2002) 1463–1465.
[1.13] J.S. Liu, A.J. Caley, M.R. Taghizadeh, Diffractive optical elements for beam shaping of monochromatic spatially incoherent light, Appl. Opt. 45 (2006) 8440–8447.

Chapter 2

ONE-DIMENSIONAL SURFACES

We begin our discussion of designer surfaces by considering one-dimensional randomly rough surfaces. The theory for such surfaces is much simpler than it is for two-dimensional randomly rough surfaces, but still provides an introduction to the approaches that have been used in dealing with the latter surfaces. Moreover, there are applications for which a one-dimensional rough surface is sufficient to their needs. We will see examples of such applications in this chapter.

We first describe an approach to the design of a one-dimensional randomly rough perfectly conducting surface that produces a specified angular dependence of the mean intensity of the scattered light. It is based on the geometrical optics limit of the Kirchhoff approximation, and the form of the surface assumed in this analysis underlies the design of many of the one-dimensional randomly rough surfaces that will be considered in this chapter. On the basis of the concepts developed in the treatment of scattering from perfectly conducting surfaces, an approach also based on the geometrical optics limit of the Kirchhoff approximation is then developed for the design of one-dimensional randomly rough surfaces, bounding finitely conducting lossy metals or dielectric media, that possess prescribed scattering properties. It is then shown how to generate a metallic surface that suppresses the conversion into bulk electromagnetic waves in the vacuum region above it when a surface electromagnetic wave propagates across it, and a surface that produces an enhanced backscattering peak for only a single, specified angle of incidence.

What the surfaces just described have in common is that they produce a scattered field with a prescribed angular dependence of its mean intensity. However, such surfaces are not the only ones of interest. Also interesting for a variety of applications are randomly rough surfaces that for fixed angles of incidence and scattering produce a scattered field with a mean intensity that has a specified dependence on its frequency. It is shown how such surfaces can be designed. As an outgrowth of this theory a method is presented for designing randomly rough surfaces with specified thermal emissivities.

The intensity of the scattered field is the limit for coinciding arguments of the correlation function of the values of the field at two different spatial points. Such a correlation function enters the definition of the spatial coherence of the scattered field, which is the property of the field that is most closely associated with interference. It is shown how a surface can be designed to produce a scattered field with a specified spectral degree of coherence in the mean scattering plane, and how to shape the incident field so that the scattered field has a specified intensity distribution in that plane. The combination of these two results shows how it is possible to design a surface that transforms an incident field with a specified intensity distribution into a scattered field with a different specified intensity distribution, a process called beam shaping.

The preceding theory is then extended to the case of the transmission of light through a thin metallic or dielectric film with a randomly rough surface, and it is shown how that surface can be designed so that the film produces a specified angular dependence of the mean intensity of the light transmitted through it.

The surfaces generated by the approaches developed in this chapter are validated by calculations of the angular, spatial, coherence, or wavelength properties of the scattered field the surfaces are designed to produce, on the basis of scattering theories that are more rigorous than those used to generate the surfaces. For a monochromatic incident field the results of the scattering calculations are averaged over the ensemble of realizations of the surface profile function, to average over the speckles that arise when a single realization of the surface profile function is illuminated by such a field. Experimentalists, however, work with just a single realization of the surface. We show how averaging over speckles in this case can be effected by the use of a broadband (polychromatic) incident field instead of a monochromatic incident field.

Finally, an approach to the fabrication on photoresist of one-dimensional randomly rough surfaces of the kind considered in this chapter is described.

2.1. Perfectly Conducting Surfaces

The scattering of light from a perfectly conducting surface is a convenient starting point for the development of an approach to the design of a randomly rough surface with specified scattering properties. Only the field in the vacuum region above the surface needs to be considered, and the boundary condition it satisfies on the surface is a simple one. Nevertheless, as we will see, the ideas that lead to a solution of this problem in this simple case can also be applied to the solution of this problem for realistic, penetrable surfaces. In addition to the mathematical simplifications the assumption of a perfectly conducting scattering medium introduces into theories of rough surface scattering, the fact that in the infrared and longer wavelength regions of the optical spectrum metals are well approximated by perfect conductors also prompts studies of the scattering of light by perfectly conducting surfaces. This section is therefore devoted to the presentation of an approach to the design of a one-dimensional randomly perfectly conducting surface that produces a prescribed angular dependence of the intensity of the field scattered from it.

2.1.1. Scattering Theory

The physical system we consider consists of vacuum in the region $x_3 > \zeta(x_1)$ and a perfect conductor in the region $x_3 < \zeta(x_1)$ (Fig. 2.1). The surface profile function $\zeta(x_1)$ is assumed

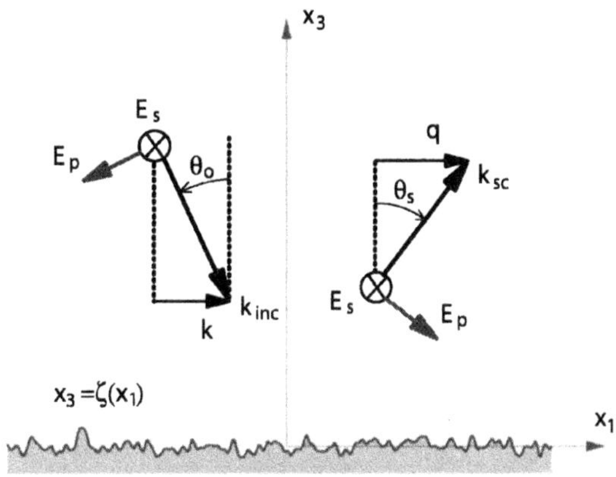

Figure 2.1. The geometry assumed in calculations of the scattering of light from a one-dimensional randomly rough perfectly conducting surface.

to be a single-valued function of x_1 that is at least twice differentiable, and constitutes a random process, but not necessarily a stationary one.

We assume that the surface $x_3 = \zeta(x_1)$ is illuminated from the vacuum region by a p- or s-polarized electromagnetic field of frequency ω, whose plane of incidence is the x_1x_3 plane. In this geometry the plane of scattering is also the x_1x_3 plane, and there is no cross-polarized scattering: the scattered field is p-polarized when the incident field is p-polarized, and s-polarized when the incident field is s-polarized. We can treat the cases of p and s polarization simultaneously by introducing the function $F_\nu(x_1, x_3|\omega)$, which is $H_2(x_1, x_3|\omega)$ when $\nu = p$ (p polarization), and is $E_2(x_1, x_3|\omega)$ when $\nu = s$ (s polarization). $H_2(x_1, x_3|\omega)$ ($E_2(x_1, x_3|\omega)$) is the single nonzero component of the total magnetic (electric) field in the vacuum region in p (s) polarization. The function $F_\nu(x_1, x_3|\omega)$ satisfies the Helmholtz equation

$$\left(\frac{\partial^2}{\partial x_1^2} + \frac{\partial^2}{\partial x_3^2} + \frac{\omega^2}{c^2}\right) F_\nu(x_1, x_3|\omega) = 0, \quad x_3 > \zeta(x_1), \tag{2.1.1}$$

where c is the speed of light in vacuum. A time dependence of the field of the form $\exp(-i\omega t)$ has been assumed, but it has not been indicated explicitly.

The field $F_\nu(x_1, x_3|\omega)$ satisfies either (a) the Dirichlet boundary condition

$$F_\nu(x_1, x_3|\omega)\Big|_{x_3=\zeta(x_1)} = 0, \tag{2.1.2}$$

when $\nu = s$ (s polarization), or (b) the Neumann boundary condition

$$\frac{\partial}{\partial n} F_\nu(x_1, x_3|\omega)\bigg|_{x_3=\zeta(x_1)} = 0, \tag{2.1.3}$$

when $\nu = p$ (p polarization), on the rough surface $x_3 = \zeta(x_1)$. In Eq. (2.1.3) $\partial/\partial n$ is the derivative along the normal to this surface at each point, directed into the vacuum,

$$\frac{\partial}{\partial n} = \frac{1}{[1 + (\zeta'(x_1))^2]^{\frac{1}{2}}}\left(-\zeta'(x_1)\frac{\partial}{\partial x_1} + \frac{\partial}{\partial x_3}\right), \tag{2.1.4}$$

where the prime denotes differentiation with respect to argument. In addition, we require that $F_\nu(x_1, x_3|\omega)$ be the sum of an incoming incident wave and outgoing scattered waves at $x_3 = \infty$.

We introduce the free space Green's function $G_0(x_1, x_3|x_1, x_3)$ as the solution of the equation

$$\left(\frac{\partial^2}{\partial x_1^2} + \frac{\partial^2}{\partial x_3^2} + \frac{\omega^2}{c^2}\right)G_0(x_1, x_3|x_1', x_3') = -4\pi\delta(x_1 - x_1')\delta(x_3 - x_3') \tag{2.1.5}$$

subject to an outgoing wave condition at infinity. It has the representations

$$G_0(x_1, x_3|x_1', x_3') = i\pi H_0^{(1)}\left((\omega/c)\left[(x_1 - x_1')^2 + (x_3 - x_3')^2\right]^{\frac{1}{2}}\right) \tag{2.1.6a}$$

$$= \int_{-\infty}^{\infty} \frac{dq}{2\pi}\frac{2\pi i}{\alpha_0(q)} \exp\left[iq(x_1 - x_1')\right]\exp\left[i\alpha_0(q)|x_3 - x_3'|\right], \tag{2.1.6b}$$

where $H_0^{(1)}(z)$ is a Hankel function of the first kind and zero order, and

$$\alpha_0(q) = \left[(\omega/c)^2 - q^2\right]^{\frac{1}{2}}, \quad \mathrm{Re}\,\alpha_0(q) > 0,\ \mathrm{Im}\,\alpha_0(q) > 0. \tag{2.1.7}$$

2.1.1.1. The Scattered Field We now invoke Green's second integral identity in the plane [2.1], which applies to a bounded, planar, singly-connected region R of the x_1x_3 plane, whose boundary is a closed regular curve C. Thus, let $u(x_1, x_3)$ and $v(x_1, x_3)$ be two arbitrary functions of x_1 and x_3 which together with their first partial derivatives are continuous in the region R and on the curve C. The theorem states that

$$\iint_R dx_1\, dx_3\left(u\nabla^2 v - v\nabla^2 u\right) = \int_C ds\left(u\frac{\partial}{\partial \nu}v - v\frac{\partial}{\partial \nu}u\right), \tag{2.1.8}$$

where $\nabla^2 = (\partial^2/\partial x_1^2 + \partial^2/\partial x_3^2)$, ds is the element of arc length along the curve C, and $\partial/\partial\nu$ is the derivative along the normal to the curve C at each point of it, directed away from the region R. We apply this theorem to the region $x_3 > \zeta(x_1)$.

In this region we set $u(x_1, x_3) = F_\nu(x_2, x_3|\omega)$ and $v(x_1, x_3) = G_0(x_1, x_3|x_1', x_3')$. For the curve C we take the union of the curve $x_3 = \zeta(x_1)$, which we denote by s, and a semicircle of infinite radius in the region $x_3 > \zeta(x_1)$, which we denote by $C^{(+\infty)}$. With the use of Eqs. (2.1.1) and (2.1.5), and the boundary conditions at infinity satisfied by $F_\nu(x_1, x_3|\omega)$ and $G_0(x_1, x_3|x_1', x_3')$, we obtain the equation

$$\begin{aligned}
&-4\pi\theta\big(x_3' - \zeta(x_1')\big)F_\nu(x_1', x_3'|\omega) \\
&\quad = -\int_s ds\left[F_\nu(x_1, x_3|\omega)\frac{\partial}{\partial n}G_0(x_1, x_3|x_1', x_3') - G_0(x_1, x_3|x_1', x_3')\frac{\partial}{\partial n}F_\nu(x_1, x_3|\omega)\right] \\
&\qquad + \int_{C^{(+\infty)}} ds\left[F_\nu(x_1, x_3|\omega)\frac{\partial}{\partial \nu}G_0(x_1, x_3|x_1', x_3') - G_0(x_1, x_3|x_1', x_3')\frac{\partial}{\partial \nu}F_\nu(x_1, x_3|\omega)\right],
\end{aligned} \tag{2.1.9}$$

where $\theta(x)$ is the Heaviside unit step function. Because the scattered field satisfies a radiation condition at infinity, its contribution to the integral over the semicircle $C^{(+\infty)}$ on the right-hand side of Eq. (2.1.9) vanishes. The latter therefore yields $-4\pi F_\nu(x_1', x_3'|\omega)_{\text{inc}}$, where $F_\nu(x_1, x_3|\omega)_{\text{inc}}$ is the incident field.

The incident field satisfies the Helmholtz equation

$$\left(\frac{\partial^2}{\partial x_1^2} + \frac{\partial^2}{\partial x_3^2} + \frac{\omega^2}{c^2}\right)F_\nu(x_1, x_3|\omega)_{\text{inc}} = 0. \tag{2.1.10}$$

We will represent it in the form of an incoming plane wave,

$$F_\nu(x_1, x_3|\omega)_{\text{inc}} = \exp\big[ikx_1 - i\alpha_0(k)x_3\big]. \tag{2.1.11}$$

We can then rewrite Eq. (2.1.9) as

$$\begin{aligned}
&\theta\big(x_3 - \zeta(x_1)\big)F_\nu(x_1, x_3|\omega) \\
&\quad = F_\nu(x_1, x_3|\omega)_{\text{inc}} + \frac{1}{4\pi}\int_s ds'\left[\left(\frac{\partial}{\partial n'}G_0(x_1, x_3|x_1', x_3')\right)F_\nu(x_1', x_3'|\omega)\right. \\
&\qquad \left. - G_0(x_1, x_3|x_1', x_3')\frac{\partial}{\partial n'}F_\nu(x_1', x_3'|\omega)\right].
\end{aligned} \tag{2.1.12}$$

In writing this equation we have used the symmetry of $G_0(x_1, x_3|x_1', x_3')$ in (x_1, x_3) and (x_1', x_3').

When $\nu = s$, then in view of the boundary condition (2.1.2) Eq. (2.1.11) becomes

$$\begin{aligned}
&\theta\big(x_3 - \zeta(x_1)\big)E_2(x_1, x_3|\omega) \\
&\quad = E_2(x_1, x_3|\omega)_{\text{inc}} - \frac{1}{4\pi}\int_s ds'\, G_0(x_1, x_3|x_1', x_3')\frac{\partial}{\partial n'}E_2(x_1', x_3'|\omega).
\end{aligned} \tag{2.1.13}$$

When $\nu = p$, then in view of the boundary condition (2.1.3) Eq. (2.1.12) becomes

$$\theta\big(x_3 - \zeta(x_1)\big) H_2(x_1, x_3|\omega)$$
$$= H_2(x_1, x_3|\omega)_{\mathrm{inc}} + \frac{1}{4\pi} \int_s ds' \left(\frac{\partial}{\partial n'} G_0(x_1, x_3|x_1', x_3') \right) H_2(x_1', x_3'|\omega). \tag{2.1.14}$$

Since we have assumed that the surface profile function $\zeta(x_1)$ is a single-valued function of x_1, we can replace integration along the curve s by integration along the x_1 axis with the aid of the relation

$$ds = \left[1 + \big(\zeta'(x_1)\big)^2\right]^{\frac{1}{2}} dx_1. \tag{2.1.15}$$

Consequently, Eq. (2.1.13) can be rewritten as

$$\theta\big(x_3 - \zeta(x_1)\big) E_2(x_1, x_3|\omega)$$
$$= E_2(x_1, x_3|\omega)_{\mathrm{inc}} - \frac{1}{4\pi} \int_{-\infty}^{\infty} dx_1' \left[G_0(x_1, x_3|x_1', x_3')\right]\Big|_{x_3'=\zeta(x_1')} F(x_1'|\omega), \tag{2.1.16}$$

where the source function $F(x_1|\omega)$ is defined by

$$F(x_1|\omega) = \frac{\partial}{\partial N} E_2(x_1, x_3|\omega)\Big|_{x_3=\zeta(x_1)}, \tag{2.1.17}$$

and

$$\frac{\partial}{\partial N} = -\zeta'(x_1) \frac{\partial}{\partial x_1} + \frac{\partial}{\partial x_3} \tag{2.1.18}$$

is the non-normalized derivative along the normal to the surface $x_3 = \zeta(x_1)$ at each point, directed into the vacuum region.

Similarly, Eq. (2.1.14) can be rewritten in the form

$$\theta\big(x_3 - \zeta(x_1)\big) H_2(x_1, x_3|\omega)$$
$$= H_2(x_1, x_3|\omega)_{\mathrm{inc}} + \frac{1}{4\pi} \int_{-\infty}^{\infty} dx_1' \left[\frac{\partial}{\partial N'} G_0(x_1, x_3|x_1', x_3')\right]\Bigg|_{x_3'=\zeta(x_1')} H(x_1'|\omega), \tag{2.1.19}$$

where the source function $H(x_1|\omega)$ is defined by

$$H(x_1|\omega) = H_2(x_1, x_3|\omega)\Big|_{x_3=\zeta(x_1)}. \tag{2.1.20}$$

The scattered field in s polarization is given by the second term on the right-hand side of Eq. (2.1.16),

$$E_2(x_1, x_3|\omega)_{\mathrm{sc}} = -\frac{1}{4\pi}\int_{-\infty}^{\infty} dx_1' \left[G_0(x_1, x_3|x_1', x_3')\right]\Big|_{x_3'=\zeta(x_1')} F(x_1'|\omega). \tag{2.1.21}$$

With the use of the representation (2.1.6b) for the Green's function $G_0(x_1, x_3|x_1', x_3')$ we find that the scattered field in the far zone is given by

$$E_2(x_1, x_3|\omega)_{\mathrm{sc}} = \int_{-\infty}^{\infty} \frac{dq}{2\pi} R_s(q|k) \exp\left[iqx_1 + i\alpha_0(q)x_3\right], \tag{2.1.22}$$

where the scattering amplitude $R_s(q|k)$ is

$$R_s(q|k) = -\frac{i}{2\alpha_0(q)}\int_{-\infty}^{\infty} dx_1\, F(x_1|\omega) \exp\left[-iqx_1 - i\alpha_0(q)\zeta(x_1)\right]. \tag{2.1.23}$$

The dependence of $R_s(q|k)$ on the wave number k arises from the dependence of the source function $F(x_1|\omega)$ on the incident field (2.1.11).

The scattered field in p polarization is given by the second term on the right-hand side of Eq. (2.1.19),

$$H_2(x_1, x_3|\omega)_{\mathrm{sc}} = \frac{1}{4\pi}\int_{-\infty}^{\infty} dx_1' \left[\frac{\partial}{\partial N'} G_0(x_1, x_3|x_1', x_3')\right]\Bigg|_{x_3'=\zeta(x_1')} H(x_1'|\omega). \tag{2.1.24}$$

With the use of Eq. (2.1.6b) the scattered field in the far zone can be represented as

$$H_2(x_1, x_3|\omega)_{\mathrm{sc}} = \int_{-\infty}^{\infty} \frac{dq}{2\pi} R_p(q|k) \exp\left[iqx_1 + i\alpha_0(q)x_3\right], \tag{2.1.25}$$

where the scattering amplitude $R_p(q|k)$ is given by

$$R_p(q|k) = -\frac{1}{2\alpha_0(q)}\int_{-\infty}^{\infty} dx_1 \left[q\zeta'(x_1) - \alpha_0(q)\right] H(x_1|\omega) \exp\left[-iqx_1 - i\alpha_0(q)\zeta(x_1)\right]. \tag{2.1.26}$$

2.1.1.2. The Mean Differential Reflection Coefficient The scattering amplitudes $R_s(q|k)$ and $R_p(q|k)$ play a central role in the present theory, because the differential reflection coefficient is expressed in terms of them. The differential reflection coefficient $(\partial R/\partial\theta_s)$ is defined in such a way that $(\partial R/\partial\theta_s)\,d\theta_s$ is the fraction of the total time-averaged flux incident on the surface $x_3 = \zeta(x_1)$ that is scattered into the angular region $(\theta_s, \theta_s + d\theta_s)$ in the limit as $d\theta_s$ tends to zero.

In calculating the total time-averaged flux in the incident and scattered fields we are interested in only the components of the Poynting vector antiparallel and parallel to the x_3 axis, respectively, not in the components parallel to the x_1x_2 plane.

Because in the present case the medium from which the field is incident, and into which it is scattered, is vacuum, we can write the real part of the 3-component of the complex Poynting vector in the region $x_3 > \zeta(x_1)$ as $(\nu = p, s)$

$$\mathrm{Re}\, S^c_{3\nu} = \mathrm{Re}\left(-i\frac{c^2}{8\pi\omega}\frac{\partial F_\nu}{\partial x_3}F^*_\nu\right). \tag{2.1.27}$$

With the form of the incident field given by Eq. (2.1.11) we obtain for the magnitude of the total time-averaged incident flux incident on the plane $x_3 = 0$,

$$P_{\mathrm{inc}} = -\int_{-\frac{L_1}{2}}^{\frac{L_1}{2}} dx_1 \int_{-\frac{L_2}{2}}^{\frac{L_2}{2}} dx_2 \left(\mathrm{Re}\, S^c_{3\nu}\right)_{\mathrm{inc}} \tag{2.1.28a}$$

$$= L_1L_2\frac{c^2\alpha_0(k)}{8\pi\omega}, \tag{2.1.28b}$$

for light of either polarization. Here L_1 and L_2 are the lengths of the x_1 and x_2 axes, respectively, covered by the rough surface, and the minus sign on the right-hand side of Eq. (2.1.28a) compensates for the fact that the 3-component of the Poynting vector of the incident field is negative.

With the scattered fields (2.1.22) and (2.1.25) written in the form $(\nu = p, s)$

$$F_\nu(x_1, x_3|\omega)_{\mathrm{sc}} = \int_{-\infty}^{\infty} \frac{dq}{2\pi} R_\nu(q|k)\exp\left[iqx_1 + i\alpha_0(q)x_3\right], \tag{2.1.29}$$

the total time-averaged scattered flux leaving the surface $x_3 = 0$ is

$$P_{\mathrm{sc}} = \int_{-\frac{L_1}{2}}^{\frac{L_1}{2}} dx_1 \int_{-\frac{L_2}{2}}^{\frac{L_2}{2}} dx_2 \left(\mathrm{Re}\, S^c_{3\nu}\right)_{\mathrm{sc}} \tag{2.1.30a}$$

$$= \mathrm{Re}\,\frac{c^2}{8\pi\omega}\int\limits_{-\frac{L_1}{2}}^{\frac{L_1}{2}} dx_1 \int\limits_{-\frac{L_2}{2}}^{\frac{L_2}{2}} dx_2 \int\limits_{-\infty}^{\infty}\frac{dq}{2\pi}\int\limits_{-\infty}^{\infty}\frac{dq'}{2\pi}\alpha_0(q)R_\nu(q|k)R_\nu^*(q'|k)$$

$$\times \exp\bigl[i(q-q')x_1\bigr]\exp\bigl\{i\bigl[\alpha_0(q)-\alpha_0^*(q')\bigr]x_3\bigr\}$$

$$= \mathrm{Re}\,L_2\frac{c^2}{8\pi\omega}\int\limits_{-\infty}^{\infty}\frac{dq}{2\pi}\alpha_0(q)\bigl|R_\nu(q|k)\bigr|^2\exp\bigl[-2\,\mathrm{Im}\,\alpha_0(q)x_3\bigr]$$

$$= L_2\frac{c^2}{8\pi\omega}\int\limits_{-\frac{\omega}{c}}^{\frac{\omega}{c}}\frac{dq}{2\pi}\alpha_0(q)\bigl|R_\nu(q|k)\bigr|^2, \tag{2.1.30b}$$

where we have allowed L_1 to tend to infinity.

We now introduce the angle of incidence θ_0, measured counterclockwise from the x_3 axis, and the scattering angle θ_s, measured clockwise from the x_3 axis (see Fig. 2.1), by

$$k = (\omega/c)\sin\theta_0, \qquad q = (\omega/c)\sin\theta_s, \tag{2.1.31}$$

so that

$$\alpha_0(k) = (\omega/c)\cos\theta_0, \qquad \alpha_0(q) = (\omega/c)\cos\theta_s, \qquad dq = (\omega/c)\cos\theta_s\,d\theta_s. \tag{2.1.32}$$

The total time-averaged incident flux, Eq. (2.1.28b) then takes the form

$$P_{\mathrm{inc}} = L_1L_2\frac{c}{8\pi}\cos\theta_0. \tag{2.1.33}$$

The total time-averaged scattered flux, Eq. (2.1.30b), becomes

$$P_{\mathrm{sc}} = \int\limits_{-\frac{\pi}{2}}^{\frac{\pi}{2}} d\theta_s\,P_{\mathrm{sc}}(\theta_s), \tag{2.1.34}$$

where

$$P_{\mathrm{sc}}(\theta_s) = L_2\frac{\omega}{16\pi^2}\cos^2\theta_s\bigl|R_\nu(q|k)\bigr|^2, \tag{2.1.35}$$

where q and k are to be expressed in terms of θ_s and θ_0 through Eqs. (2.1.31).

By definition, the differential reflection coefficient is given by

$$\left(\frac{\partial R_\nu}{\partial\theta_s}\right) = \frac{P_{\mathrm{sc}}(\theta_s)}{P_{\mathrm{inc}}} = \frac{1}{L_1}\left(\frac{\omega}{2\pi c}\right)\frac{\cos^2\theta_s}{\cos\theta_0}\bigl|R_\nu(q|k)\bigr|^2. \tag{2.1.36}$$

Because we are concerned with scattering from a randomly rough surface defined by $x_3 = \zeta(x_1)$, it is the mean differential reflection coefficient that we must calculate. It is given by

$$\left\langle \frac{\partial R_\nu}{\partial \theta_s} \right\rangle = \frac{1}{L_1}\left(\frac{\omega}{2\pi c}\right)\frac{\cos^2\theta_s}{\cos\theta_0}\left\langle \left|R_\nu(q|k)\right|^2\right\rangle, \tag{2.1.37}$$

where q and k are given by Eqs. (2.1.31). Here, and in all that follows, the angle brackets denote an average over the ensemble of realizations of the surface profile function.

2.1.1.3. The Equations Satisfied by the Source Functions From the results expressed by Eqs. (2.1.23) and (2.1.26) we see that the determination of the scattering amplitudes $R_s(q|k)$ and $R_p(q|k)$ that enter the expression for the mean differential reflection coefficient given by Eq. (2.1.37) requires knowledge of the source functions $F(x_1|\omega)$ and $H(x_1|\omega)$. These functions satisfy integral equations that can be obtained in the following way.

We return to Eq. (2.1.16), set $x_3 = \zeta(x_1) + \eta$ and $x_3 = \zeta(x_1) - \eta$ in turn in it, where η is a positive infinitesimal, and add the resulting two equations. Keeping in mind the boundary condition (2.1.2), we obtain

$$2E_2(x_1, x_3|\omega)\big|_{x_3=\zeta(x_1)} = \frac{1}{2\pi}\int_{-\infty}^{\infty} dx_1' \left[G_0(x_1, x_3|x_1', x_3')\right]_{\substack{x_3'=\zeta(x_1')\\ x_3=\zeta(x_1)}} F(x_1'|\omega), \tag{2.1.38}$$

a Fredholm equation of the first kind for $F(x_1|\omega)$. In obtaining this equation we have used the result that [2.2]

$$\begin{aligned}\int_{-\infty}^{\infty} dx_1' \left[G_0(x_1, x_3|x_1', x_3')\right]_{\substack{x_3'=\zeta(x_1')\\ x_3=\zeta(x_1)\pm\eta}} F(x_1'|\omega) \\ = \int_{-\infty}^{\infty} dx_1' \left[G_0(x_1, x_3|x_1', x_3')\right]_{\substack{x_3'=\zeta(x_1')\\ x_3=\zeta(x_1)}} F(x_1'|\omega),\end{aligned} \tag{2.1.39}$$

i.e., that the singularity of the Green's function $[G_0(x_1, x_3|x_1', x_3')]_{x_3'=\zeta(x_1'), x_3=\zeta(x_1)}$ at $x_1 = x_1'$ is integrable.

An equation for $F(x_1|\omega)$ that is more useful for certain purposes is obtained if we assume that $x_3 > \zeta(x_1)$ and apply the operator $\partial/\partial N$, Eq. (2.1.18), to both sides of Eq. (2.1.16). In this way we obtain

$$\begin{aligned}&\theta\big(x_3 - \zeta(x_1)\big)\big(\partial E_2(x_1, x_3|\omega)/\partial N\big) \\ &\quad = \big(\partial E_2(x_1, x_3|\omega)_{\mathrm{inc}}/\partial N\big)\end{aligned}$$

$$-\frac{1}{4\pi}\int_{-\infty}^{\infty} dx_1' \left[\partial G_0(x_1, x_3|x_1', x_3')/\partial N\right]_{x_3'=\zeta(x_1')} F(x_1'|\omega). \tag{2.1.40}$$

We next set x_3 equal to $\zeta(x_1)+\eta$ and $\zeta(x_1)-\eta$ in turn, where η is a positive infinitesimal, and add the two resulting equations. The result is the equation

$$F(x_1|\omega) = 2F(x_1|\omega)_{\text{inc}} - \frac{1}{2\pi} P \int_{-\infty}^{\infty} dx_1' \left[\partial G_0(x_1, x_3|x_1', x_3')/\partial N\right]_{\substack{x_3'=\zeta(x_1') \\ x_3=\zeta(x_1)}} F(x_1'|\omega), \tag{2.1.41}$$

where P denotes the Cauchy principal value, and

$$F(x_1|\omega)_{\text{inc}} = \left[\partial E_2(x_1, x_3|\omega)_{\text{inc}}/\partial N\right]_{x_3=\zeta(x_1)}. \tag{2.1.42}$$

Equation (2.1.41) is an inhomogeneous Fredholm equation of the second kind for $F(x_1|\omega)$. In obtaining Eq. (2.1.41) we have used the result that

$$\begin{aligned}
&\frac{\partial}{\partial N} G_0(x_1, x_3|x_1', x_3')\Big|_{\substack{x_3'=\zeta(x_1') \\ x_3=\zeta(x_1)+\eta}} + \frac{\partial}{\partial N} G_0(x_1, x_3|x_1', x_3')\Big|_{\substack{x_3'=\zeta(x_1') \\ x_3=\zeta(x_1)-\eta}} \\
&= \begin{cases} 2\frac{\partial}{\partial N} G_0(x_1, x_3|x_1', x_3')\Big|_{\substack{x_3'=\zeta(x_1') \\ x_3=\zeta(x_1)}} & x_1 \neq x_1', \\ 0 & x_1 = x_1'. \end{cases}
\end{aligned} \tag{2.1.43}$$

We obtain an integral equation satisfied by the source function $H(x_1|\omega)$ by setting x_3 equal to $\zeta(x_1)+\eta$ and $\zeta(x_1)-\eta$ in turn in Eq. (2.1.19) and adding the two resulting equations. In this way we obtain the equation

$$H(x_1|\omega) = 2H(x_1|\omega)_{\text{inc}} + \frac{1}{2\pi} P \int_{-\infty}^{\infty} dx_1' \left[\partial G_0(x_1, x_3|x_1', x_3')/\partial N'\right]_{\substack{x_3'=\zeta(x_1') \\ x_3=\zeta(x_1)}} H(x_1'|\omega), \tag{2.1.44}$$

where

$$H(x_1|\omega)_{\text{inc}} = H_2(x_1, x_3|\omega)_{\text{inc}}\Big|_{x_3=\zeta(x_1)}. \tag{2.1.45}$$

Equation (2.1.44) is an inhomogeneous Fredholm equation of the second kind for $H(x_1|\omega)$. In obtaining Eq. (2.1.44) we have used the result that

$$\frac{\partial}{\partial N'} G_0(x_1, x_3|x_1', x_3')\Big|_{\substack{x_3'=\zeta(x_1') \\ x_3=\zeta(x_1)+\eta}} + \frac{\partial}{\partial N'} G_0(x_1, x_3|x_1', x_3')\Big|_{\substack{x_3'=\zeta(x_1') \\ x_3=\zeta(x_1)-\eta}}$$

$$
= \begin{cases} 2\frac{\partial}{\partial N'} G_0(x_1, x_3|x_1', x_3')\Big|_{\substack{x_3'=\zeta(x_1') \\ x_3=\zeta(x_1)}} & x_1 \neq x_1', \\ 0 & x_1 = x_1'. \end{cases} \tag{2.1.46}
$$

We will return to Eqs. (2.1.41) and (2.1.44) later in this chapter.

2.1.1.4. The Kirchhoff Approximation and its Geometrical Optics Limit In the preceding section of this chapter we have presented a rigorous theory of the scattering of p- and s-polarized light, incident from vacuum, from a one-dimensional, rough, perfectly conducting surface. The very rigor of these results makes them ill-suited for the determination of a surface profile function that produces a scattered field with a prescribed dependence of the mean differential reflection coefficient on the scattering angle θ_s. No solutions of Eqs. (2.1.41) and (2.1.44) for the source functions exist, apart from purely numerical ones. We therefore use an approximate version of the theory developed in the preceding sections as the basis for the solution of this inverse problem. The approximate version of the rigorous theory we use is the Kirchhoff approximation or, more accurately, the tangent plane approximation. In fact, a major part of the theoretical and numerical analysis in this book is based on this approximation due to its simplicity, and because its use avoids some difficulties that can arise when other approximate theories are used. The rigorous results obtained in the preceding sections of this chapter will be used later in this chapter to validate the results obtained on the basis of the Kirchhoff approximation. We now turn to a derivation of the equations underlying this approximation, adapting the approach presented by Voronovich [2.3] to the present case of scattering from a one-dimensional rough surface.

We begin by assuming that the plane wave (2.1.11) is incident on the plane $x_3 = 0$ bounding the scattering medium. The total field in the vacuum region $x_3 > 0$ evaluated on the plane $x_3 = 0$ is then given by

$$
F_\nu^>(x_1, 0|\omega) = \big[1 + R_\nu\big(\alpha_0(k)\big)\big] F_\nu(x_1, 0|\omega)_{\text{inc}}, \tag{2.1.47}
$$

while its normal derivative, evaluated on the same plane, is

$$
\frac{\partial}{\partial x_3} F_\nu^>(x_1, 0|\omega) = -i\alpha_0(k)\big[1 - R_\nu\big(\alpha_0(k)\big)\big] F_\nu(x_1, 0)_{\text{inc}}, \tag{2.1.48}
$$

where $R_\nu(\alpha_0(k))$ is the Fresnel reflection amplitude. We now approximate the rough surface $x_3 = \zeta(x_1)$ at each point on it by the plane tangent to it at this point, and assume that the total field and its normal derivative at this point are given locally by Eqs. (2.1.47) and (2.1.48). We substitute these expressions into the second term on the right-hand side of Eq. (2.1.12) and obtain for the scattered field

$$F_\nu(x_1, x_3|\omega)_{sc} = \frac{1}{4\pi} \int_S ds' \Big\{ \big[1 + R_\nu\big(\alpha_\ell(k|x')\big)\big] F_\nu(x_1', x_3'|\omega)_{inc} \frac{\partial}{\partial n'} G_0(x_1, x_3|x_1', x_3') + i\alpha_\ell(k|x_1')\big[1 - R_\nu\big(\alpha_\ell(k|x_1')\big)\big] F_\nu(x_1', x_3'|\omega)_{inc} G_0(x_1, x_3|x_1', x_3') \Big\}. \tag{2.1.49}$$

In this equation $\alpha_\ell(k|x_1)$ is the modulus of the projection of the wave vector $(k, 0, -\alpha_0(k))$ of the incident field on a local normal to the surface,

$$\hat{\mathbf{n}} = \frac{(-\zeta'(x_1), 0, 1)}{[1 + (\zeta'(x_1))^2]^{\frac{1}{2}}}, \tag{2.1.50}$$

so that

$$\alpha_\ell(k|x_1) = \frac{\alpha_0(k) + k\zeta'(x_1)}{[1 + (\zeta'(x_1))^2]^{\frac{1}{2}}}. \tag{2.1.51}$$

We now replace integration along the surface profile function s in Eq. (2.1.49) by integration along the x_1 axis by the use of Eq. (2.1.15), use the representation (2.1.6b) for the Green's function $G_0(x_1, x_3|x_1', x_3')$ in the case that $x_3 \gg x_3'$, and substitute the expression (2.1.11) for $F_\nu(x_1, x_3|\omega)_{inc}$. In this way we obtain the scattered field in the form

$$\begin{aligned} F_\nu(x_1, x_3|\omega)_{sc} = -\int_{-\infty}^{\infty} \frac{dq}{2\pi} \exp\big[iqx_1 + i\alpha_0(q)x_3\big] \\ \times \frac{1}{2\alpha_0(q)} \int_{-\infty}^{\infty} dx_1' \exp\big\{-i(q-k)x_1' - i\big[\alpha_0(q) + \alpha_0(k)\big]\zeta(x_1')\big\} \\ \times \big\{\big[1 + R_\nu\big(\alpha_\ell(kx_1')\big)\big]\big[q\zeta'(x_1') - \alpha_0(q)\big] \\ + \big[1 - R_\nu\big(\alpha_\ell(k|x_1')\big)\big]\big[k\zeta'(x_1') + \alpha_0(k)\big]\big\}. \end{aligned} \tag{2.1.52}$$

On comparing this result with the expression for the scattered field given by Eqs. (2.1.22) and (2.1.25) we find that the scattering amplitude $R_\nu(q|k)$ is given by

$$\begin{aligned} R_\nu(q|k) = -\frac{1}{2\alpha_0(q)} \int_{-\infty}^{\infty} dx_1 \exp\big\{-i(q-k)x_1 - i\big[\alpha_0(q) + \alpha_0(k)\big]\zeta(x_1)\big\} \\ \times \big\{\big[1 + R_\nu\big(\alpha_\ell(k|x_1)\big)\big]\big[q\zeta'(x_1) - \alpha_0(q)\big] \\ + \big[1 - R_\nu\big(\alpha_\ell(k|x_1)\big)\big]\big[k\zeta'(x_1) + \alpha_0(k)\big]\big\}. \end{aligned} \tag{2.1.53}$$

A particularly convenient expression for the scattering amplitude is obtained if we evaluate the integral in Eq. (2.1.53) by the method of stationary phase. The stationary points are obtained from the requirement that the derivative of the exponent vanish, with the result that the ith stationary point is obtained from

$$\zeta'(x_{1i}) = -\frac{q-k}{\alpha_0(q)+\alpha_0(k)}. \tag{2.1.54}$$

We see that at each stationary point $\zeta'(x_1)$ has the value given by the right-hand side of Eq. (2.1.54), and $\alpha_\ell(k|x_1)$ has the value

$$\alpha_\ell(k|x_{1i}) = \left[\frac{(\omega/c)^2 + \alpha_0(q)\alpha_0(k) - qk}{2}\right]^{\frac{1}{2}} \tag{2.1.55a}$$

$$= \frac{\omega}{c}\cos\left(\frac{\theta_s+\theta_0}{2}\right), \tag{2.1.55b}$$

where we have used Eqs. (2.1.31) and (2.1.32). We approximate the factor multiplying the exponential in the integrand on the right-hand side of Eq. (2.1.53) by its value calculated at the stationary points, and find that

$$R_\nu(q|k) = R_\nu\big(\alpha_\ell(k|x_{1i})\big)\frac{(\omega/c)^2 + \alpha_0(q)\alpha_0(k) - qk}{\alpha_0(q)[\alpha_0(q)+\alpha_0(k)]} \times \int_{-\infty}^{\infty} dx_1 \exp\{-i(q-k)x_1 - i[\alpha_0(q)+\alpha_0(k)]\zeta(x_1)\}. \tag{2.1.56}$$

This is the tangent plane or Kirchhoff approximation to the scattering amplitude $R_\nu(q|k)$.

In the case of interest here, namely a perfectly conducting surface, the Fresnel reflection amplitudes $R_s(\alpha_0(k))$ and $R_p(\alpha_0(k))$ entering Eqs. (2.1.47) and (2.1.48) are readily found to be

$$R_s\big(\alpha_0(k)\big) = -1, \tag{2.1.57a}$$

$$R_p\big(\alpha_0(k)\big) = 1, \tag{2.1.57b}$$

independent of $\alpha_0(k)$. It follows that the reflection amplitudes $R_s(\alpha_\ell(k|x_{1i}))$ and $R_p(\alpha_\ell(k|x_{1i}))$ entering Eq. (2.1.56) have the same values. We thus obtain that the Kirchhoff approximation to the scattering amplitude $R_\nu(q|k)$ takes the form

$$R_s(q|k) = -\frac{(\omega/c)^2 + \alpha_0(q)\alpha_0(k) - qk}{\alpha_0(q)[\alpha_0(q)+\alpha_0(k)]} \times \int_{-\infty}^{\infty} dx_1 \exp\{-i(q-k)x_1 - i[\alpha_0(q)+\alpha_0(k)]\zeta(x_1)\} \tag{2.1.58a}$$

$$= -R_p(q|k). \tag{2.1.58b}$$

It should be noted that the results given by Eqs. (2.1.58) are precisely the results that are obtained from Eqs. (2.1.23) and (2.1.25) when the source functions $F(x_1|\omega)$ and $H(x_1|\omega)$ entering them are given by the inhomogeneous term on the right-hand side of Eqs. (2.1.41)–(2.1.42) and (2.1.44)–(2.1.45), respectively.

We see from Eqs. (2.1.58) that within the Kirchhoff approximation the amplitudes for the scattering of s- and p-polarized light from a one-dimensional rough perfectly conducting surface differ only in sign. These amplitudes enter the mean differential reflection coefficients (2.1.37) only through the square of their moduli. Consequently, within the Kirchhoff approximation the mean differential reflection coefficient is independent of the polarization of the incident light, and can be written

$$\left\langle\frac{\partial R}{\partial\theta_s}\right\rangle = \frac{1}{L_1}\left(\frac{\omega}{2\pi c}\right)\frac{[1+\cos(\theta_s+\theta_0)]^2}{\cos\theta_0(\cos\theta_s+\cos\theta_0)^2} \times \int_{-\infty}^{\infty} dx_1 \int_{-\infty}^{\infty} dx_1' \exp\left[-i(q-k)(x_1-x_1')\right] \times \left\langle\exp\left[-i\left(\alpha_0(q)+\alpha_0(k)\right)\left(\zeta(x_1)-\zeta(x_1')\right)\right]\right\rangle. \tag{2.1.59}$$

As we will see later, if one goes beyond the Kirchhoff approximation, and includes contributions from multiple-scattering processes, it is no longer the case that $\langle\partial R_s/\partial\theta_s\rangle$ equals $\langle\partial R_p/\partial\theta_s\rangle$, so that these functions have to be calculated separately.

Our goal in this work is to determine the surface profile function $\zeta(x_1)$ that produces a specified form of $\langle\partial R/\partial\theta_s\rangle$ as a function of θ_s for a given θ_0. As it stands, the expression for $\langle\partial R/\partial\theta_s\rangle$ as a function of $\zeta(x_1)$ given by Eq. (2.1.59) is still too complicated for us to invert to obtain $\zeta(x_1)$ in terms of $\langle\partial R/\partial\theta_s\rangle$. To simplify Eq. (2.1.59) we pass to the geometrical optics limit of the Kirchhoff approximation. This limit is obtained if we make the change of variable $x_1' = x_1 + u$ in Eq. (2.1.59), expand the difference $\zeta(x_1)-\zeta(x_1+u)$ in powers of u, and retain only the leading nonzero term:

$$\left\langle\frac{\partial R}{\partial\theta_s}\right\rangle = \frac{1}{L_1}\left(\frac{\omega}{2\pi c}\right)\frac{1}{\cos\theta_0}\left[\frac{1+\cos(\theta_s+\theta_0)}{\cos\theta_s+\cos\theta_0}\right]^2 \times \int_{-\infty}^{\infty} dx_1 \int_{-\infty}^{\infty} du \exp\left[i(q-k)u\right]\left\langle\exp\left[iau\zeta'(x_1)\right]\right\rangle. \tag{2.1.60}$$

To simplify the notation, in writing Eq. (2.1.60) we have introduced the function

$$a(\theta_s,\theta_0) = \alpha_0(q)+\alpha_0(k) = (\omega/c)(\cos\theta_s+\cos\theta_0). \tag{2.1.61}$$

Equation (2.1.60) is the starting point for the determination of the surface profile function $\zeta(x_1)$ that produces a specified form of $\langle\partial R/\partial\theta_s\rangle$.

2.1.2. A Surface that Produces a Prescribed Angular Dependence of the Mean Intensity of the Field Scattered from It

Our first goal in this chapter is to show how a one-dimensional randomly rough perfectly conducting surface can be designed to produce a scattered field whose mean intensity has a prescribed dependence on the scattering angle θ_s for a given value of the angle of incidence θ_0. In tackling this problem, it is good to know at the outset what kinds of surfaces can produce a mean scattered intensity of a prescribed form or, what is equally important, what kinds of surfaces cannot produce a mean scattered intensity of a desired form, so as to avoid starting the solution of the inverse problem on the basis of invalid assumptions about the nature of the surface we seek. In this section we address these questions.

As noted in Chapter 1, an optical element that has a variety of potential applications is a band-limited uniform diffuser. In the present context this is a one-dimensional random surface that produces a scattered field whose mean intensity is constant within a specified range of scattering angles and vanishes outside this range. In Section 2.1.2.1 we show that a surface whose surface profile function is a stationary Gaussian random process—a commonly used model for a randomly rough surface—cannot act as a band-limited uniform diffuser. In Section 2.1.2.2 we present a representation of a one-dimensional surface profile function that describes a perfectly conducting surface that can produce a scattered field whose mean intensity can have essentially any specified (single-valued) dependence on the scattering angle, including one that corresponds to a band-limited uniform diffuser. This representation will be the basis for many of the results that will be obtained in the rest of this chapter.

2.1.2.1. A Stationary Surface Let us assume for the moment that the surface profile function $\zeta(x_1)$ is a stationary random process. The average in Eq. (2.1.60) is then independent of x_1. The integral over x_1 therefore gives a factor of L_1, the length of the x_1 axis covered by the rough surface. The integral over u then leads to the result

$$\left\langle \frac{\partial R}{\partial \theta_s} \right\rangle = \left(\frac{\omega}{2\pi c}\right) \frac{1}{\cos\theta_0} \left[\frac{1+\cos(\theta_s+\theta_0)}{\cos\theta_s+\cos\theta_0}\right]^2 \times \int_{-\infty}^{\infty} du\, g(u) \exp\bigl[i(\omega/c)(\sin\theta_s - \sin\theta_0)u\bigr], \tag{2.1.62}$$

where

$$g(u) = \left\langle \exp\left[i\frac{\omega}{c}(\cos\theta_s + \cos\theta_0)u\zeta'(x_1)\right]\right\rangle. \tag{2.1.63}$$

To simplify the following discussion we assume that the angles of incidence and scattering are sufficiently small that the function of θ_s and θ_0 multiplying the integral in Eq. (2.1.62) can be replaced by unity. This is not a very restrictive assumption. For example, if $\theta_s = \theta_0 = 15°$, this function departs from unity by less than 3.5%. In this limit we have

$$\left\langle \frac{\partial R}{\partial \theta_s} \right\rangle \cong \left(\frac{\omega}{2\pi c} \right) \int_{-\infty}^{\infty} du\, g(u) \exp\left[i(\omega/c)(\theta_s - \theta_0)u\right], \tag{2.1.64}$$

where now

$$g(u) = \left\langle \exp\left[i2(\omega/c)u\zeta'(x_1)\right]\right\rangle. \tag{2.1.65}$$

If we are interested, in particular, in obtaining a uniform distribution of the mean scattered intensity in the range $-\theta_m < \theta_s - \theta_0 < \theta_m$, and a vanishing mean scattered intensity outside this range, so that

$$\left\langle \frac{\partial R}{\partial \theta_s} \right\rangle = K\theta\left(\theta_m - |\theta_s - \theta_0|\right), \tag{2.1.66}$$

where K is a constant, we obtain from Eqs. (2.1.64) and (2.1.66) that

$$g(u) = 2K\theta_m \operatorname{sinc}\left[(\omega/c)\theta_m u\right], \tag{2.1.67}$$

where $\operatorname{sinc} x = \sin x/x$.

We can rewrite Eqs. (2.1.64) and (2.1.65) as

$$\begin{aligned} \left\langle \frac{\partial R}{\partial \theta_s} \right\rangle &= \left(\frac{\omega}{2\pi c} \right) \left\langle \int_{-\infty}^{\infty} du \exp\left\{i(\omega/c)\left[\theta_s - \theta_0 + 2\zeta'(x_1)\right]u\right\} \right\rangle \\ &= \frac{\omega}{c} \left\langle \delta\left(\frac{\omega}{c}\left[\theta_s - \theta_0 + 2\zeta'(x_1)\right]\right)\right\rangle \\ &= \frac{1}{2} \left\langle \delta\left(\frac{\theta_0 - \theta_s}{2} - \zeta'(x_1)\right)\right\rangle. \end{aligned} \tag{2.1.68}$$

Now, the probability density function (pdf) of slopes on the surface, $p(s)$, is defined such that $p(s)\,ds$ is the probability that the slope $\zeta'(x_1)$ of the surface at an arbitrary point on the x_1 axis lies between s and $s + ds$. It is given formally by

$$p(s) = \left\langle \delta\left(s - \zeta'(x_1)\right)\right\rangle. \tag{2.1.69}$$

The fact that it is independent of x_1 is a consequence of the assumed stationarity of the surface profile function $\zeta(x_1)$. On equating the right-hand sides of Eqs. (2.1.66) and (2.1.68), and keeping the definition (2.1.69) in mind, we find that

$$p(s) = 2K\theta\left(\frac{1}{2}\theta_m - |s| \right). \tag{2.1.70}$$

The normalization of $p(s)$,

$$\int_{-\infty}^{\infty} ds\, p(s) = 1, \tag{2.1.71}$$

yields the result that $K = 1/(2\theta_m)$, so that finally

$$p(s) = \frac{\theta(\frac{1}{2}\theta_m - |s|)}{\theta_m}. \tag{2.1.72}$$

This result shows that Eq. (2.1.66) will be satisfied if the pdf of slopes has a rectangular shape.

Let us now assume that the surface profile function $\zeta(x_1)$ is a zero-mean Gaussian random process in addition to being a stationary random process. The resulting random process is defined by a two-point surface height correlation function of the form

$$\langle \zeta(x_1)\zeta(x_1')\rangle = \delta^2 W(|x_1 - x_1'|), \tag{2.1.73}$$

where $\delta = \langle \zeta^2(x_1)\rangle^{\frac{1}{2}}$ is the root-mean-square (rms) height of the surface. The precise form of the surface height autocorrelation function $W(|x_1|)$ is not required here. However, it is required to possess several general properties. By definition

$$W(0) = 1. \tag{2.1.74}$$

From the inequalities

$$\langle (\zeta(x_1) \pm \zeta(x_1'))^2 \rangle \geqslant 0 \tag{2.1.75}$$

and the definition (2.1.73) it follows that

$$-1 < W(|x_1 - x_1'|) < 1. \tag{2.1.76}$$

Since the heights of the surface at two widely separated points on a random surface can be assumed to be uncorrelated, it follows that $W(|x_1|)$ tends to zero as $|x_1| \to \infty$. Finally, for small values of $|x_1|$ it is assumed that $W(|x_1|)$ has the form

$$W(|x_1|) = 1 - \frac{x_1^2}{a^2} + O(x_1^4), \tag{2.1.77}$$

where the characteristic length a is called the transverse correlation length of the surface roughness. The absence of a term linear in $|x_1|$ in the expansion (2.1.77) means that we are omitting fractal surfaces from the present discussion. It also means that $W(|x_1|)$ possesses a second derivative at $x_1 = 0$.

The property of a Gaussian random process $\zeta(x_1)$ that will be most useful in the remainder of this section is expressed by [2.4]

$$\begin{aligned}\left\langle \exp\left\{-i\left[a\zeta(x_1) - b\zeta(x_1')\right]\right\}\right\rangle &= \exp\left\{-\frac{1}{2}\left\langle\left[a\zeta(x_1) - b\zeta(x_1')\right]^2\right\rangle\right\} \\ &= \exp\left\{-\frac{1}{2}\delta^2\left[a^2 + b^2 - 2abW\left(|x_1 - x_1'|\right)\right]\right\},\end{aligned} \tag{2.1.78}$$

where a and b are arbitrary coefficients.

Since the random process that is obtained by differentiating a Gaussian random process is itself a Gaussian random process [2.5], it follows that if $\zeta(x_1)$ is assumed to be a Gaussian random process, then $\zeta'(x_1)$ is also a Gaussian random process. The pdf of slopes (2.1.69) in this case is given by

$$\begin{aligned}p(s)_G &= \lim_{\epsilon\to 0}\left\langle \delta\left(s - \frac{\zeta(x_1+\epsilon) - \zeta(x_1)}{\epsilon}\right)\right\rangle \\ &= \lim_{\epsilon\to 0}\int_{-\infty}^{\infty}\frac{dk}{2\pi}\exp(iks)\left\langle\exp\left\{-ik\left[\zeta(x_1+\epsilon) - \zeta(x_1)\right]/\epsilon\right\}\right\rangle \\ &= \lim_{\epsilon\to 0}\int_{-\infty}^{\infty}\frac{dk}{2\pi}\exp(iks)\exp\left\{-(\delta k/\epsilon)^2\left[1 - W\left(|\epsilon|\right)\right]\right\} \\ &= \int_{-\infty}^{\infty}\frac{dk}{2\pi}\exp(iks)\exp\left(-\delta^2k^2/a^2\right) \\ &= \frac{a}{2\sqrt{\pi}\delta}\exp\left[-\frac{a^2s^2}{4\delta^2}\right],\end{aligned} \tag{2.1.79}$$

where we have used the Fourier integral representation of the Dirac delta function,

$$\delta(x) = \int_{-\infty}^{\infty}\frac{dk}{2\pi}\exp(ikx), \tag{2.1.80}$$

and the small argument representation of $W(|x_1|)$, Eq. (2.1.77), in obtaining Eq. (2.1.79).

A comparison of the expressions for $p(s)$ given by Eqs. (2.1.72) and (2.1.79) shows that if we wish to design a one-dimensional random surface that acts as a band-limited uniform diffuser, the corresponding surface profile function $\zeta(x_1)$ cannot be assumed to be a stationary zero-mean Gaussian random process.

Although this result was obtained on the basis of the geometrical optics limit of the Kirchhoff approximation, it has a more general validity. For, if we return to the Kirchhoff approximation to the mean differential reflection coefficient given by Eq. (2.1.59) we find that the function $g(u)$ appearing in Eq. (2.1.62) is now given by

$$g(u) = \langle \exp[-i(\omega/c)(\cos\theta_s + \cos\theta_0)(\zeta(x_1) - \zeta(x_1 + u))] \rangle \tag{2.1.81}$$

instead of by Eq. (2.1.63). Due to the assumption that $\zeta(x_1)$ is a stationary random process, the function defined by Eq. (2.1.81) is independent of x_1. By the use of Eq. (2.1.78) we find that Eq. (2.1.81) becomes

$$\begin{aligned} g(u) &= \exp\{-(\omega\delta/c)^2(\cos\theta_s + \cos\theta_0)^2[1 - W(|u|)]\} \\ &\to \exp\{-(2\omega\delta/c)^2[1 - W(|u|)]\}, \end{aligned} \tag{2.1.82}$$

where the second expression obtains for small angles of incidence and scattering. Since the magnitude of $W(|u|)$ is always smaller or equal to unity, Eq. (2.1.76), the right-hand side of Eq. (2.1.82) is positive for all u, so that Eq. (2.1.67) cannot be satisfied, even in the Kirchhoff approximation, by the assumption that $\zeta(x_1)$ is a stationary, zero-mean, Gaussian random process.

It should be kept in mind that the failure of a surface profile function, assumed to constitute a stationary, zero-mean Gaussian random process, to generate a scattered field with a specified angular dependence of its mean intensity, has been demonstrated here only for a surface intended to act as a band-limited uniform diffuser. Indeed, later in this book we will show that a surface defined by a profile function that is a stationary, zero-mean, Gaussian random process can generate a scattered field with certain specified spatial correlation properties. Nevertheless, the conclusion to be drawn from the analysis presented in this section is that in seeking to design a random surface that has prescribed scattering properties it is best to begin with as few assumptions about the statistical properties of the surface sought as possible. We now turn to a discussion of a representation of a random surface that has sufficient flexibility to be able to produce scattering patterns of great variety.

2.1.2.2. A Nonstationary Surface We now relax the assumption of the stationarity of the surface. In this case because $\zeta(x_1)$ has been assumed to be a nonstationary random process, we cannot assume that $\zeta'(x_1)$ is a stationary random process. The average $\langle \exp[i(\omega/c)(\cos\theta_s + \cos\theta_0)u\zeta'(x_1)]\rangle$ in Eq. (2.1.60) therefore has to be assumed to be a function of x_1, and we cannot carry out the integral over x_1 to yield L_1, as we did when $\zeta(x_1)$ was assumed to be a stationary random process.

To evaluate the double integral in Eq. (2.1.60) we proceed as follows. We define a set of equally-spaced points along the x_1-axis by $x_n = nb$, where b is a characteristic length and $n = 0, \pm 1, \pm 2, \ldots$. Within the interval $nb \leqslant x_1 \leqslant (n+1)b$ the surface profile function

$\zeta(x_1)$ is assumed to have the linear form

$$\zeta(x_1) = a_n x_1 + b_n, \quad nb \leqslant x_1 \leqslant (n+1)b. \tag{2.1.83}$$

The slopes $\{a_n\}$ are assumed to be independent identically distributed random deviates. Therefore the pdf of a_n,

$$f(\gamma) = \langle \delta(\gamma - a_n) \rangle, \tag{2.1.84}$$

is independent of n. In order that the surface be continuous at, say, $x_1 = (n+1)b$, the relation

$$b_{n+1} = b_n - (n+1)(a_{n+1} - a_n)b \tag{2.1.85}$$

must be satisfied. From this recurrence relation the $\{b_n\}$ can be determined from a knowledge of the $\{a_n\}$, provided that an initial value, e.g., that of b_0, is specified. With no loss of generality we can assume that $b_0 = 0$, and in what follows we do so. The solution of Eq. (2.1.85) is then

$$b_n = (a_0 + a_1 + \cdots + a_{n-1} - na_n)b, \quad n \geqslant 1, \tag{2.1.86a}$$

$$b_{-n} = -\big(a_{-1} + a_{-2} + \cdots + a_{-(n-1)} - (n-1)a_{-n}\big)b, \quad n \geqslant 1. \tag{2.1.86b}$$

The double integral in Eq. (2.1.60) can now be evaluated. Thus, we have that

$$\begin{aligned}
&\int_{-\infty}^{\infty} dx_1 \int_{-\infty}^{\infty} du \exp\big[i(q-k)u\big]\big\langle \exp\big[iau\zeta'(x_1)\big]\big\rangle \\
&\quad = \int_{-\infty}^{\infty} du \exp\big[i(q-k)u\big] \sum_{n=-N_s}^{N_s-1} \int_{nb}^{(n+1)b} dx_1 \big\langle \exp(iaua_n)\big\rangle \\
&\quad = \int_{-\infty}^{\infty} du \exp\big[i(q-k)u\big] \sum_{n=-N_s}^{N_s-1} \int_{nb}^{(n+1)b} dx_1 \int_{-\infty}^{\infty} d\gamma\, f(\gamma) \exp(iau\gamma) \\
&\quad = 2N_s b \int_{-\infty}^{\infty} d\gamma\, f(\gamma) \int_{-\infty}^{\infty} du \exp\big[i(q-k+a\gamma)\big] \\
&\quad = 2N_s b \int_{-\infty}^{\infty} d\gamma\, f(\gamma) 2\pi\delta(q-k+a\gamma) \\
&\quad = 2N_s b \frac{2\pi}{a} f\left(\frac{k-q}{a}\right)
\end{aligned}$$

$$= 2N_s b \frac{2\pi}{(\omega/c)(\cos\theta_0 + \cos\theta_s)} f\left(\frac{\sin\theta_0 - \sin\theta_s}{\cos\theta_0 + \cos\theta_s}\right). \tag{2.1.87}$$

If we now use the fact that the length L_1 is equal to $2N_s b$, then on combining Eqs. (2.1.60) and (2.1.87) we obtain

$$\left\langle\frac{\partial R}{\partial\theta_s}\right\rangle(\theta_s, \theta_0) = \frac{[1 + \cos(\theta_0 + \theta_s)]^2}{\cos\theta_0(\cos\theta_0 + \cos\theta_s)^3} f\left(\frac{\sin\theta_0 - \sin\theta_s}{\cos\theta_0 + \cos\theta_s}\right), \tag{2.1.88}$$

where we have indicated explicitly that the mean differential reflection coefficient depends on both θ_s and θ_0. We now make the change of variable

$$\frac{\sin\theta_0 - \sin\theta_s}{\cos\theta_0 + \cos\theta_s} = -\gamma. \tag{2.1.89}$$

With the aid of the identities

$$\sin A - \sin B = 2\cos\frac{1}{2}(A + B)\sin\frac{1}{2}(A - B), \tag{2.1.90a}$$

$$\cos A + \cos B = 2\cos\frac{1}{2}(A + B)\cos\frac{1}{2}(A - B) \tag{2.1.90b}$$

we can rewrite Eq. (2.1.89) as

$$\gamma = \tan\frac{1}{2}(\theta_s - \theta_0), \tag{2.1.91}$$

so that

$$\theta_s = \theta_0 + 2\tan^{-1}\gamma. \tag{2.1.92}$$

From Eq. (2.1.89) we obtain the useful results

$$\cos\theta_s = \frac{(1-\gamma^2)\cos\theta_0 - 2\gamma\sin\theta_0}{1+\gamma^2}, \qquad \sin\theta_s = \frac{(1-\gamma^2)\sin\theta_0 + 2\gamma\cos\theta_0}{1+\gamma^2}. \tag{2.1.93}$$

It follows from these results and Eq. (2.1.88) that

$$f(-\gamma) = \frac{\cos\theta_0(\cos\theta_0 + \cos\theta_s)^3}{[1 + \cos(\theta_0 + \theta_s)]^2}\left\langle\frac{\partial R}{\partial\theta_s}\right\rangle(\theta_0 + 2\tan^{-1}\gamma, \theta_0). \tag{2.1.94}$$

With the use of Eqs. (2.1.93) we find that

$$\frac{\cos\theta_0(\cos\theta_0 + \cos\theta_s)^3}{[1 + \cos(\theta_s + \theta_0)]^2} = \frac{2}{1+\gamma^2}\frac{\cos\theta_0}{\cos\theta_0 - \gamma\sin\theta_0}. \tag{2.1.95}$$

Therefore, we can rewrite Eq. (2.1.94) as

$$f(\gamma) = \frac{2}{1+\gamma^2} \frac{\cos\theta_0}{\cos\theta_0 + \gamma \sin\theta_0} \left\langle \frac{\partial R}{\partial \theta_s} \right\rangle \left(\theta_0 - 2\tan^{-1}\gamma, \theta_0\right). \tag{2.1.96}$$

Thus, if we wish to design a one-dimensional, randomly rough, perfectly conducting surface which, when illuminated by s- or p-polarized light at an angle of incidence θ_0, produces a specified form of $\langle \partial R/\partial \theta_s \rangle(\theta_s, \theta_0)$, we must first replace θ_s by $\theta_0 - 2\tan^{-1}\gamma$ and then determine $f(\gamma)$ from Eq. (2.1.96). A long sequence of $\{a_n\}$ is determined from $f(\gamma)$, e.g., by the rejection method [2.6], and the corresponding sequence of $\{b_n\}$ is obtained from Eqs. (2.1.85) or (2.1.86). A realization of the surface profile function $\zeta(x_1)$ is then constructed on the basis of Eq. (2.1.83).

The surface profile functions generated in this fashion possess the following general properties. The mean value of $\zeta(x_1)$ is given by

$$\langle \zeta(x_1) \rangle = \langle a_n x_1 + b_n \rangle, \quad nb \leqslant x_1 \leqslant (n+1)b. \tag{2.1.97}$$

The average of a_n is $\langle a \rangle$, independent of n. The average of b_n is therefore zero, according to Eqs. (2.1.86). Consequently, we find that

$$\langle \zeta(x_1) \rangle = \langle a \rangle x_1, \tag{2.1.98}$$

independent of n. This result means that the mean surface profile function will grow linearly with x_1 unless $\langle a \rangle = 0$. This condition will be satisfied if $f(\gamma)$ is an even function of γ.

The mean square height of the surface is given by

$$\begin{aligned} \langle \zeta^2(x_1) \rangle &= \langle (a_n x_1 + b_n)^2 \rangle \\ &= \langle a_n^2 \rangle x_1^2 + 2\langle a_n b_n \rangle x_1 + \langle b_n^2 \rangle, \quad nb \leqslant x_1 \leqslant (n+1)b. \end{aligned} \tag{2.1.99}$$

The value of $\langle a_n^2 \rangle$ is $\langle a^2 \rangle$, independent of n. From Eqs. (2.1.86) we find that $\langle a_n b_n \rangle = -n\langle a^2 \rangle b$ for $n \geqslant 0$, while $\langle b_n^2 \rangle = n(n+1)\langle a^2 \rangle b^2$ for $n \geqslant 0$. Thus, we find that

$$\langle \zeta^2(x_1) \rangle = \langle a^2 \rangle \left[x_1^2 - 2nbx_1 + n(n+1)b^2 \right], \quad nb \leqslant x_1 \leqslant (n+1)b, \tag{2.1.100a}$$

for $n \geqslant 0$. A similar analysis shows that

$$\langle \zeta^2(x_1) \rangle = \langle a^2 \rangle \left[x_1^2 + 2nbx_1 + n(n+1)b^2 \right], \quad -(n+1)b \leqslant x_1 \leqslant -nb, \tag{2.1.100b}$$

for $n \geqslant 0$. These results show that $\zeta(x_1)$ is not a stationary function of x_1. In addition they show that $\langle \zeta^2(x_1) \rangle$ is an even function of x_1.

In the common case of normal incidence, $\theta_0 = 0$, Eq. (2.1.96) simplifies to

$$f(\gamma) = \frac{2}{1+\gamma^2}\left(\frac{\partial R}{\partial \theta_s}\right)(-2\tan^{-1}\gamma, 0), \tag{2.1.101}$$

with

$$\cos\theta_s = \frac{1-\gamma^2}{1+\gamma^2}, \qquad \sin\theta_s = \frac{2\gamma}{1+\gamma^2}. \tag{2.1.102}$$

2.1.3. Solution of the Scattering Problem

How well the random surfaces generated in the manner described in the preceding section produce a calculated scattered field for which the mean differential reflection coefficient matches the input mean differential reflection coefficient, can be determined in the following way. A large number N_p of realizations of the random surface profile function is generated in this manner, and the scattering problem for a plane wave of ν polarization incident on it is solved for each of these realizations. The value of $|R_\nu(q|k)|^2$ is calculated from each of these solutions, and an arithmetic average of the N_p results for $|R_\nu(q|k)|^2$ thus obtained yields the average indicated in Eq. (2.1.37). A comparison of the result obtained in this manner with the input mean differential reflection coefficient reveals the quality of the random surface generated.

The approach to the design of randomly rough surfaces with specified scattering properties described in Section 2.1.2.2 is based on the geometrical optics limit of the Kirchhoff approximation for the mean differential reflection coefficient, a single-scattering approximation that does not distinguish between incident light of p and s polarization. For the comparison of the calculated and input mean differential reflection coefficients to be meaningful, the former should be calculated by an approach that is more rigorous than the geometrical optics limit of the Kirchhoff approximation. Otherwise only the numerical accuracy with which the mean differential reflection coefficient in the geometrical limit of the Kirchhoff approximation is calculated is being tested, a circular process.

One such approach is to calculate the mean differential reflection coefficient by a rigorous computer simulation approach that takes multiple scattering processes of all orders into account, and distinguishes between p- and s-polarized incident light. A second approach is to calculate the mean differential reflection coefficient on the basis of the Kirchhoff approximation, but without passing to the geometrical optics limit of it. This is a simple and computationally rapid approach that, however, is a single-scattering approximation, and moreover does not distinguish between incident light of p and s polarization. However, its use permits calculations of the mean differential reflection coefficient to be carried out for much larger values of the characteristic length b and for longer randomly rough surfaces than is possible at the present time in calculations based on the rigorous computer simulation approach. It will be seen below that the larger the value of b used in the solution of the scattering problems the better is the agreement between the calculated and input mean differential reflection coefficients. In this section we consider each of these approaches in turn.

2.1.3.1. Computer Simulations The Kirchhoff approximation is a single-scattering approximation, and it is of considerable interest to determine how well surfaces designed on the basis of the geometrical optics limit of this approximation, also a single-scattering approximation, reproduce the input mean differential reflection coefficient in calculations in which multiple scattering processes are taken into account. Such a determination can be made on the basis of the solution of the scattering equations by rigorous computer simulation calculations.

In going beyond the single-scattering approximation we can no longer assume that the scattering amplitudes $R_p(q|k)$ and $R_s(q|k)$ differ only in sign. The scattering amplitude for each polarization has to be calculated separately.

In the case of s polarization a convenient starting point for this calculation is Eq. (2.1.40). When we set $x_3 = \zeta(x_1) + \eta$ in it, where η is a positive infinitesimal, and recall the definitions (2.1.17) and (2.1.42), we obtain as the equation for the source function $F(x_1|\omega)$,

$$F(x_1|\omega) = F(x_1|\omega)_{\text{inc}} - \int_{-\infty}^{\infty} dx_1'\, H^{(0)}(x_1|x_1')F(x_1'|\omega), \tag{2.1.103}$$

where the kernel $H^{(0)}(x_1|x_1')$ is given by

$$H^{(0)}(x_1|x_1') = \frac{1}{4\pi}\frac{\partial}{\partial N} G_0(x_1, x_3|x_1', x_3')\Big|_{\substack{x_3'=\zeta(x_1') \\ x_3=\zeta(x_1)+\eta}} \tag{2.1.104a}$$

$$= -\frac{i}{4}\frac{\omega^2}{c^2}\frac{H_1^{(1)}\left(\frac{\omega}{c}[(x_1-x_1')^2 + (\zeta(x_1)-\zeta(x_1')+\eta)^2]^{\frac{1}{2}}\right)}{\frac{\omega}{c}[(x_1-x_1')^2 + (\zeta(x_1)-\zeta(x_1')+\eta)^2]^{\frac{1}{2}}}$$
$$\times\left[-\zeta'(x_1)(x_1-x_1') + \left(\zeta(x_1)-\zeta(x_1')+\eta\right)\right], \tag{2.1.104b}$$

where $H_1^{(1)}(z)$ is the Hankel function of the first kind and first order.

For the incident field we assume the plane wave given by Eq. (2.1.11), so that the function $F(x_1|\omega)_{\text{inc}}$ is obtained from Eq. (2.1.42) as

$$F(x_1|\omega)_{\text{inc}} = -i(\omega/c)\left[\zeta'(x_1)\sin\theta_0 + \cos\theta_0\right]\exp\left\{i(\omega/c)\left[x_1\sin\theta_0 - \zeta(x_1)\cos\theta_0\right]\right\}. \tag{2.1.105}$$

Equation (2.1.103) is solved by converting it into a matrix equation that is then solved numerically. Since the procedure involved in carrying out this conversion will be used in several places later on in this chapter, we describe it in some detail here.

As a first step, the infinite range of integration is replaced by the finite range $(-L/2, L/2)$. This range is divided into N segments of equal length, whose midpoints are given by the points $\{x_n\}$ defined by

$$x_n = -\frac{L}{2} + \left(n - \frac{1}{2}\right)\Delta x, \quad n = 1, 2, \ldots, N, \tag{2.1.106}$$

where $\Delta x = L/N$. Equation (2.1.103) is then rewritten as

$$F(x_1|\omega) = F(x_1|\omega)_{\text{inc}} - \sum_{n=1}^{N} \int_{x_n - \frac{1}{2}\Delta x}^{x_n + \frac{1}{2}\Delta x} dx_1' \, H^{(0)}(x_1|x_1') F(x_1'|\omega). \tag{2.1.107}$$

On the assumption that $F(x_1|\omega)$ is a slowly varying function of x_1 in each interval $(x_n - \frac{1}{2}\Delta x, x_n + \frac{1}{2}\Delta x)$, we evaluate it at the midpoint of each interval and remove it from the integral. In this way we obtain the equation

$$F(x_1|\omega) = F(x_1|\omega)_{\text{inc}} - \sum_{n=1}^{N} \left\{ \int_{-\frac{1}{2}\Delta x}^{\frac{1}{2}\Delta x} du \, H^{(0)}(x_1|x_n + u) \right\} F(x_n|\omega). \tag{2.1.108}$$

We next set $x_1 = x_m$ $(m = 1, 2, \ldots, N)$ and obtain the matrix equation

$$F(x_m|\omega) = F(x_m|\omega)_{\text{inc}} - \sum_{n=1}^{N} H_{mn}^{(0)} F(x_n|\omega), \quad m = 1, 2, \ldots, N, \tag{2.1.109}$$

where

$$H_{mn}^{(0)} = \int_{-\frac{1}{2}\Delta x}^{\frac{1}{2}\Delta x} du \, H^{(0)}(x_m|x_n + u). \tag{2.1.110}$$

The Hankel function appearing in the expression (2.1.104b) for the kernel $H^{(0)}(x_1|x_1')$ is singular when its argument vanishes, so that the kernel is also singular when $x_1 = x_1'$. This has to be taken into account in the evaluation of the matrix elements $H_{mn}^{(0)}$. For the off-diagonal elements of this matrix the integrand in Eq. (2.1.110) is never singular. We can therefore expand it in powers of u and integrate the result term-by-term. In this way we obtain the result that to the leading nonzero order in Δx

$$H_{mn}^{(0)} = \Delta x H^{(0)}(x_m|x_n) = \Delta x\left(-\frac{i}{4}\right)\frac{\omega^2}{c^2}\frac{H_1^{(1)}(\frac{\omega}{c}[(x_m-x_n)^2+(\zeta(x_m)-\zeta(x_n))^2]^{\frac{1}{2}})}{\frac{\omega}{c}[(x_m-x_n)^2+(\zeta(x_m)-\zeta(x_n))^2]^{\frac{1}{2}}} \times\left[-\zeta'(x_m)(x_m-x_n)+\big(\zeta(x_m)-\zeta(x_n)\big)\right],\quad m\neq n. \tag{2.1.111}$$

The evaluation of the diagonal elements requires some care. From Eqs. (2.1.104) and (2.1.110) the expression for the diagonal element $H_{mm}^{(0)}$ becomes

$$H_{mm}^{(0)} = -\frac{i}{4}\frac{\omega^2}{c^2}\int_{-\frac{1}{2}\Delta x}^{\frac{1}{2}\Delta x} du\frac{H_1^{(1)}(\frac{\omega}{c}[u^2+(\zeta(x_m)-\zeta(x_m+u)+\eta)^2]^{\frac{1}{2}})}{\frac{\omega}{c}[u^2+(\zeta(x_m)-\zeta(x_m+u)+\eta)^2]^{\frac{1}{2}}} \times\left[\zeta'(x_m)u+\big(\zeta(x_m)-\zeta(x_m+u)+\eta\big)\right] \tag{2.1.112a}$$

$$\cong -\frac{i}{4}\frac{\omega^2}{c^2}\int_{-\frac{1}{2}\Delta x}^{\frac{1}{2}\Delta x} du\frac{H_1^{(1)}(\frac{\omega}{c}\psi_m)}{\frac{\omega}{c}\psi_m}\left[\eta-\frac{1}{2}u^2\zeta''(x_m)-\frac{1}{6}u^3\zeta'''(x_m)+\cdots\right], \tag{2.1.112b}$$

where

$$\psi_m = \left[u^2\phi_m^2-2u\eta\zeta'(x_m)+\eta^2-\eta u^2\zeta''(x_m)+u^3\zeta'(x_m)\zeta''(x_m)\right]^{\frac{1}{2}}, \tag{2.1.113a}$$

and

$$\phi_m = \left[1+\big(\zeta'(x_m)\big)^2\right]^{\frac{1}{2}}. \tag{2.1.113b}$$

With the aid of the small argument expansion of $H_1^{(1)}(z)/z$ [2.7],

$$H_1^{(1)}(z)/z = -\frac{2i}{\pi}\frac{1}{z^2}+\frac{i}{\pi}\ell n\left(\frac{z}{2}\right)+\frac{i}{\pi}\left(\gamma-\frac{1}{2}\right)+\frac{1}{2}+O\big(z^2\ell n z\big), \tag{2.1.114}$$

where $\gamma = 0.5772157$ is Euler's constant, the leading nonzero contribution in Δx to $H_{mm}^{(0)}$ is obtained from the first term on the right-hand side of Eq. (2.1.114),

$$H_{mm}^{(0)} = -\frac{1}{2\pi}\int_{-\frac{1}{2}\Delta x}^{\frac{1}{2}\Delta x} du\frac{\eta}{\phi_m^2u^2-2u\eta\zeta'(x_m)+\eta^2}+\frac{1}{2\pi}\int_{-\frac{1}{2}\Delta x}^{\frac{1}{2}\Delta x} du\frac{\zeta''(x_m)}{2\phi_m^2}, \tag{2.1.115}$$

since the terms omitted either vanish with vanishing η or are of higher order in Δx. On carrying out the integrations we obtain finally

$$H_{mm}^{(0)} = -\frac{1}{2} + \Delta x \frac{\zeta''(x_m)}{4\pi\phi_m^2}. \tag{2.1.116}$$

The presence of a contribution independent of Δx to this matrix element is a consequence of the singularity of the kernel $H^{(0)}(x_1|x_1')$ for coinciding arguments.

When these results are used in Eq. (2.1.109), we can rewrite it as

$$F(x_m|\omega) = 2F(x_m|\omega)_{\text{inc}} - \sum_{n=1}^{N} \mathcal{H}_{mn}^{(0)} F(x_n|\omega), \tag{2.1.117}$$

where

$$\mathcal{H}_{mn}^{(0)} = \Delta x \left(-\frac{i}{2}\right) \frac{\omega^2}{c^2} \frac{H_1^{(1)}\left(\frac{\omega}{c}[(x_m - x_n)^2 + (\zeta(x_m) - \zeta(x_n))^2]^{\frac{1}{2}}\right)}{\frac{\omega}{c}[(x_m - x_n)^2 + (\zeta(x_m) - \zeta(x_n))^2]^{\frac{1}{2}}}$$

$$\times \left[-\zeta'(x_m)(x_m - x_n) + \left(\zeta(x_m) - \zeta(x_n)\right)\right] \quad m \neq n \tag{2.1.118a}$$

$$= \Delta x \frac{\zeta''(x_m)}{2\pi\phi_m^2} \quad m = n. \tag{2.1.118b}$$

The convergence of the solution of Eq. (2.1.117) is tested by increasing L and N until the change in the magnitude of $F(x_m|\omega)$ for each m is smaller than some preassigned value.

We note that Eq. (2.1.117) has the form of the matrix equation that would have been obtained if we had applied the method described here to the solution of Eq. (2.1.41), except that in the latter case the Cauchy principal value of the integral is usually dealt with by setting the diagonal matrix element $\mathcal{H}_{mm}^{(0)}$ equal to zero, rather than by assigning it the nonzero value given by Eq. (2.1.118b). However, it has been argued by Toporkov et al. [2.8] that the retention of the term given by Eq. (2.1.118b) significantly improves the convergence of the numerical solution for the source function $F(x_1|\omega)$, in comparison with calculations where this term is neglected. It is for this reason that we derived Eqs. (2.1.117)–(2.1.118) by starting from Eq. (2.1.40) rather than by starting directly from Eq. (2.1.41).

When the source function $F(x_m|\omega)$ has been obtained from the solution of Eq. (2.1.117), the mean differential reflection coefficient, Eq. (2.1.37), is given by

$$\left\langle \frac{\partial R_s}{\partial \theta_s} \right\rangle = \frac{1}{L} \frac{c}{8\pi\omega} \frac{\langle |r_s(\theta_s|\theta_0)|^2 \rangle}{\cos\theta_0}, \tag{2.1.119}$$

where

$$r_s(\theta_s|\theta_0) = \int_{-\infty}^{\infty} dx_1\, F(x_1|\omega)\exp\{-i(\omega/c)[x_1\sin\theta_s + \zeta(x_1)\cos\theta_s]\} \quad (2.1.120a)$$

$$\cong \Delta x \sum_{n=1}^{N} F(x_n|\omega)\exp\{-i(\omega/c)[x_n\sin\theta_s + \zeta(x_n)\cos\theta_s]\}. \quad (2.1.120b)$$

The dependence of $r_s(\theta_s|\theta_0)$ on θ_0 arises from the fact that the source function $F(x_1|\omega)$ is a functional of the incident field through $F(x_1|\omega)_{\mathrm{inc}}$, Eq. (2.1.105).

Turning now to the case that the incident field is p-polarized, a convenient starting point for a rigorous numerical solution of the scattering problem is Eq. (2.1.19). We set $x_3 = \zeta(x_1) + \eta$ in it and obtain as the equation for the source function $H(x_1|\omega)$,

$$H(x_1|\omega) = H(x_1|\omega)_{\mathrm{inc}} + \int_{-\infty}^{\infty} dx_1'\, \tilde{H}^{(0)}(x_1|x_1')H(x_1'|\omega), \quad (2.1.121)$$

where the kernel $\tilde{H}^{(0)}(x_1|x_1')$ is

$$\tilde{H}^{(0)}(x_1|x_1') = \frac{1}{4\pi}\frac{\partial}{\partial N'}G_0(x_1, x_3|x_1', x_3')\Big|_{\substack{x_3'=\zeta(x_1')\\ x_3=\zeta(x_1)+\eta}}$$

$$= -\frac{i}{4}\frac{\omega^2}{c^2}\frac{H_1^{(1)}\left(\frac{\omega}{c}[(x_1 - x_1')^2 + (\zeta(x_1) - \zeta(x_1') + \eta)^2]^{\frac{1}{2}}\right)}{\frac{\omega}{c}[(x_1 - x_1')^2 + (\zeta(x_1) - \zeta(x_1') + \eta)^2]^{\frac{1}{2}}}$$

$$\times \left[(x_1 - x_1')\zeta'(x_1') - (\zeta(x_1) - \zeta(x_1') + \eta)\right]. \quad (2.1.122)$$

The incident field is again taken to be the plane wave given by Eq. (2.1.11), so that the function $H(x_1|\omega)_{\mathrm{inc}}$ obtained from Eq. (2.1.20) is

$$H(x_1|\omega)_{\mathrm{inc}} = \exp\{i(\omega/c)[x_1\sin\theta_0 - \zeta(x_1)\cos\theta_0]\}. \quad (2.1.123)$$

In exactly the same way as Eq. (2.1.109) was obtained from Eq. (2.1.103), we can derive the following matrix equation for $H(x_m|\omega)$:

$$H(x_m|\omega) = H(x_m|\omega)_{\mathrm{inc}} + \sum_{n=1}^{N} \tilde{H}_{mn}^{(0)} H(x_n|\omega), \quad m = 1, 2, \ldots, N, \quad (2.1.124)$$

where $\tilde{H}^{(0)}_{mn}$ is given by

$$\tilde{H}^{(0)}_{mn} = \Delta x\left(-\frac{i}{4}\right)\frac{\omega^2}{c^2}\frac{H_1^{(1)}(\frac{\omega}{c}[(x_m - x_n)^2 + (\zeta(x_m) - \zeta(x_n))^2]^{\frac{1}{2}})}{\frac{\omega}{c}[(x_m - x_n)^2 + (\zeta(x_m) - \zeta(x_n))^2]^{\frac{1}{2}}}$$

$$\times\left[(x_m - x_n)\zeta'(x_n) - (\zeta(x_m) - \zeta(x_n))\right] \quad m \neq n \tag{2.1.125a}$$

$$= \frac{1}{2} + \Delta x\frac{\zeta''(x_m)}{4\pi\phi_m^2} \qquad m = n. \tag{2.1.125b}$$

In view of the result given by Eq. (2.1.125b), we rewrite Eq. (2.1.124) as

$$H(x_m|\omega) = 2H(x_m|\omega)_{\text{inc}} + \sum_{n=1}^{N}\tilde{\mathcal{H}}^{(0)}_{mn}H(x_n|\omega), \quad m = 1, 2, \ldots, N, \tag{2.1.126}$$

where

$$\tilde{\mathcal{H}}^{(0)}_{mn} = \Delta x\left(-\frac{i}{2}\right)\frac{\omega^2}{c^2}\frac{H_1^{(1)}(\frac{\omega}{c}[(x_m - x_n)^2 + (\zeta(x_m) - \zeta(x_n))^2]^{\frac{1}{2}})}{\frac{\omega}{c}[(x_m - x_n)^2 + (\zeta(x_m) - \zeta(x_n))^2]^{\frac{1}{2}}}$$

$$\times\left[(x_m - x_n)\zeta'(x_n) - (\zeta(x_m) - \zeta(x_n))\right] \quad m \neq n \tag{2.1.127a}$$

$$= \Delta x\frac{\zeta''(x_m)}{2\pi\phi_m^2} \qquad m = n. \tag{2.1.127b}$$

The matrix $\tilde{\mathcal{H}}^{(0)}_{mn}$ is just the transpose of the matrix $\mathcal{H}^{(0)}_{mn}$ defined by Eq. (2.1.118).

Once the source function $H(x_m|\omega)$ has been obtained from the solution of Eq. (2.1.126), the mean differential reflection coefficient, Eq. (2.1.37) is given by

$$\left\langle\frac{\partial R_p}{\partial\theta_s}\right\rangle = \frac{1}{L_1}\frac{c}{8\pi\omega}\frac{\langle|r_p(\theta_s|\theta_0)|^2\rangle}{\cos\theta_0}, \tag{2.1.128}$$

where

$$r_p(\theta_s|\theta_0) = \frac{\omega}{c}\int_{-\infty}^{\infty} dx_1\, H(x_1|\omega)\left[\zeta'(x_1)\sin\theta_s - \cos\theta_s\right]$$

$$\times \exp\left\{-i(\omega/c)\left[x_1\sin\theta_s + \zeta(x_1)\cos\theta_s\right]\right\} \tag{2.1.129a}$$

$$\cong \Delta x\frac{\omega}{c}\sum_{n=1}^{N} H(x_n|\omega)\left[\zeta'(x_n)\sin\theta_s - \cos\theta_s\right]$$

$$\times \exp\left\{-i(\omega/c)\left[x_n\sin\theta_s + \zeta(x_n)\cos\theta_s\right]\right\}. \tag{2.1.129b}$$

2.1.3.2. The Kirchhoff Approximation The simplest approach to the solution of the scattering problem for a given realization of the surface profile function is arguably the Kirchhoff approximation. The expression for the scattering amplitude $R(q|k)$ given by Eq. (2.1.58a) can be rewritten in the light of Eqs. (2.1.31), (2.1.32), and (2.1.83) as

$$R(q|k) = -\frac{[1+\cos(\theta_s+\theta_0)]}{\cos\theta_s(\cos\theta_s+\cos\theta_0)} \sum_{n=-N_s}^{N_s-1} \exp\left[-i\frac{\omega}{c}(\cos\theta_s+\cos\theta_0)b_n\right]$$
$$\times \int_{nb}^{(n+1)b} dx_1 \exp\left\{-i\frac{\omega}{c}\left[\sin\theta_s - \sin\theta_0 + (\cos\theta_s+\cos\theta_0)a_n\right]x_1\right\}$$
$$= -b\frac{[1+\cos(\theta_s+\theta_0)]}{\cos\theta_s(\cos\theta_s+\cos\theta_0)} r(\theta_s|\theta_0), \tag{2.1.130}$$

where

$$r(\theta_s|\theta_0) = \sum_{n=-N_s}^{N_s-1} \exp\left(-i\frac{\omega}{c}\left\{(\cos\theta_s+\cos\theta_0)b_n\right.\right.$$
$$\left.\left. + \left[\sin\theta_s - \sin\theta_0 + (\cos\theta_s+\cos\theta_0)a_n\right]\left(n+\frac{1}{2}\right)b\right\}\right)$$
$$\times \operatorname{sinc}\left(\frac{\omega}{c}\left[\sin\theta_s - \sin\theta_0 + (\cos\theta_s+\cos\theta_0)a_n\right]\frac{b}{2}\right). \tag{2.1.131}$$

If we define L_1 as $2N_s b$, the expression for the mean differential reflection coefficient given by Eq. (2.1.37) becomes

$$\left\langle \frac{\partial R}{\partial \theta_s} \right\rangle = \frac{b}{2N_s} \frac{\omega}{2\pi c} \frac{[1+\cos(\theta_s+\theta_0)]^2}{\cos\theta_0(\cos\theta_s+\cos\theta_0)^2} \left\langle \left| r(\theta_s|\theta_0) \right|^2 \right\rangle. \tag{2.1.132}$$

The procedure now is to generate a large number N_p of realizations of the random surface, i.e., of sets of the slopes $\{a_n\}$ and the corresponding $\{b_n\}$, and for each realization to calculate the corresponding value of $|r(\theta_s|\theta_0)|^2$ by the use of Eq. (2.1.131). An arithmetic average of the N_p results for $|r(\theta_s|\theta_0)|^2$ obtained in this way yields the average indicated in Eq. (2.1.132). It should be noted that $\langle |r(\theta_s|\theta_0)|^2 \rangle$ must grow linearly with N_s if this result for $\langle \partial R/\partial\theta_s \rangle$ is to be independent of N_s. This is indeed what occurs.

2.1.4. Results

To illustrate the approach developed in the preceding sections to the design of one-dimensional randomly rough surfaces that produce a mean differential reflection coefficient with a specified dependence on the scattering angle for a given angle of incidence, we will use it to design a surface that a) acts as a band-limited uniform diffuser; or b) acts as a Lambertian diffuser; or c) suppresses single-scattering processes. These examples give an indication of the versatility of the present approach.

2.1.4.1. A Band-Limited Uniform Diffuser If we wish to design a randomly rough surface that produces a mean differential reflection coefficient that is constant in the angular interval $|\theta_s| < \theta_m$, where $\theta_m < \pi/2$, and vanishes outside this interval, the mean differential reflection coefficient we seek to reproduce has the form

$$\left\langle \frac{\partial R}{\partial \theta_s} \right\rangle = A\theta(\theta_s + \theta_m)\theta(\theta_m - \theta_s), \tag{2.1.133}$$

where $\theta(z)$ is the Heaviside unit step function, and A is a coefficient to be determined. On substituting Eq. (2.1.133) into Eq. (2.1.96) we find that the probability density function of a_n is given by

$$f(\gamma) = \frac{2A}{1+\gamma^2} \frac{\cos\theta_0}{\cos\theta_0 + \gamma \sin\theta_0} \theta(\theta_0 - 2\tan^{-1}\gamma + \theta_m)\theta(\theta_m - \theta_0 + 2\tan^{-1}\gamma). \tag{2.1.134}$$

Because $\tan^{-1}\gamma$ is a monotonically increasing function of γ for $-\infty \leqslant \gamma \leqslant \infty$, we can rewrite Eq. (2.1.134) in the more convenient form

$$f(\gamma) = \frac{2A}{1+\gamma^2} \frac{\cos\theta_0}{\cos\theta_0 + \gamma \sin\theta_0} \theta\left(\tan\frac{\theta_m + \theta_0}{2} - \gamma\right)\theta\left(\tan\frac{\theta_m - \theta_0}{2} + \gamma\right). \tag{2.1.135}$$

The coefficient A is obtained from the normalization condition for $f(\gamma)$,

$$\int_{-\infty}^{\infty} d\gamma\, f(\gamma) = 1, \tag{2.1.136}$$

and is given by

$$A = \frac{1}{2\cos\theta_0}\left[\theta_m \cos\theta_0 + \sin\theta_0 \ell n \frac{\cos(\frac{\theta_m - \theta_0}{2})}{\cos(\frac{\theta_m + \theta_0}{2})}\right]^{-1}. \tag{2.1.137}$$

The fact that $f(\gamma)$ depends explicitly on the angle of incidence θ_0 means that a different ensemble of randomly rough surfaces has to be generated for each value of θ_0 in order for the mean differential reflection coefficient to have the form of a band-limited uniform diffuser in the angular range $|\theta_s| < \theta_m$ described by Eqs. (2.1.133) and (2.1.137).

This is illustrated in Fig. 2.2, where we have plotted a segment of one realization of the random surface defined by Eqs. (2.1.135) and (2.1.137) for $\theta_m = 20°$, and angles of

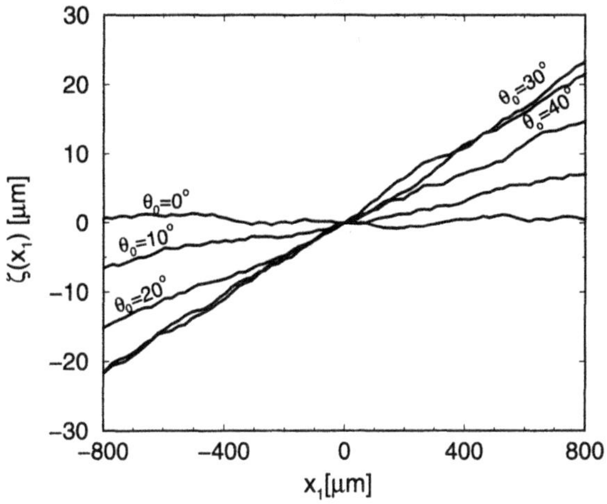

Figure 2.2. A segment of a single realization of the surface profile function $\zeta(x_1)$ designed to act as a band-limited uniform diffuser, for which $f(\gamma)$ is given by Eqs. (2.1.135) and (2.1.137), for each of five angles of incidence θ_0. The values of the parameters employed are $\theta_m = 20°$ and $b = 22$ μm.

incidence equal to $\theta_0 = 0°, 10°, 20°, 30°$, and $40°$. The value of the characteristic length b assumed in calculating these surface profile functions was $b = 22$ μm. It is seen that as the angle of incidence increases the mean surface tilts in such a way as to keep it close to being perpendicular to the incident wave. However, we emphasize that the surface calculated for a nonzero value of θ_0 is not simply the surface calculated for $\theta_0 = 0°$ rotated through the angle θ_0. In Fig. 2.3 we present a plot of the mean differential reflection coefficient $\langle \partial R/\partial\theta_s \rangle$ as a function of θ_s for the band-limited uniform diffusers defined by Eqs. (2.1.133), (2.1.135), and (2.1.137) for the same five angles of incidence that were used in plotting Fig. 2.2. The parameters used in obtaining these results, which were calculated on the basis of the Kirchhoff approximation, Eqs. (2.1.131) and (2.1.132), were $\theta_m = 20°$, $b = 220$ μm, $\lambda = 632.8$ nm, $N_s = 200$. The results for $N_p = 120\,000$ realizations of the surface profile function were used in calculating the average in Eq. (2.1.132). (The calculations of $\langle \partial R/\partial\theta_s \rangle$ in the Kirchhoff approximation are very fast, so that a large number of realizations can be used in calculating this average, if desired.) We see that the random surface generated by our approach indeed acts as a band-limited uniform diffuser within the range $|\theta_s| < 20°$ for each angle of incidence: there is virtually no scattered intensity for values of $|\theta_s|$ greater than 20°, and for values of $|\theta_s|$ smaller than 20° the scattered intensity is very nearly uniform. However, the magnitude of the scattered intensity for $|\theta_s|$ smaller than 20° increases with increasing θ_0. This is due to the increase of the amplitude A in Eq. (2.1.133) as θ_0 increases, as can be seen from Eq. (2.1.137).

The derivation of the pdf (2.1.135) was based on the geometrical optics limit of the Kirchhoff approximation, a single-scattering approximation, that yields the same result for $f(\gamma)$ for incident light of p and s polarization. It is of interest to examine whether the random surfaces generated by its use retain the scattering properties for which they were

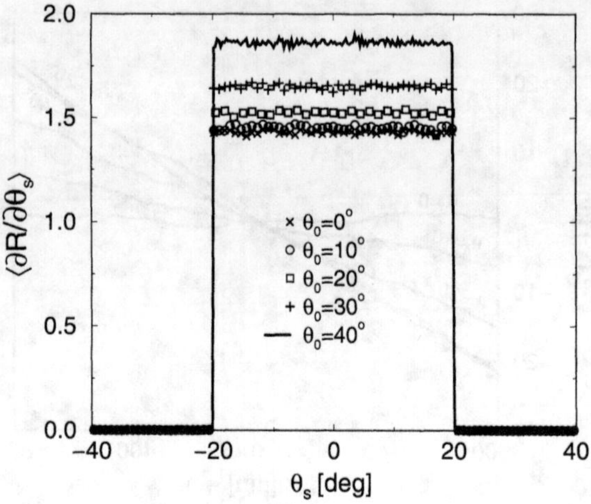

Figure 2.3. The mean differential reflection coefficient $\langle \partial R/\partial\theta_s \rangle$ estimated from $N_p = 120\,000$ realizations of the surface profile function for the case of a band-limited uniform diffuser for which $f(\gamma)$ is given by Eqs. (2.1.135) and (2.1.137). The calculations have been carried out for the same five angles of incidence that were used in plotting Fig. 2.2, on the basis of the Kirchhoff approximation for scattering from a one-dimensional perfectly conducting surface. The values of the parameters employed were $\theta_m = 20°$, $b = 220$ μm, $\lambda = 632.8$ nm, and $N_s = 200$.

designed when multiple scattering is taken into account, in which case, as we have seen in Section 2.1.3.1, different expressions for $\langle \partial R/\partial\theta_s \rangle$ are obtained for the same surface profile function for incident light of p and s polarization. In Fig. 2.4 we present results for the mean differential reflection coefficient $\langle \partial R_s/\partial\theta_s \rangle$ as a function of θ_s for the band-limited diffusers defined by Eqs. (2.1.133), (2.1.135), and (2.1.137), for the same five angles of incidence that were used in obtaining the results plotted in Figs. 2.2 and 2.3. The parameters used in obtaining these results, which were calculated by the rigorous computer simulation approach for the scattering of s-polarized light from a one-dimensional perfectly conducting surface, developed in Section 2.1.3.1, that takes multiple-scattering processes of all orders into account, were $\theta_m = 20°$, $b = 22$ μm, $\lambda = 632.8$ nm, and $N_s = 6$. The results obtained for $N_p = 30\,000$ realizations of the surface profile function were used in calculating the average in Eq. (2.1.119).[1] From a comparison of the results presented in Figs. 2.3 and 2.4 we see that not only do the random surfaces generated by our approach act as band-limited uniform diffusers for $|\theta_s| < \theta_m$ for each angle of incidence, but, also that the results obtained by the rigorous computer simulation approach, for the values of the experimental and roughness parameters assumed, are in complete agreement with those obtained in the Kirchhoff approximation.

[1] The results presented in Fig. 2.4 correct the corresponding results presented as Fig. 1 of Ref. [2.9], which were calculated with an incorrect normalization of $f(\gamma)$.

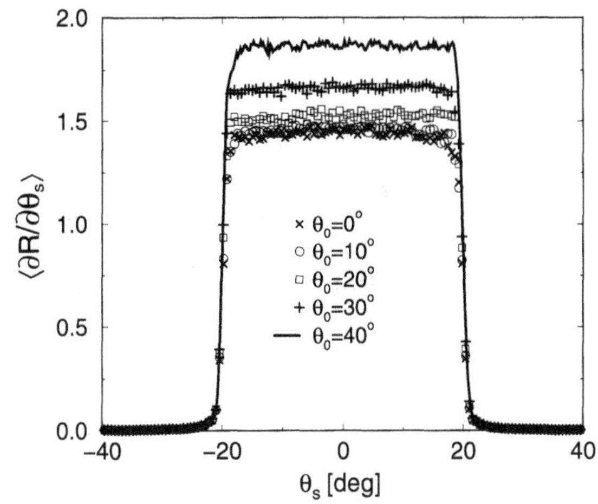

Figure 2.4. The mean differential reflection coefficient $\langle \partial R_s/\partial \theta_s \rangle$ estimated from $N_p = 30\,000$ realizations of the surface profile function for the case of a band-limited uniform diffuser for which $f(\gamma)$ is given by Eqs. (2.1.135) and (2.1.137). The calculations have been carried out for the same five angles of incidence that were used in plotting Figs. 2.2 and 2.3, on the basis of a rigorous computer simulation approach to the scattering of s-polarized light from a one-dimensional perfectly conducting surface. The values of the parameters employed were $\theta_m = 20°$, $b = 22$ μm, $\lambda = 632.8$ nm, and $N_s = 6$.

We now turn to a consideration of some results obtained at normal incidence. In this case the mean differential reflection coefficient (2.1.133) takes the form

$$\left\langle \frac{\partial R}{\partial \theta_s} \right\rangle = \frac{\theta(\theta_m - |\theta_s|)}{2\theta_m}, \tag{2.1.138}$$

so that the pdf becomes

$$f(\gamma) = \frac{1}{2\tan^{-1}\gamma_m} \frac{\theta(\gamma_m - |\gamma|)}{1+\gamma^2}, \tag{2.1.139}$$

where $\gamma_m = \tan(\theta_m/2)$.

A segment of one realization of the surface profile function $\zeta(x_1)$ and its derivative, calculated by the approach presented here, are plotted in Figs. 2.5(a) and 2.5(b), respectively, for the case of a band-limited uniform diffuser, for which $f(\gamma)$ is given by Eq. (2.1.139). The values of the parameters used in generating these functions were $\theta_m = 20°$ and $b = 22$ μm.

In Fig. 2.6 we present plots of $\langle \partial R/\partial \theta_s \rangle$ as a function of θ_s, calculated on the basis of the Kirchhoff approximation, Eqs. (2.1.131)–(2.1.132) for the band-limited uniform diffuser defined by Eqs. (2.1.138) and (2.1.139). The values of the parameters used in obtaining the result plotted in Fig. 2.6(a) were $\theta_m = 20°$, $b = 22$ μm, $\lambda = 632.8$ nm, and $N_s = 2000$. The values of the parameters used in obtaining the results plotted in Fig. 2.6(b) were

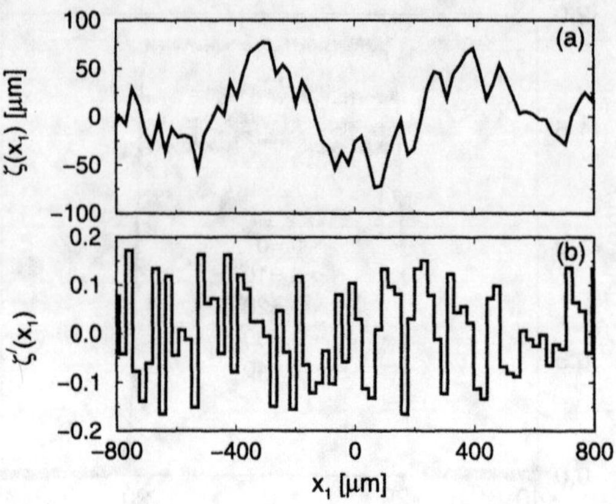

Figure 2.5. (a) A segment of one realization of the surface profile function $\zeta(x_1)$ designed to act as a band-limited uniform diffuser, for which $f(\gamma)$ is given by Eq. (2.1.139). The values of the parameters employed in obtaining this surface are $\theta_m = 20°$ and $b = 22$ μm. (b) The derivative $\zeta'(x_1)$ of this surface profile function.

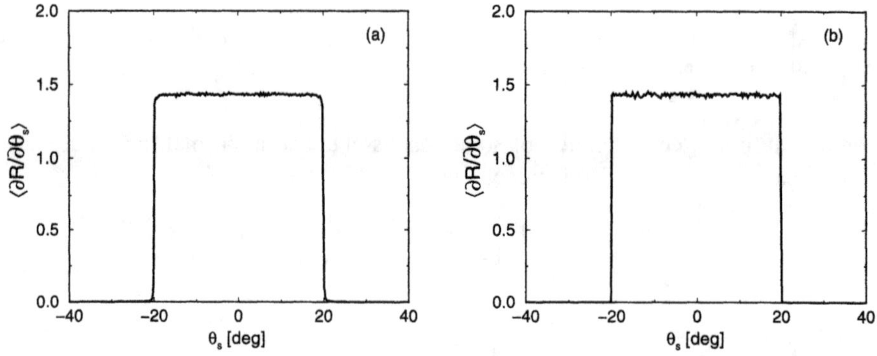

Figure 2.6. The mean differential reflection coefficient $\langle \partial R/\partial\theta_s \rangle$ estimated from $N_p = 100\,000$ realizations of the surface profile function for the case of a band-limited uniform diffuser for which $f(\gamma)$ is given by Eq. (2.1.139). The calculations have been carried out on the basis of the Kirchhoff approximation for scattering from a one-dimensional perfectly conducting surface. The values of the parameters employed are: (a) $\theta_m = 20°$, $b = 22$ μm, $\lambda = 632.8$ nm, and $N_s = 2000$; (b) $\theta_m = 20°$, $b = 220$ μm, $\lambda = 632.8$ nm, and $N_s = 200$.

$\theta_m = 20°$, $b = 220$ μm, $\lambda = 632.8$ nm, and $N = 200$. Thus, although the characteristic length b was significantly larger in the calculations that led to Fig. 2.6(b) than it was in obtaining Fig. 2.6(a), the overall length of the surface, $L_1 = 2N_s b$, was the same in both cases. The results obtained for $N_p = 100\,000$ realizations of the surface profile function were used in calculating the average in Eq. (2.1.132).

From the results presented in Fig. 2.6 we see that the random surface generated from the pdf $f(\gamma)$ given by Eq. (2.1.139) behaves as a band-limited uniform diffuser. We also see that the larger the characteristic length b is, the sharper are the corners of $\langle \partial R/\partial\theta_s \rangle$ and the

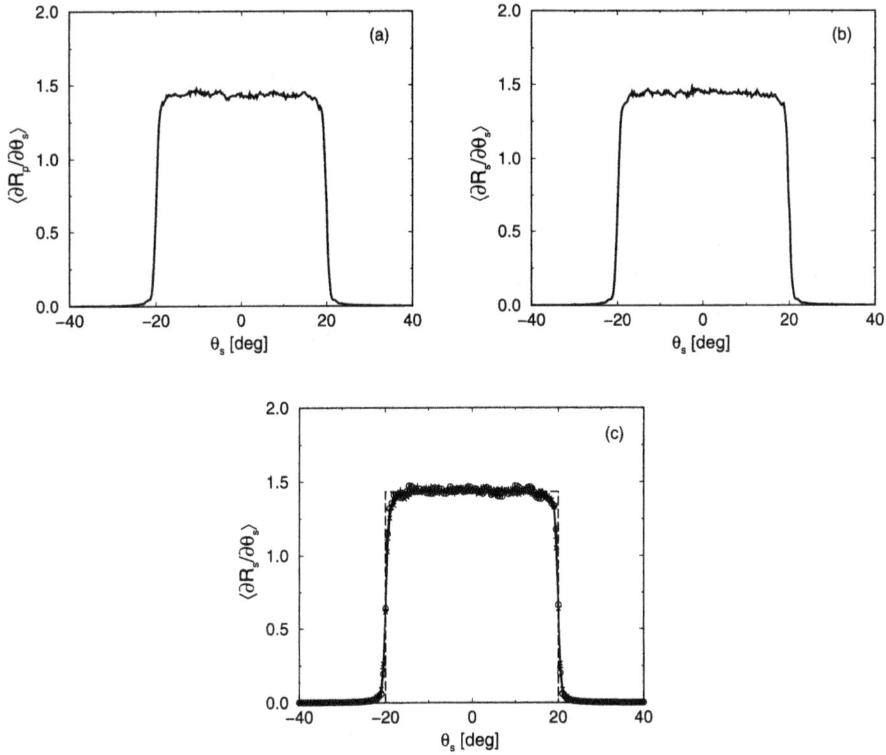

Figure 2.7. The mean differential reflection coefficient $\langle \partial R/\partial\theta_s\rangle$ estimated from $N_p = 20\,000$ realizations of the surface profile function for the case of a band-limited uniform diffuser for which $f(\gamma)$ is given by Eq. (2.1.139). The calculations have been carried out on the basis of a rigorous computer simulation approach to the scattering of p- (a) and s-polarized (b) light from a one-dimensional perfectly conducting surface. The values of the parameters employed are $\theta_m = 20°$, $b = 22$ μm, and $\lambda = 632.8$ nm. (c) The mean differential reflection coefficient $\langle \partial R_s/\partial\theta_s\rangle$ calculated for three different wavelengths of the incident light: $\lambda = 457.9$ nm ($\circ$), $\lambda = 533$ nm ($+$), and $\lambda = 632.8$ nm ($\times$).

more uniform is the scattered intensity in the region $|\theta_s| < 20°$. This appears to be a general result: the larger b is, the closer the calculated $\langle \partial R/\partial\theta_s\rangle$ is to the input mean differential reflection coefficient. Additional calculations, whose results are not shown here, suggest that, for a given value of b, increasing the length L_1 of the surface also brings the calculated $\langle \partial R/\partial\theta_s\rangle$ closer to the input $\langle \partial R/\partial\theta_s\rangle$.

In Fig. 2.7 we present results for the mean differential reflection coefficient $\langle \partial R/\partial\theta_s\rangle$ as a function of θ_s for the scattering of both p- (a) and s-polarized (b) light from a one-dimensional randomly rough perfectly conducting surface designed to act as a band-limited uniform diffuser. These results were obtained on the basis of the rigorous numerical simulation approach described in Section 2.1.3.1, which takes into account multiple-scattering processes of all orders. The values of the parameters assumed in carrying out these calculations were $\theta_0 = 0°, \theta_m = 20°$, $b = 22$ μm, and $\lambda = 632.8$ nm. Results obtained for $N_p = 20\,000$ realizations of the surface profile functions were averaged to obtain the mean differential reflection coefficient. We see that at least for the values of the roughness and experimental parameters assumed in carrying out the computer simulations, there is virtually no difference between the mean differential reflection coefficients calculated for p- and s-polarized incident light, and that both reproduce the input result (2.1.138) for $\langle \partial R/\partial\theta_s \rangle$ quite closely. This indicates that multiple-scattering processes are not significant in scattering from this surface. From a comparison of the results presented in Fig. 2.7 with the result plotted in Fig. 2.6(a), we see that the Kirchhoff approximation provides an accurate solution of the scattering problem for this surface.

In Fig. 2.7(c) we present results for the mean differential reflection coefficients $\langle \partial R_s/\partial\theta_s \rangle$ obtained for scattering from the surface designed to act as the uniform diffuser defined by Eqs. (2.1.138) and (2.1.139), calculated for the same parameters used in obtaining the results presented in Figs. 2.7(a) and 2.7(b) but for three different wavelengths of the incident light: $\lambda = 457.9$ nm (○), $\lambda = 533$ nm (+), and $\lambda = 632.8$ nm (×). The three results plotted in Fig. 2.7(c) are practically indistinguishable. These results were obtained on the basis of the rigorous numerical simulation approach, which takes into account multiple-scattering processes of all orders. It is clearly seen from the plots that the achromatic properties of the designed surfaces are not destroyed by multiple-scattering processes.

Due to limitations on the memory and speed of present day computers, rigorous computational simulation calculations of the scattering from one-dimensional randomly rough surface of the kind considered here and characterized by a length $b = 220$ μm, when the incident light has a wavelength $\lambda = 632.8$ nm, are extremely difficult to carry out with good accuracy: too few segments of length b would be contained within the length of the x_1-axis covered by the random surface that can be used in such simulations. It is for this reason that the results for $\langle \partial R/\partial\theta_s \rangle$ presented in Figs. 2.4 and 2.7 were all calculated for a value of $b = 22$ μm. Even with this value of b only a few segments (5 or 6) could be taken into account.

In concluding this section we note that the surface profile function $\zeta(x_1)$ within the interval $nb \leqslant x_1 \leqslant (n+1)b$ is characterized by only a single slope a_n, unlike the surface profile function assumed in Refs. [2.10–2.13]. The latter was fabricated from equally spaced identical trapezoidal grooves, each of which was characterized by the slopes of its two walls. As a result, the expression for $\langle \partial R/\partial\theta_s \rangle$ in terms of the pdf $f(\gamma)$ of the random amplitudes of these grooves, obtained in the geometrical optics approximation, was the sum of two contributions, one arising from each of the two slopes of the walls of the trapezoidal groove. In the case that it was desired for $\langle \partial R/\partial\theta_s \rangle$ to have the rectangular form defined by Eq. (2.1.138), this produced a scattering distribution that consisted of two adjacent rectangular distributions. Diffraction effects rounded the corners of these two contributions, and led to a small peak in the retroreflection direction due to the overlap of the tails of these two rectangular distributions. An approximately uniform scattering distribution could be obtained by replacing $f(\gamma)$ by $f(\gamma - \epsilon)$, where ϵ is a small constant. In the present case, where the surface profile function has only a single slope within each interval $nb \leqslant x_1 \leqslant (n+1)b$, $n = 0, \pm 1, \pm 2, \ldots$, the expression for $\langle \partial R/\partial \theta_s \rangle$ obtained in the geometrical optics approximation, Eq. (2.1.88) consists of only a single contribution, associated with that single slope, and the results obtained by the present approach do not possess the small peak in the retroreflection direction seen in the results obtained in Refs. [2.10–2.13]. This is an attractive feature of the method developed in this section.

2.1.4.2. A Lambertian Diffuser Optical diffusers that produce a scattered field for which the mean differential reflection coefficient is proportional to the cosine of the polar scattering angle are widely used in the calibration of scatterometers [2.14]. Such diffusers have the property that their radiance or luminance is the same in all scattering directions. Due to this angular dependence such devices are often referred to as *Lambertian diffusers*. In the visible region of the optical spectrum volume disordered media such as compacted powdered barium sulfate or freshly smoked magnesium oxide [2.15] are used as Lambertian diffusers. However, this type of diffuser is inapplicable in the infrared region due to its strong absorption and the presence of a specular component of the scattered radiation. The design of a random surface that acts as a Lambertian diffuser, especially in the infrared region of the optical spectrum, is therefore a desirable goal, and one that has been regarded as difficult to achieve [2.16].

The mean differential reflection coefficient we seek to reproduce in this case has the form, at normal incidence $\theta_0 = 0°$,

$$\left\langle \frac{\partial R}{\partial \theta_s} \right\rangle = A \cos\theta_s, \quad -\pi/2 \leqslant \theta_s \leqslant \pi/2. \tag{2.1.140}$$

On combining Eqs. (2.1.101), (2.1.102), and (2.1.140) we find that the probability density function of a_n is

$$f(\gamma) = 2A \frac{1-\gamma^2}{(1+\gamma^2)^2} \theta(1-|\gamma|). \tag{2.1.141}$$

The normalization condition for $f(\gamma)$ yields the result that $A = 1/2$, so that in this case

$$\left\langle \frac{\partial R}{\partial \theta_s} \right\rangle = \frac{1}{2} \cos\theta_s, \quad -\pi/2 \leqslant \theta_s \leqslant \pi/2, \tag{2.1.142}$$

$$f(\gamma) = \frac{1-\gamma^2}{(1+\gamma^2)^2} \theta(1-|\gamma|). \tag{2.1.143}$$

In Fig. 2.8 we present a plot of a segment of a single realization of the surface profile function and its derivative, calculated by the approach presented in this chapter, for the case of a Lambertian diffuser for which $f(\gamma)$ is given by Eq. (2.1.143). The roughness parameter used in generating these functions was $b = 22$ µm.

The parameters used in calculating the corresponding mean differential reflection coefficients $\langle \partial R/\partial \theta_s \rangle$, plotted in Figs. 2.9(a) and 2.9(b), which were carried out on the basis of the Kirchhoff approximation, were the same as were used in obtaining the results plotted

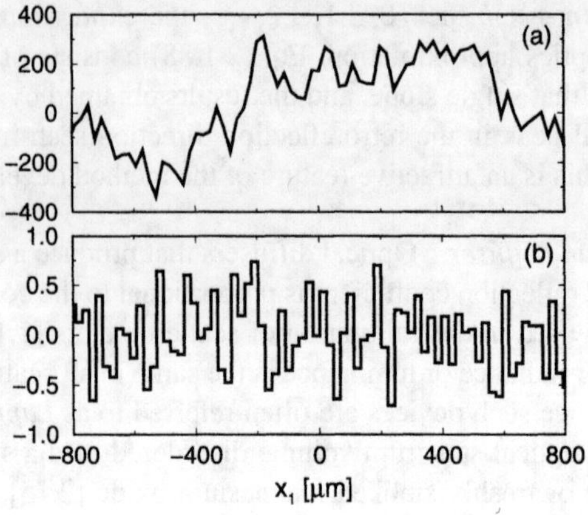

Figure 2.8. (a) A segment of a single realization of the surface profile function $\zeta(x_1)$ designed to act as a Lambertian diffuser, for which $f(\gamma)$ is given by Eq. (2.1.143). The values of the parameter b employed is $b = 22$ μm. (b) The derivative $\zeta'(x_1)$ of this surface profile function.

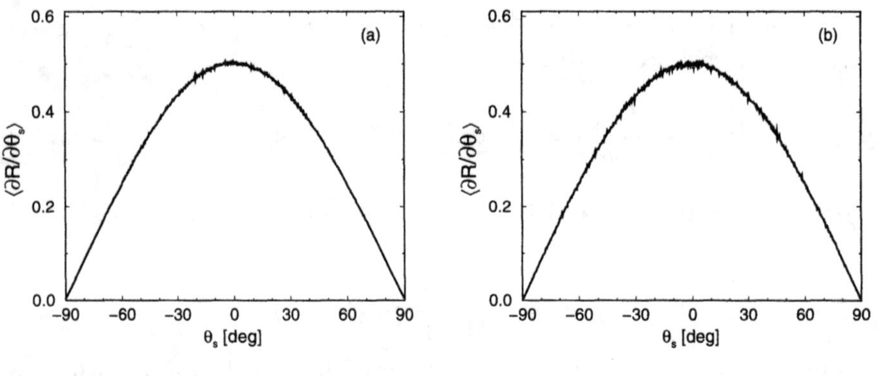

Figure 2.9. The mean differential reflection coefficient $\langle \partial R/\partial \theta_s \rangle$ estimated from $N_p = 100\,000$ realizations of the surface profile function for the case of a Lambertian diffuser for which $f(\gamma)$ is given by Eq. (2.1.143). The calculations have been carried out on the basis of the Kirchhoff approximation for scattering from a one-dimensional perfectly conducting surface. The values of the parameters employed are: (a) $\theta_0 = 0°$, $b = 22$ μm, $\lambda = 632.8$ nm, and $N_s = 2000$, (b) $\theta_0 = 0°$, $b = 200$ μm, $\lambda = 632.8$ nm, and $N_s = 200$. The input distribution given by Eq. (2.1.142) is also plotted for comparison.

in Figs. 2.6(a), and 2.6(b), respectively. Results obtained for $N_p = 100\,000$ realizations of the surface profile function were used in calculating the average in Eq. (2.1.132). The input distribution given by Eq. (2.1.142) is also plotted for comparison. From the results presented in Figs. 2.9(a) and 2.9(b) we see that the calculated mean differential reflection coefficient $\langle \partial R/\partial \theta_s \rangle$ follows the cosine law (2.1.142) very closely. In this case increasing the value of the characteristic length b does not bring the calculated $\langle \partial R/\partial \theta_s \rangle$ significantly closer to the input $\langle \partial R/\partial \theta_s \rangle$, but it makes it a little less noisy.

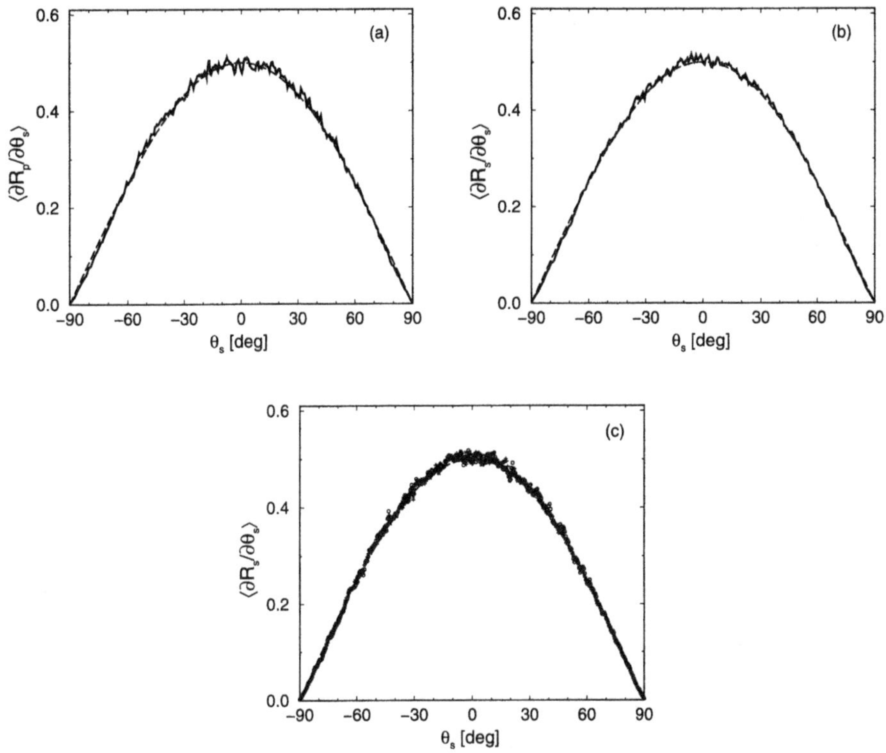

Figure 2.10. The mean differential reflection coefficients $\langle \partial R_p/\partial\theta_s \rangle$ (a) and $\langle \partial R_s/\partial\theta_s \rangle$ (b), estimated from $N_p = 30\,000$ realizations of the surface profile function for the case of a Lambertian diffuser for which $f(\gamma)$ is given by Eq. (2.1.143). The calculations have been carried out on the basis of a rigorous computer simulation approach to the scattering of p- and s-polarized light from a one-dimensional perfectly conducting surface. The values of the parameters employed are $\theta_0 = 0°$, $b = 22$ μm, $\lambda = 632.8$ nm, and $N_s = 6$. The input distribution given by Eq. (2.1.142) is also plotted for comparison. (c) The mean differential reflection coefficient $\langle \partial R_s/\partial\theta_s \rangle$ calculated for three different wavelengths of the incident light: $\lambda = 457.9$ nm ($\circ$), $\lambda = 533$ nm ($+$), and $\lambda = 632.8$ nm ($\times$).

In Figs. 2.10(a) and 2.10(b) we present results for the mean differential reflection coefficients $\langle \partial R_p/\partial\theta_s \rangle$ and $\langle \partial R_s/\partial\theta_s \rangle$ obtained for scattering from the surface designed to act as the Lambertian diffuser defined by Eqs. (2.1.142) and (2.1.143), calculated on the basis of the rigorous computer simulation approach described in Section 2.1.3.1. The values of the parameters assumed in carrying out these calculations were $\theta_0 = 0°$, $b = 22$ μm, $\lambda = 632.8$ nm, $N_s = 6$, and $N_p = 30\,000$. The input distribution given by Eq. (2.1.142) is also plotted for comparison. The results plotted in Figs. 2.10(a) and 2.10(b) follow the cosine law (2.1.142) very closely and, at least for the experimental and roughness parameters assumed, validate the use of the Kirchhoff approximation for the solution of the scattering problem.

In Fig. 2.10(c) we present results for the mean differential reflection coefficients $\langle \partial R_s/\partial \theta_s \rangle$ obtained for scattering from the surface designed to act as the Lambertian diffuser defined by Eqs. (2.1.142) and (2.1.143), calculated for the same parameters used in obtaining the results presented in Figs. 2.10(a) and 2.10(b) but for three different wavelength of the incident light: $\lambda = 457.9$ nm (○), $\lambda = 533$ nm (+), and $\lambda = 632.8$ nm (×). The three results plotted in Fig. 2.10(c) are practically indistinguishable. These results were obtained on the basis of the rigorous numerical simulation approach, which takes into account multiple-scattering processes of all orders. It is clearly seen from the plots that the achromatic properties of the designed surfaces are not destroyed by multiple-scattering processes.

2.1.4.3. A Surface that Suppresses Single-Scattering Processes Such effects as enhanced backscattering, which is the presence of a well-defined peak in the retroreflection direction $\theta_s = -\theta_0$ in the dependence of the mean differential reflection coefficient on the scattering angle, and arises due to the multiple scattering of light from a randomly rough surface, are often superimposed on a large background due to single scattering processes. Such effects can become more readily observable if the background due to single-scattering process can be suppressed in the angular regions in which one wishes to study them, so that only multiple-scattering processes contribute to the mean differential reflection coefficient in these regions.

West and O'Donnell [2.17] have designed and fabricated one-dimensional weakly rough random metal surfaces that enhance the backscattering phenomenon, by coupling the incident electromagnetic field strongly to the surface electromagnetic waves supported by the vacuum–metal interface. At the same time, these surfaces suppress single-scattering processes within a range of scattering angles centered at $\theta_s = 0°$, which makes the enhanced backscattering peak easy to see. These surfaces are defined by surface profile functions that are stationary, zero-mean, Gaussian random processes.

In this section we show how a different kind of one-dimensional random surface on a perfect conductor can be designed that suppresses the single-scattering contribution to the mean differential reflection coefficient within a specified range of scattering angles. Such a surface can be designed on the basis of the following considerations [2.11]. Let us consider the scattering of an s-polarized plane wave of frequency ω from a one-dimensional, randomly rough, perfectly conducting, surface, when the plane of incidence is perpendicular to the generators of the surface, and its angle of incidence is θ_0. The inhomogeneous integral equation for the normal derivative of the single nonzero component of the electric field in the vacuum, evaluated on the surface, is given by Eq. (2.1.41). If we solve this equation by iteration,

$$F(x_1|\omega) = 2F(x_1|\omega)_{\text{inc}} - \frac{1}{2\pi} P \int_{-\infty}^{\infty} dx_1' \left[\partial G_0(x_1, x_3|x_1', x_3')/\partial N\right]\Big|_{\substack{x_3'=\zeta(x_1') \\ x_3=\zeta(x_1)}} 2F(x_1'|\omega)_{\text{inc}} + \cdots, \tag{2.1.144}$$

the inhomogeneous term, as we have seen, yields the Kirchhoff approximation to the mean differential reflection coefficient, a single scattering approximation, the first iterate yields the pure double-scattering contribution, and so on [2.18]. (Strictly speaking, this is rigorously the case only in the high frequency limit, but is a very good approximation in the visible region of the optical spectrum.) Consequently, if a surface can be designed with the property that the Kirchhoff approximation to the mean differential reflection coefficient of the light scattered from it vanishes for the scattering angle θ_s in the interval $(-\theta_m, \theta_m)$, say, all of the intensity of the light scattered into this range of scattering angles will be due to multiple-scattering processes. The aim, therefore, is to design a one-dimensional, perfectly conducting, randomly rough surface for which the Kirchhoff approximation to the mean differential reflection coefficient vanishes identically for θ_s in the interval $(-\theta_m, \theta_m)$. The analysis is greatly simplified by working in the geometrical optics limit of the Kirchhoff approximation. The results obtained in this limit still display the behavior sought.

Thus, let us assume that $\langle \partial R/\partial\theta_s \rangle$ has the form

$$\left\langle \frac{\partial R}{\partial \theta_s} \right\rangle = \begin{cases} 0 & |\theta_s| < \theta_m, \\ A\cos\theta_s & \theta_m < |\theta_s| < \pi/2. \end{cases} \tag{2.1.145}$$

By the use of Eqs. (2.1.92) and (2.1.93) we can rewrite this expression as

$$\begin{aligned} &\left\langle \frac{\partial R}{\partial \theta_s} \right\rangle (\theta_0 + 2\tan^{-1}\gamma, \theta_0) \\ &\quad = \frac{A}{1+\gamma^2}\left[(1-\gamma^2)\cos\theta_0 - 2\gamma\sin\theta_0\right] \\ &\qquad \times \left[\theta\left(-\gamma - \tan\frac{1}{2}(\theta_m + \theta_0)\right)\theta\left(\gamma + \tan\frac{1}{2}\left(\frac{\pi}{2} + \theta_0\right)\right)\right. \\ &\qquad \left. + \theta\left(\gamma - \tan\frac{1}{2}(\theta_m - \theta_0)\right)\theta\left(\tan\frac{1}{2}\left(\frac{\pi}{2} - \theta_0\right) - \gamma\right)\right]. \end{aligned} \tag{2.1.146}$$

It follows from Eq. (2.1.96) that the pdf of a_n is given by

$$\begin{aligned} f(\gamma) = &\frac{2A}{(1+\gamma^2)^2}\frac{\cos\theta_0}{\cos\theta_0 + \gamma\sin\theta_0}\left[(1-\gamma^2)\cos\theta_0 + 2\gamma\sin\theta_0\right] \\ &\times \left[\theta\left(\tan\frac{1}{2}\left(\frac{\pi}{2} + \theta_0\right) - \gamma\right)\theta\left(\gamma - \tan\frac{1}{2}(\theta_m + \theta_0)\right)\right. \\ &\left. + \theta\left(\gamma + \tan\frac{1}{2}\left(\frac{\pi}{2} - \theta_0\right)\right)\theta\left(-\gamma - \tan\frac{1}{2}(\theta_m - \theta_0)\right)\right]. \end{aligned} \tag{2.1.147}$$

The coefficient A is obtained from the normalization condition for $f(\gamma)$.

A segment of a single realization of the surface profile function $\zeta(x_1)$ and its derivative calculated by the approach described in this section, are plotted in Figs. 2.11(a) and 2.11(b),

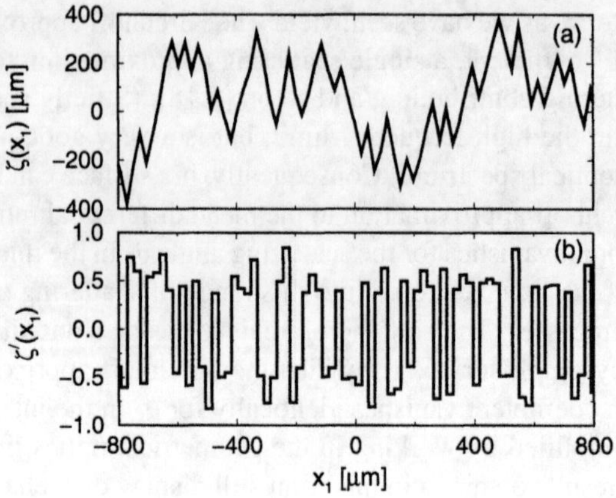

Figure 2.11. A segment of a single realization of the surface profile function $\zeta(x_1)$ designed to suppress the single scattering contribution to the mean differential reflection coefficient within a range of scattering angles $|\theta_s| < 40°$, for which $f(\gamma)$ is given by Eq. (2.1.147). The value of the parameter b employed is $b = 22$ μm. (b) The derivative $\zeta'(x_1)$ of this surface profile function.

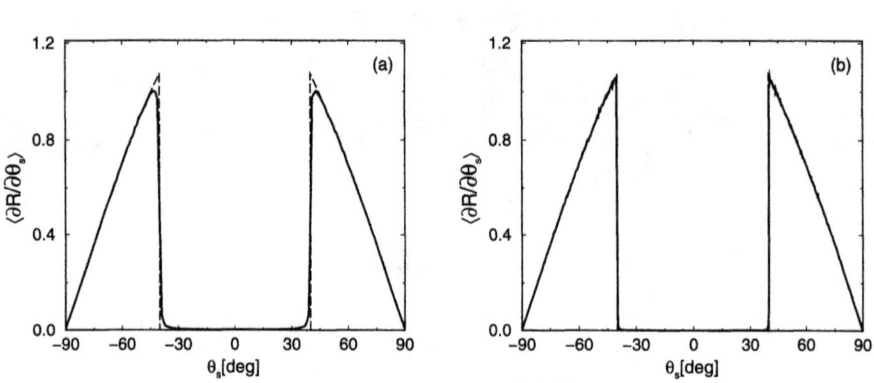

Figure 2.12. The mean differential reflection coefficient $\langle \partial R/\partial\theta_s \rangle$ estimated from $N_p = 100000$ realizations of the surface profile function for which $f(\gamma)$ is given by Eq. (2.1.147). The calculations have been carried out on the basis of the Kirchhoff approximation for scattering from a one-dimensional perfectly conducting surface. The values of the parameters employed are: (a) $\theta_0 = 0°$, $\theta_m = 40°$, $b = 22$ μm, $\lambda = 632.8$ nm, and $N_s = 1000$; (b) $\theta_0 = 0°$, $\theta_m = 40°$, $b = 200$ μm, $\lambda = 632.8$ nm, and $N_s = 100$. The input distribution given by Eq. (2.1.142) is also plotted for comparison.

respectively, for the case of a surface that suppresses single-scattering processes for which $f(\gamma)$ is given by Eq. (2.1.147). The parameters used in generating the surface profile and its derivative were $\theta_0 = 0°$, $\theta_m = 40°$, $b = 22$ μm.

In obtaining the mean differential reflection coefficients calculated on the basis of the Kirchhoff approximation described in Section 3.1.3.1, that are plotted in Figs. 2.12(a) and 2.12(b), the same values of θ_0 and θ_m were assumed, and the results obtained

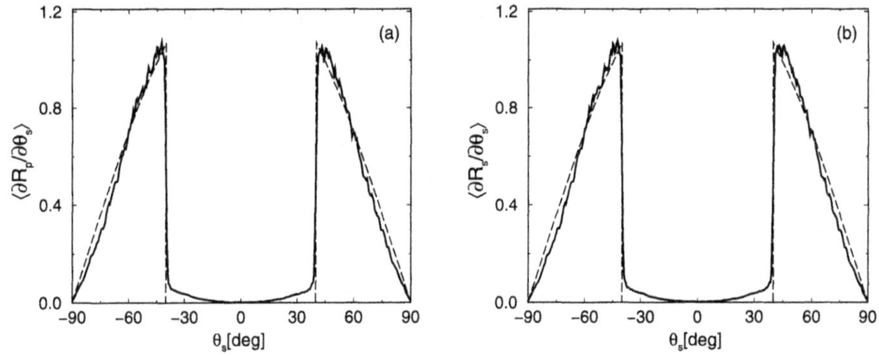

Figure 2.13. The mean differential reflection coefficients $\langle \partial R_p/\partial \theta_s \rangle$ (a) and $\langle \partial R_s/\partial \theta_s \rangle$ (b) estimated from $N_p = 30\,000$ realizations of the surface profile function for which $f(\gamma)$ is given by Eq. (2.1.147). The calculations have been carried out on the basis of the rigorous computer simulation approach for scattering from a one-dimensional perfectly conducting surface. The values of the parameters employed are: (a) $\theta_0 = 0°$, $\theta_m = 40°$, $b = 22$ μm, $\lambda = 632.8$ nm, and $N_s = 6$.

for $N_p = 100\,000$ realizations were used in calculating the average appearing in Eq. (2.1.132). The values $b = 22$ μm and $N_s = 1000$ were used in obtaining the result plotted in Fig. 2.12(a), while the values $b = 220$ μm and $N_s = 100$ were used in obtaining Fig. 2.12(b). It is seen from Figs. 2.12(a) and 2.12(b) that $\langle \partial R/\partial \theta_s \rangle$ vanishes for θ_s in the interval $(-40°, 40°)$ for both values of b assumed. However, the use of the larger value of b sharpens the corners of the mean differential reflection coefficient, and brings the calculated $\langle \partial R/\partial \theta_s \rangle$ into closer agreement with the input function. Since the results plotted in Figs. 2.12(a) and 2.12(b) were calculated on the basis of a single-scattering approximation, the vanishing of $\langle \partial R/\partial \theta_s \rangle$ for $|\theta_s| < 40°$ demonstrates that the surfaces used in obtaining these results indeed suppress single scattering for θ_s in this interval.

The situation is somewhat different when we calculate $\langle \partial R/\partial \theta_s \rangle$ as a function of θ_s by means of the rigorous computer simulation approach described in Section 2.1.3.1. The parameters used in obtaining the result plotted in Fig. 2.13 were $\theta_0 = 0°$, $\theta_m = 40°$, $b = 22$ μm, $\lambda = 632.8$ nm, $N_s = 100$, and $N_p = 30000$. The input distribution (2.1.145) is plotted in this figure as well. The result plotted in Fig. 2.13 shows a significant reduction of the mean differential reflection coefficient, but not a total suppression of it, for $|\theta_s| < 40°$. The result that $\langle \partial R/\partial \theta_s \rangle$ is not identically zero in this angular interval is due to the fact that the rigorous computer simulation approach takes multiple-scattering processes of all orders into account, and it is these processes that produce the small background in $\langle \partial R_s/\partial \theta_s \rangle$ in the interval $|\theta_s| < 40°$. There are no traces of the enhanced backscattering peak in these results because the surface is too "smooth".

2.2. Penetrable Surfaces

Up to now we have presented an account of how a one-dimensional randomly rough surface can be designed to scatter light in a specified manner that has been based on the assumption that the scattering medium is a perfect conductor, in other words an impenetrable medium. The experience gained from the preceding treatment enables extending the method described there to the design of a one-dimensional randomly rough surface bounding a finitely conducting metal or a dielectric that scatters light in a prescribed fashion. We present this extension here to demonstrate the versatility of the original approach.

2.2.1. Scattering Theory

The scattering system we consider now consists of vacuum in the region $x_3 > \zeta(x_1)$ and the scattering medium characterized by a dielectric constant ϵ in the region $x_3 < \zeta(x_1)$. If ϵ is real and positive, the scattering medium is a dielectric; if ϵ is complex with a negative real part and a positive imaginary part, the scattering medium is a lossy reflecting metal.

The surface profile function $\zeta(x_1)$ has the properties described in Section 2.1.1. We assume that the surface $x_3 = \zeta(x_1)$ is illuminated from the vacuum region by a p- or s-polarized plane wave of frequency ω, given by Eq. (2.1.11), whose plane of incidence is the x_1x_3 plane.

To solve the scattering problem for this system we will use the Kirchhoff approximation developed in Section 2.1.1.4, due to its simplicity. In order to apply it we first need to obtain the scattering amplitude $R_\nu(\alpha_0(k))$ when light of p or s polarization is incident on the planar surface $x_3 = 0$ separating the dielectric medium of incidence from the dielectric scattering medium. We can deal with the case of p- and s-polarized incident light simultaneously by working with the function $F_\nu(x_1, x_3|\omega)$ introduced in Section 2.2.1. This function is $H_2(x_1, x_3|\omega)$ when $\nu = p$, and is $E_2(x_1, x_3|\omega)$ when $\nu = s$. The field $F_\nu^>(x_1, x_3|\omega)$ in the vacuum region $x_3 > 0$ satisfies the Helmholtz equation

$$\left(\frac{\partial^2}{\partial x_1^2} + \frac{\partial^2}{\partial x_3^2} + \frac{\omega^2}{c^2}\right) F_\nu^>(x_1, x_3|\omega) = 0. \tag{2.2.1}$$

Its solution that satisfies the boundary conditions at $x_3 = \infty$ can be written as the sum of an incident plane wave and a reflected plane wave,

$$F_\nu^>(x_1, x_3|\omega) = \exp\left[ikx_1 - i\alpha_0(k)x_3\right] + R_\nu\left(\alpha_0(k)\right)\exp\left[ikx_1 + i\alpha_0(k)x_3\right], \tag{2.2.2}$$

where $\alpha_0(k)$ is defined in Eq. (2.1.7). The field in the dielectric region $x_3 < 0$ satisfies the Helmholtz equation

$$\left(\frac{\partial^2}{\partial x_1^2} + \frac{\partial^2}{\partial x_3^2} + \epsilon\frac{\omega^2}{c^2}\right) F_\nu^<(x_1, x_3|\omega) = 0. \tag{2.2.3}$$

Its solution that satisfies the boundary condition at $x_3 = -\infty$ can be written as a transmitted refracted plane wave,

$$F_\nu^<(x_1, x_3|\omega) = T_\nu\left(\alpha_0(k)\right)\exp\left[ikx_1 - i\alpha(k)x_3\right], \tag{2.2.4}$$

where

$$\alpha(k) = \left[\epsilon(\omega/c)^2 - k^2\right]^{\frac{1}{2}}, \quad \mathrm{Re}\,\alpha(k) > 0,\ \mathrm{Im}\,\alpha(k) > 0. \tag{2.2.5}$$

The boundary conditions at the interface $x_3 = 0$ require the continuity of the tangential components of the electric and magnetic fields across it, and can be written as

$$F_\nu^>(x_1, x_3|\omega)\big|_{x_3=0} = F_\nu^<(x_1, x_3|\omega)\big|_{x_3=0}, \tag{2.2.6}$$

$$\frac{\partial}{\partial x_3} F_\nu^>(x_1, x_3|\omega)\bigg|_{x_3=0} = \frac{1}{\kappa_\nu}\frac{\partial}{\partial x_3} F_\nu^<(x_1, x_3|\omega)\bigg|_{x_3=0}, \tag{2.2.7}$$

where $\kappa_p = \epsilon$ and $\kappa_s = 1$. On substituting Eqs. (2.2.2) and (2.2.4) into Eqs. (2.2.6) and (2.2.7) we obtain a pair of coupled equations for the reflection and transmission amplitudes R_ν and T_ν:

$$1 + R_\nu = T_\nu, \tag{2.2.8a}$$

$$-i\alpha_0(k) + i\alpha_0(k)R_\nu = -\frac{i}{\kappa_\nu}\alpha(k)T_\nu, \tag{2.2.8b}$$

whose solutions are

$$R_\nu\big(\alpha_0(k)\big) = \frac{\kappa_\nu\alpha_0(k) - \alpha(k)}{\kappa_\nu\alpha_0(k) + \alpha(k)}, \tag{2.2.9}$$

$$T_\nu\big(\alpha_0(k)\big) = \frac{2\kappa_\nu\alpha_0(k)}{\kappa_\nu\alpha_0(k) + \alpha(k)}. \tag{2.2.10}$$

If we rewrite $\alpha(k)$ as

$$\alpha(k) = \left[(\epsilon - 1)(\omega/c)^2 + \alpha_0^2(k)\right]^{\frac{1}{2}} \tag{2.2.11}$$

and replace $\alpha_0(k)$ here and in Eq. (2.2.9) by $\alpha_\ell(k|x_{1i})$, Eq. (2.1.55b), we find that $R_\nu(\alpha_\ell(k|x_{1i}))$ is given by

$$R_\nu\big(\alpha_\ell(k|x_{1i})\big) = \frac{\kappa_\nu\cos(\frac{\theta_s+\theta_0}{2}) - [\epsilon - \sin^2(\frac{\theta_s+\theta_0}{2})]^{\frac{1}{2}}}{\kappa_\nu\cos(\frac{\theta_s+\theta_0}{2}) + [\epsilon - \sin^2(\frac{\theta_s+\theta_0}{2})]^{\frac{1}{2}}} \tag{2.2.12a}$$

$$\equiv R_\nu\left(\frac{\theta_s + \theta_0}{2}\right). \tag{2.2.12b}$$

The scattering amplitude $R_\nu(q|k)$ obtained in the Kirchhoff approximation for scattering from a one-dimensional rough surface, and given by Eq. (2.1.56), then becomes

$$R_\nu(q|k) = R_\nu\left(\frac{\theta_s + \theta_0}{2}\right)\frac{1}{\cos\theta_s}\frac{\cos\frac{1}{2}(\theta_s + \theta_0)}{\cos\frac{1}{2}(\theta_s - \theta_0)}$$

$$\times \int_{-\infty}^{\infty} dx_1 \exp[-i(q-k)x_1] \exp[-ia\zeta(x_1)], \tag{2.2.13}$$

where $a(\theta_s, \theta_0)$ is defined in Eq. (2.1.61)

A derivation identical to the one carried out in Section 2.1.1.2 yields the result that the mean differential reflection coefficient for incident light of either polarization is given in terms of the scattering amplitude $R_\nu(q|k)$ by

$$\left\langle \frac{\partial R_\nu}{\partial \theta_s} \right\rangle = \frac{1}{L_1}\left(\frac{\omega}{2\pi c}\right)\frac{\cos^2\theta_s}{\cos\theta_0}\left\langle \left|R_\nu(q|k)\right|^2 \right\rangle. \tag{2.2.14}$$

In the present case this becomes

$$\left\langle \frac{\partial R_\nu}{\partial \theta_s} \right\rangle = \frac{1}{L_1}\left(\frac{\omega}{2\pi c}\right)\frac{1}{\cos\theta_0}\left|R_\nu\left(\frac{\theta_s+\theta_0}{2}\right)\right|^2 \frac{\cos^2\frac{1}{2}(\theta_s+\theta_0)}{\cos^2\frac{1}{2}(\theta_s-\theta_0)}$$
$$\times \int_{-\infty}^{\infty} dx_1 \int_{-\infty}^{\infty} dx_1' \exp[-i(q-k)(x_1-x_1')]\left\langle \exp\left[-ia\left(\zeta(x_1)-\zeta(x_1')\right)\right]\right\rangle. \tag{2.2.15}$$

On passing to the geometrical optics limit of this expression, as this was done in Section 2.1.1.4, we obtain

$$\left\langle \frac{\partial R_\nu}{\partial \theta_s} \right\rangle = \frac{1}{L_1}\left(\frac{\omega}{2\pi c}\right)\frac{1}{\cos\theta_0}\left|R_\nu\left(\frac{\theta_s+\theta_0}{2}\right)\right|^2 \frac{\cos^2\frac{1}{2}(\theta_s+\theta_0)}{\cos^2\frac{1}{2}(\theta_s-\theta_0)}$$
$$\times \int_{-\infty}^{\infty} du \exp[i(q-k)u] \int_{-\infty}^{\infty} dx_1 \left\langle \exp[iau\zeta'(x_1)]\right\rangle. \tag{2.2.16}$$

The double integral in Eqs. (2.2.16) is the same one that appears in Eq. (2.1.60), and we evaluate it in the same way this was done in Section 2.1.2.2. With the use of the representation for the surface profile function given by Eqs. (2.1.83) and (2.1.85), and the result given by Eq. (2.1.87), Eq. (2.2.16) becomes

$$\left\langle \frac{\partial R_\nu}{\partial \theta_s} \right\rangle(\theta_s, \theta_0) = \left|R_\nu\left(\frac{\theta_s+\theta_0}{2}\right)\right|^2 \frac{1}{2\cos\theta_0}\frac{\cos\frac{1}{2}(\theta_s+\theta_0)}{\cos^3\frac{1}{2}(\theta_s-\theta_0)} f\left(-\tan\frac{1}{2}(\theta_s-\theta_0)\right). \tag{2.2.17}$$

The change of variable (2.1.91) yields the useful results

$$\cos\frac{1}{2}(\theta_s-\theta_0)=\frac{1}{\sqrt{1+\gamma^2}},\qquad \cos\frac{1}{2}(\theta_s+\theta_0)=\frac{1}{\sqrt{1+\gamma^2}}[\cos\theta_0-\gamma\sin\theta_0], \tag{2.2.18}$$

$$\sin\frac{1}{2}(\theta_s-\theta_0)=\frac{\gamma}{\sqrt{1+\gamma^2}},\qquad \sin\frac{1}{2}(\theta_s+\theta_0)=\frac{1}{\sqrt{1+\gamma^2}}[\sin\theta_0+\gamma\cos\theta_0]. \tag{2.2.19}$$

It follows from these results that Eq. (2.2.17) can be rewritten as

$$\left\langle\frac{\partial R_\nu}{\partial\theta_s}\right\rangle\!\left(\theta_0+2\tan^{-1}\gamma,\theta_0\right)=\left|r_\nu(\gamma|\theta_0)\right|^2\frac{1+\gamma^2}{2}\frac{\cos\theta_0-\gamma\sin\theta_0}{\cos\theta_0}f(-\gamma), \tag{2.2.20}$$

where we have introduced the definition

$$R_\nu\left(\frac{\theta_s+\theta_0}{2}\right)=\frac{\kappa_\nu(\cos\theta-\gamma\sin\theta_0)-[\epsilon(1+\gamma^2)-(\sin\theta_0+\gamma\cos\theta_0)^2]^{\frac{1}{2}}}{\kappa_\nu(\cos\theta_0-\gamma\sin\theta_0)+[\epsilon(1+\gamma^2)-(\sin\theta_0+\gamma\cos\theta_0)^2]^{\frac{1}{2}}} \tag{2.2.21a}$$

$$\equiv r_\nu(\gamma|\theta_0). \tag{2.2.21b}$$

We invert Eq. (2.2.20) and obtain for the pdf of the slope a_n

$$f_\nu(\gamma)=\frac{1}{|r_\nu(-\gamma|\theta_0)|^2}\frac{2}{1+\gamma^2}\frac{\cos\theta_0}{\cos\theta_0+\gamma\sin\theta_0}\left\langle\frac{\partial R_\nu}{\partial\theta_s}\right\rangle\!\left(\theta_0-2\tan^{-1}\gamma,\theta_0\right). \tag{2.2.22}$$

In the case of normal incidence, $\theta_0=0$, Eq. (2.2.22) simplifies to

$$f_\nu(\gamma)=\frac{2}{1+\gamma^2}\left|\frac{\kappa_\nu+\sqrt{\epsilon+(\epsilon-1)\gamma^2}}{\kappa_\nu-\sqrt{\epsilon+(\epsilon-1)\gamma^2}}\right|^2\left\langle\frac{\partial R_\nu}{\partial\theta_s}\right\rangle\!\left(-2\tan^{-1}\gamma,0\right). \tag{2.2.23}$$

2.2.2. *Solution of the Scattering Problem*

We again validate the quality of a surface designed on the basis of the geometrical optics limit of a single-scattering approximation, in the present case the Kirchhoff approximation, by calculating the mean differential reflection coefficient of the light scattered from it by means of a more rigorous theory. In this section we describe how such calculations are carried out on the basis of a rigorous computer simulation approach, and on the basis of the Kirchhoff approximation, but without passing to the geometrical optics limit of the latter theory.

2.2.2.1. Computer Simulations In obtaining a rigorous computer simulation solution to the problem of the scattering of light from a one-dimensional randomly rough interface between vacuum and a penetrable medium, such as a metal or a dielectric, when the plane of incidence is normal to the generators of the surface, one can deal with the cases of p- and s-polarized incident light simultaneously by working with the functions $F_{\nu}^{\gtrless}(x_1, x_3|\omega)$ introduced in Section 2.2.1. They now satisfy Eqs. (2.2.1) and (2.2.3) in the regions $x_3 > \zeta(x_1)$ and $x_3 < \zeta(x_1)$, respectively. We also introduce the Green's functions $G_0(x_1, x_3|x_1', x_3')$ and $G_\epsilon(x_1, x_3|x_1, x_3)$ as the solutions of the equations

$$\left(\frac{\partial^2}{\partial x_1^2} + \frac{\partial^2}{\partial x_3^2} + \frac{\omega^2}{c^2}\right) G_0(x_1, x_3|x_1', x_3') = -4\pi\delta(x_1 - x_1')\delta(x_3 - x_3') \tag{2.2.24}$$

and

$$\left(\frac{\partial^2}{\partial x_1^2} + \frac{\partial^2}{\partial x_3^2} + \epsilon(\omega)\frac{\omega^2}{c^2}\right) G_\epsilon(x_1, x_3|x_1', x_3') = -4\pi\delta(x_1 - x_1')\delta(x_3 - x_3') \tag{2.2.25}$$

respectively, subject to outgoing wave and vanishing boundary conditions at infinity. Explicit expressions for these functions are

$$G_0(x_1, x_3|x_1', x_3') = i\pi H_0^{(1)}\left((\omega/c)\left[(x_1 - x_1')^2 + (x_3 - x_3')^2\right]^{\frac{1}{2}}\right) \tag{2.2.26a}$$

$$= \int_{-\infty}^{\infty} \frac{dq}{2\pi} \frac{2\pi i}{\alpha_0(q)} \exp\left[iq(x_1 - x_1') + i\alpha_0(q)|x_3 - x_3'|\right] \tag{2.2.26b}$$

and

$$G_\epsilon(x_1, x_3|x_1', x_3') = i\pi H_0^{(1)}\left(n_c(\omega/c)\left[(x_1 - x_1')^2 + (x_3 - x_3')^2\right]^{\frac{1}{2}}\right) \tag{2.2.27a}$$

$$= \int_{-\infty}^{\infty} \frac{dq}{2\pi} \frac{2\pi i}{\alpha(q)} \exp\left[iq(x_1 - x_1') + i\alpha(q)|x_3 - x_3'|\right], \tag{2.2.27b}$$

where $H_0^{(1)}(z)$ is a Hankel function of the first kind and zero order, and $n_c(\omega)$ is the complex index of refraction of the scattering medium,

$$n_c(\omega) = \left[\epsilon(\omega)\right]^{\frac{1}{2}}, \quad \mathrm{Re}\, n_c(\omega) > 0,\ \mathrm{Im}\, n_c(\omega) > 0. \tag{2.2.28}$$

We now apply the results given by Eqs. (2.2.1) and (2.2.24) and Green's second integral identity in the plane, Eq. (2.1.8), to the region $x_3 > \zeta(x_1)$ and obtain the following equation for $F_\nu^>(x_1, x_3|\omega)$:

$$\begin{aligned}\theta\big(x_3 - \zeta(x_1)\big)&F_\nu^>(x_1, x_3|\omega)\\ &= F_\nu^>(x_1, x_3|\omega)_{\text{inc}} + \frac{1}{4\pi}\int_S ds'\bigg[\bigg(\frac{\partial}{\partial n'}G_0(x_1, x_3|x_1', x_3')\bigg)F_\nu^>(x_1', x_3'|\omega)\\ &\quad - G_0(x_1, x_3|x_1', x_3')\frac{\partial}{\partial n'}F_\nu^>(x_1', x_3'|\omega)\bigg].\end{aligned} \tag{2.2.29}$$

We next apply the results given by Eqs. (2.2.3) and (2.2.25) and Green's second integral identity in the plane, Eq. (2.1.8), to the region $x_3 < \zeta(x_1)$ and obtain the following equation for $F_\nu^<(x_1, x_3|\omega)$:

$$\begin{aligned}\theta\big(\zeta(x_1) - x_3\big)&F_\nu^<(x_1, x_3|\omega)\\ &= -\frac{1}{4\pi}\int_S ds'\bigg[\bigg(\frac{\partial}{\partial n'}G_\epsilon(x_1, x_3|x_1', x_3')\bigg)F_\nu^<(x_1', x_3'|\omega)\\ &\quad - G_\epsilon(x_1, x_3|x_1', x_3')\frac{\partial}{\partial n'}F_\nu^<(x_1', x_3'|\omega)\bigg].\end{aligned} \tag{2.2.30}$$

Since we have assumed that the surface profile function $\zeta(x_1)$ is a single-valued function of x_1, we can use Eqs. (2.1.4), (2.1.15), (2.1.18), and the boundary conditions (2.2.6)–(2.2.7) to rewrite Eqs. (2.2.29) and (2.2.30) in the forms

$$\begin{aligned}\theta\big(x_3 - \zeta(x_1)\big)&F_\nu^>(x_1, x_3|\omega)\\ &= F_\nu^>(x_1, x_3|\omega)_{\text{inc}} + \frac{1}{4\pi}\int_{-\infty}^{\infty} dx_1'\bigg\{\bigg[\frac{\partial}{\partial N'}G_0(x_1, x_3|x_1', x_3')\bigg]_{x_3'=\zeta(x_1')}H_\nu(x_1'|\omega)\\ &\quad - \big[G_0(x_1, x_3|x_1', x_3')\big]_{x_3'=\zeta(x_1')}L_\nu(x_1'|\omega)\bigg\},\end{aligned} \tag{2.2.31}$$

$$\begin{aligned}\theta\big(\zeta(x_1) - x_3\big)&F_\nu^<(x_1, x_3|\omega)\\ &= -\frac{1}{4\pi}\int_{-\infty}^{\infty} dx_1'\bigg\{\bigg[\frac{\partial}{\partial N'}G_\epsilon(x_1, x_3|x_1', x_3')\bigg]_{x_3'=\zeta(x_1')}H_\nu(x_1'|\omega)\\ &\quad - \kappa_\nu(\omega)\big[G_\epsilon(x_1, x_3|x_1', x_3')\big]_{x_3'=\zeta(x_1')}L_\nu(x_1'|\omega)\bigg\},\end{aligned} \tag{2.2.32}$$

where the source functions $H_\nu(x_1|\omega)$ and $L_\nu(x_1|\omega)$ are defined by

$$H_\nu(x_1|\omega) = F_\nu^>(x_1, x_3|\omega)\big|_{x_3=\zeta(x_1)}, \tag{2.2.33a}$$

$$L_\nu(x_1|\omega) = \frac{\partial}{\partial N} F_\nu^>(x_1, x_3|\omega)\bigg|_{x_3=\zeta(x_1)}. \tag{2.2.33b}$$

The scattered field $F_\nu^>(x_1, x_3\omega)_{\text{sc}}$ is given by the integral term on the right-hand side of Eq. (2.2.31). When we use the representation (2.2.26b) for the Green's function $G_0(x_1, x_3|x_1', x_3')$ in the case that $x_3 > \zeta(x_1)_{\max}$, we can write the scattered field in the form

$$F_\nu^>(x_1, x_3|\omega)_{\text{sc}} = \int_{-\infty}^{\infty} \frac{dq}{2\pi} R_\nu(q|k) \exp\bigl[iqx_1 + i\alpha_0(q)x_3\bigr], \tag{2.2.34}$$

where the scattering amplitude $R_\nu(q|k)$ is given by

$$R_\nu(q|k) = \frac{i}{2\alpha_0(q)} \int_{-\infty}^{\infty} dx_1 \exp\bigl[-iqx_1 - i\alpha_0(q)\zeta(x_1)\bigr]$$
$$\times \bigl\{i\bigl[q\zeta'(x_1) - \alpha_0(q)\bigr]H_\nu(x_1|\omega) - L_\nu(x_1|\omega)\bigr\}. \tag{2.2.35}$$

The dependence of the scattering amplitude on the wave number k arises from the dependence of the source functions $H_\nu(x_1|\omega)$ and $L_\nu(x_1|\omega)$ on the incident field, which we assume is given by Eq. (2.1.11).

Because the scattered field given by Eq. (2.2.34) has exactly the same form as the expressions for the scattered fields given by Eqs. (2.1.22) and (2.1.25), we can conclude that the mean differential reflection coefficient in the present case is still given by Eq. (2.1.37) together with Eq. (2.1.31). It is convenient for what follows to write it in the form

$$\left\langle \frac{\partial R_\nu}{\partial \theta_s} \right\rangle = \frac{1}{L_1} \frac{c}{8\pi\omega} \frac{\langle |r_\nu(\theta_s|\theta_0)|^2 \rangle}{\cos\theta_0}, \tag{2.2.36}$$

where

$$r_\nu(\theta_s|\theta_0) = \int_{-\infty}^{\infty} dx_1 \exp\bigl\{-i(\omega/c)\bigl[x_1 \sin\theta_s + \zeta(x_1)\cos\theta_s\bigr]\bigr\}$$
$$\times \bigl\{i(\omega/c)\bigl[\zeta'(x_1)\sin\theta_s - \cos\theta_s\bigr]H_\nu(x_1|\omega) - L_\nu(x_1|\omega)\bigr\}. \tag{2.2.37}$$

The source functions $H_\nu(x_1|\omega)$ and $L_\nu(x_1|\omega)$ satisfy a pair of coupled inhomogeneous integral equations. These equations can be obtained from Eqs. (2.2.31) and (2.2.32) by evaluating them at $x_3 = \zeta(x_1) + \eta$, where η is a positive infinitesimal that we eventually let go to zero:

$$H_\nu(x_1|\omega) = H_\nu(x_1|\omega)_{\mathrm{inc}} + \int_{-\infty}^{\infty} dx_1' \left[H^{(0)}(x_1|x_1')H_\nu(x_1'|\omega) - L^{(0)}(x_1|x_1')L_\nu(x_1'|\omega)\right], \tag{2.2.38}$$

$$0 = -\int_{-\infty}^{\infty} dx_1' \left[H^{(\epsilon)}(x_1|x_1')H_\nu(x_1'|\omega) - \kappa_\nu(\omega)L^{(\epsilon)}(x_1|x_1')L_\nu(x_1'|\omega)\right], \tag{2.2.39}$$

where

$$H_\nu(x_1|\omega)_{\mathrm{inc}} = F_\nu^>(x_1, x_3|\omega)_{\mathrm{inc}}\Big|_{x_3=\zeta(x_1)}, \tag{2.2.40}$$

and

$$\begin{aligned} H^{(\epsilon)}(x_1|x_1') &= \frac{1}{4\pi}\frac{\partial}{\partial N'}G_\epsilon(x_1, x_3|x_1', x_3')\Bigg|_{\substack{x_3'=\zeta(x_1')| \\ x_3=\zeta(x_1)+\eta}} \\ &= \left(-\frac{i}{4}\right)n_c^2\frac{\omega^2}{c^2}\frac{H_1^{(1)}(n_c(\omega/c)[(x_1-x_1')^2 + \left(\zeta(x_1)-\zeta(x_1')+\eta\right)^2]^{\frac{1}{2}})}{n_c(\omega/c)[(x_1-x_1')^2 + (\zeta(x_1)-\zeta(x_1')+\eta)^2]^{\frac{1}{2}}} \\ &\quad \times \left[(x_1-x_1')\zeta'(x_1') - \left(\zeta(x_1)-\zeta(x_1')+\eta\right)\right], \end{aligned} \tag{2.2.41}$$

$$\begin{aligned} L^{(\epsilon)}(x_1|x_1') &= \frac{1}{4\pi}G_\epsilon(x_1, x_3|x_1', x_3')\Bigg|_{\substack{x_3'=\zeta(x_1') \\ x_3=\zeta(x_1)+\eta}} \\ &= \frac{i}{4}H_0^{(1)}\left(n_c(\omega/c)\left[(x_1-x_1')^2 + \left(\zeta(x_1)-\zeta(x_1')+\eta\right)^2\right]^{\frac{1}{2}}\right). \end{aligned} \tag{2.2.42}$$

The kernels $H^{(0)}(x_1|x_1')$ and $L^{(0)}(x_1|x_1')$ are obtained by setting $n_c = 1$ in $H^{(\epsilon)}(x_1|x_1')$ and $L^{(\epsilon)}(x_1|x_1')$, respectively.

Equations (2.2.38) and (2.2.39) are solved by converting them into matrix equations that are then solved numerically, exactly as this was done in Section 2.1.3.1. The infinite range of integration is replaced by the finite range $(-L/2, L/2)$, which is then subdivided into N equal segments each of length $\Delta x = L/n$, whose midpoints are at the points defined by $x_n = -\frac{L}{2} + (n - \frac{1}{2})\Delta x$, $n = 1, 2, \ldots, N$. On the assumption that $H_\nu(x_1|\omega)$ and $L_\nu(x_1|\omega)$ are slowly varying functions of x_1 in each of these intervals, one obtains the pair of matrix equations

$$H_\nu(x_m|\omega) = H_\nu(x_m|\omega)_{\rm inc} + \sum_{n=1}^{N}\big[H_{mn}^{(0)} H_\nu(x_n|\omega) - L_{mn}^{(0)} L_\nu(x_n|\omega)\big], \quad m = 1, 2, \ldots, N, \tag{2.2.43}$$

$$\sum_{n=1}^{N}\big[H_{mn}^{(\epsilon)} H_\nu(x_n|\omega) - \kappa_\nu(\omega) L_{mn}^{(\epsilon)} L_\nu(x_n|\omega)\big] = 0, \quad m = 1, 2, \ldots, N, \tag{2.2.44}$$

where

$$H_{mn}^{(0,\epsilon)} = \int_{-\frac{1}{2}\Delta x}^{\frac{1}{2}\Delta x} du\, H^{(0,\epsilon)}(x_m|x_n + u), \tag{2.2.45}$$

$$L_{mn}^{(0,\epsilon)} = \int_{-\frac{1}{2}\Delta x}^{\frac{1}{2}\Delta x} du\, L^{(0,\epsilon)}(x_m|x_n + u). \tag{2.2.46}$$

The kernels $H^{(\epsilon)}(x_1|x_1')$ and $L^{(\epsilon)}(x_1|x_1')$ are singular when $x_1 = x_1'$, due to the singularities of the Hankel functions entering them for vanishing arguments. This has to be taken into account in evaluating the matrix elements $H_{mn}^{(\epsilon)}$ and $L_{mn}^{(\epsilon)}$. For the off-diagonal elements of these matrices the integrands in Eqs. (2.2.45) and (2.2.46) never vanish. We can therefore expand them in powers of u and integrate the resulting expressions term-by-term. In this way we find that to leading nonzero order in Δx

$$\begin{aligned} H_{mn}^{(\epsilon)} &= \Delta x H^{(\epsilon)}(x_m|x_n) \\ &= \Delta x\left(-\frac{i}{4}\right) n_c^2 \left(\frac{\omega}{c}\right)^2 \\ &\quad\times \frac{H_1^{(1)}(n_c(\omega/c)[(x_m - x_n)^2 + (\zeta(x_m) - \zeta(x_n))^2]^{\frac{1}{2}})}{n_c(\omega/c)[(x_m - x_n)^2 + (\zeta(x_m) - \zeta(x_n))^2]^{\frac{1}{2}}} \\ &\quad\times \big[(x_m - x_n)\zeta'(x_n) - \big(\zeta(x_m) - \zeta(x_n)\big)\big], \quad m \neq n, \end{aligned} \tag{2.2.47}$$

$$\begin{aligned} L_{mn}^{(\epsilon)} &= \Delta x L^{(\epsilon)}(x_m|x_n) \\ &= \Delta x\left(\frac{i}{4}\right) H_0^{(1)}\big(n_c(\omega/c)\big[(x_m - x_n)^2 + \big(\zeta(x_m) - \zeta(x_n)\big)^2\big]^{\frac{1}{2}}\big), \quad m \neq n. \end{aligned} \tag{2.2.48}$$

The evaluation of the diagonal element $L_{mm}^{(\epsilon)}$ is carried out in the following way. From Eqs. (2.2.42) and (2.2.46) we obtain

$$L_{mm}^{(\epsilon)} = \frac{i}{4} \int_{-\frac{1}{2}\Delta x}^{\frac{1}{2}\Delta x} du \, H_0^{(1)}\left(n_c(\omega/c)\left[u^2 + \left(\zeta(x_m) - \zeta(x_m + u)\right)^2\right]^{\frac{1}{2}}\right)$$

$$= \frac{i}{4} \int_{-\frac{1}{2}\Delta x}^{\frac{1}{2}\Delta x} du \, H_0^{(1)}\left(n_c(\omega/c)\left[\phi_m^2 u^2 + \zeta'(x_m)\zeta''(x_m)u^2 + \cdots\right]^{\frac{1}{2}}\right), \quad (2.2.49)$$

where

$$\phi_m = \left[1 + \left(\zeta'(x_m)\right)^2\right]^{\frac{1}{2}}, \quad (2.2.50)$$

and where we have passed to the limit $\eta \to 0+$ since the integral is convergent in this limit. With the aid of the small argument expansion of the Hankel function $H_0^{(1)}(z)$ [2.17],

$$H_0^{(1)}(z) = \frac{2i}{\pi} \ell n \frac{z}{2} + i\gamma + 1 + O\left(z^2 \ell n z\right), \quad (2.2.51)$$

where $\gamma = 0.5772157$ is Euler's constant, we obtain from Eq. (2.2.49) the result that

$$L_{mm}^{(\epsilon)} = \frac{i}{2} \int_0^{\frac{1}{2}\Delta x} du \left\{ \frac{2i}{\pi} \ell n\left(n_c(\omega/c)(\phi_m/2)u\right) + i\gamma + 1 \right\} + O\left(\Delta x^3 \ell n \Delta x\right)$$

$$= \frac{i}{4} \Delta x \left\{ \frac{2i}{\pi} \ell n\left(\frac{1}{2} \Delta x n_c(\omega/c)(\phi_m/2e) \right) + i\gamma + 1 \right\}$$
$$+ O\left(\Delta x^3 \ell n \Delta x\right). \quad (2.2.52)$$

A second use of Eq. (2.2.51) yields this result in a compact form:

$$L_{mm}^{(\epsilon)} = \Delta x \left(\frac{i}{4} \right) H_0^{(1)}\left(n_c \frac{\omega}{c} \frac{\phi_m \Delta x}{2e} \right). \quad (2.2.53)$$

The result that $L_{mm}^{(\epsilon)}$ is of $O(\Delta x \ell n \Delta x)$ for small Δx, rather than of $O(\Delta x)$, is a reflection of the (integrable) singularity of the kernel $L^{(\epsilon)}(x_1|x_1')$ for coinciding arguments.

From Eqs. (2.2.41) and (2.2.45) we obtain for the diagonal element $H_{mm}^{(\epsilon)}$

$$H_{mm}^{(\epsilon)} = \left(-\frac{i}{4}\right) n_c^2 \left(\frac{\omega}{c}\right)^2$$

$$\times \int_{-\frac{1}{2}\Delta x}^{\frac{1}{2}\Delta x} du \frac{H_1^{(1)}(n_c(\omega/c)[u^2 + (\zeta(x_m) - \zeta(x_m + u) + \eta)^2]^{\frac{1}{2}})}{n_c(\omega/c)[u^2 + (\zeta(x_m) - \zeta(x_m + u) + \eta)^2]^{\frac{1}{2}}}$$

$$\times \left[-u\zeta'(x_m + u) - (\zeta(x_m) - \zeta(x_m + u) + \eta)\right]$$

$$= \frac{i}{4} n_c^2 \left(\frac{\omega}{c}\right)^2 \int_{-\frac{1}{2}\Delta x}^{\frac{1}{2}\Delta x} du \frac{H_1^{(1)}(n_c(\omega/c)\psi_m)}{n_c(\omega/c)\psi_m}$$

$$\times \left[\eta + \frac{1}{2}u^2\zeta''(x_m) + \frac{1}{3}u^3\zeta'''(x_m) + \cdots\right], \qquad (2.2.54)$$

where ψ_m is given by Eq. (2.1.113a). With the aid of the small argument expansion of $H_1^{(1)}(z)/z$, Eq. (2.1.114), the leading nonzero contribution in Δx to $H_{mm}^{(0)}$ is given by

$$H_{mm}^{(\epsilon)} = \lim_{\eta \to 0+} \frac{1}{2\pi} \int_{-\frac{1}{2}\Delta x}^{\frac{1}{2}\Delta x} du \frac{\eta}{\phi_m^2 u^2 - 2\eta\zeta'(x_m)u + \eta^2} + \frac{1}{2\pi} \int_{-\frac{1}{2}\Delta x}^{\frac{1}{2}\Delta x} du \frac{1}{2} \frac{\zeta''(x_m)}{\phi_m^2}, \qquad (2.2.55)$$

because terms omitted either vanish with vanishing η or are of higher order in Δx. On carrying out the integrations we obtain finally

$$H_{mm}^{(\epsilon)} = \frac{1}{2} + \Delta x \frac{\zeta''(x_m)}{4\pi\phi_m^2}. \qquad (2.2.56)$$

The presence of a contribution independent of Δx to this matrix element is a consequence of the singularity of the kernel $H^{(\epsilon)}(x_1|x_1')$ for coinciding arguments.

The expressions for the matrix elements $H_{mn}^{(0)}$ and $L_{mn}^{(0)}$ are obtained by setting $n_c = 1$ in the expressions for $H_{mn}^{(\epsilon)}$ and $L_{mn}^{(\epsilon)}$, respectively.

When these results are used in Eqs. (2.2.43) and (2.2.44), we can rewrite these equations as

$$H_\nu(x_m|\omega) = 2H_\nu(x_m|\omega)_{\text{inc}} + \sum_{n=1}^{N} \left[\mathcal{H}_{mn}^{(0)} H_\nu(x_n|\omega) - \mathcal{L}_{mn}^{(0)} L_\nu(x_n|\omega)\right], \quad m = 1, 2, \ldots, N, \qquad (2.2.57)$$

$$H_\nu(x_m|\omega) + \sum_{n=1}^{N} \left[\mathcal{H}_{mn}^{(\epsilon)} H_\nu(x_n|\omega) - \kappa_\nu(\omega)\mathcal{L}_{mn}^{(\epsilon)} L_\nu(x_n|\omega)\right] = 0, \quad m = 1, 2, \ldots, N, \qquad (2.2.58)$$

where

$$\mathcal{H}_{mn}^{(\epsilon)} = \Delta x\left(-\frac{i}{2}\right)n_c^2\left(\frac{\omega}{c}\right)^2$$

$$\times \frac{H_1^{(1)}\left(n_c(\omega/c)[(x_m-x_n)^2+\left(\zeta(x_m)-\zeta(x_n)\right)^2]^{\frac{1}{2}}\right)}{n_c(\omega/c)[(x_m-x_n)^2+\left(\zeta(m)-\zeta(x_n)\right)^2]^{\frac{1}{2}}}$$

$$\times\left[(x_m-x_n)\zeta'(x_n)-\left(\zeta(x_m)-\zeta(x_n)\right)\right] \quad m\neq n \tag{2.2.59a}$$

$$= \Delta x\frac{\zeta''(x_m)}{2\pi\phi_m^2} \quad m=n, \tag{2.2.59b}$$

$$\mathcal{L}_{mn}^{(\epsilon)} = \Delta x\left(\frac{i}{2}\right)H_0^{(1)}\left(n_c(\omega/c)\left[(x_m-x_n)^2+\left(\zeta(x_m)-\zeta(x_n)\right)^2\right]^{\frac{1}{2}}\right) \quad m\neq n \tag{2.2.60a}$$

$$= \Delta x\left(\frac{i}{2}\right)H_0^{(1)}\left(n_c\frac{\omega}{c}\frac{\phi_m\Delta x}{2e}\right) \quad m=n. \tag{2.2.60b}$$

The expressions for $\mathcal{H}_{mn}^{(0)}$ and $\mathcal{L}_{mn}^{(0)}$ are obtained from Eqs. (2.2.59) and (2.2.60), respectively, by replacing n_c by unity.

When $H_\nu(x_n|\omega)$ and $L_\nu(x_n|\omega)$ have been obtained by solving Eqs. (2.2.57) and (2.2.58), the amplitude $r_\nu(\theta_s|\theta_0)$ that appears in the expression (2.2.36) for the differential reflection coefficient is calculated from Eq. (2.2.37) according to

$$r_\nu(\theta_s|\theta_0) = \Delta x\sum_{n=1}^{N}\exp\left\{-i(\omega/c)\left[x_n\sin\theta_s+\zeta(x_n)\cos\theta_s\right]\right\}$$
$$\times\left\{i(\omega/c)\left[\zeta'(x_n)\sin\theta_s-\cos\theta_s\right]H_\nu(x_n|\omega)-L_\nu(x_n|\omega)\right\}. \tag{2.2.61}$$

2.2.2.2. The Kirchhoff Approximation A simple and rapid solution to the scattering problem is provided by the Kirchhoff approximation, but without passing to the geometrical optics limit of this approximation.

The scattering amplitude $R_\nu(q|k)$ in the Kirchhoff approximation is given by Eq. (2.2.13). If we represent the surface profile function $\zeta(x_1)$ in the form given by Eq. (2.1.83), Eq. (2.2.13) becomes

$$R_\nu(q|k) = R_\nu\left(\frac{\theta_s+\theta_0}{2}\right)\frac{1}{\cos\theta_s}\frac{\cos\frac{1}{2}(\theta_s+\theta_0)}{\cos\frac{1}{2}(\theta_s-\theta_0)}$$
$$\times\sum_{n=-N_s}^{N_s-1}\exp\left[-i(\omega/c)(\cos\theta_s+\cos\theta_0)b_n\right]$$

$$\times \int_{nb}^{(n+1)b} dx_1 \exp\{-i(\omega/c)[\sin\theta_s - \sin\theta_0] + (\cos\theta_s + \cos\theta_0)a_n\}x_1$$

$$= b\,R_\nu\left(\frac{\theta_s+\theta_0}{2}\right)\frac{1}{\cos\theta_s}\frac{\cos\frac{1}{2}(\theta_s+\theta_0)}{\cos\frac{1}{2}(\theta_s-\theta_0)}r(\theta_s|\theta_0), \tag{2.2.62}$$

where $r(\theta_s|\theta_0)$ is defined by Eq. (2.1.131). If, again, we define L_1 as $2N_s b$, the mean differential reflection coefficient, Eq. (2.2.14), becomes

$$\left\langle\frac{\partial R_\nu}{\partial\theta_s}\right\rangle = \frac{b}{2N_s}\left(\frac{\omega}{2\pi c}\right)\frac{1}{\cos\theta_0}\left|R_\nu\left(\frac{\theta_s+\theta_0}{2}\right)\right|^2\left[\frac{\cos\frac{1}{2}(\theta_s+\theta_0)}{\cos\frac{1}{2}(\theta_s-\theta_0)}\right]^2\langle|r(\theta_s|\theta_0)|^2\rangle. \tag{2.2.63}$$

A large number N_p of random surface profile functions is then generated, and for each realization the corresponding function $|r(\theta_s|\theta_0)|^2$ is calculated by the use of Eq. (2.1.131). The arithmetic average of the N_p results for $|r(\theta_s|\theta_0)|^2$ obtained yields the average appearing in (2.2.63).

2.2.3. *Results*

To illustrate the approach developed in Sections 2.2.1 for the design of one-dimensional randomly rough penetrable surfaces that produce a mean differential reflection coefficient with a specified dependence on the scattering angle for a given angle of incidence, as in Section 2.1.4 we will apply them in this section to the design of a surface that a) acts as a band-limited uniform diffuser; or b) acts as a Lambertian diffuser; or c) suppresses single-scattering processes. The surfaces generated will bound both a semi-infinite metal (silver) and a semi-infinite dielectric (photoresist). Each surface will be illuminated by light of p and s polarization whose wavelength is $\lambda = 612.7$ nm. The dielectric function of silver at this wavelength is $\epsilon(\omega) = -17.2 + i0.498$, and the dielectric constant of photoresist will be taken to be $\epsilon = 2.69$. To test the quality of the surfaces generated, the mean differential reflection coefficients $\langle\partial R_p/\partial\theta_s\rangle$ and $\langle\partial R_s/\partial\theta_s\rangle$ will be calculated for the scattering of p- and s-polarized light from them by means of the rigorous numerical simulation method presented in Section 2.2.2.1, and will then be compared with the expressions for these coefficients that served as inputs to the design of these surfaces.

2.2.3.1. A Band-Limited Uniform Diffuser The mean differential reflection coefficient we wish the surface to produce when illuminated at normal incidence by ν-polarized light is

$$\left\langle\frac{\partial R_\nu}{\partial\theta_s}\right\rangle = A_\nu\theta(\theta_s+\theta_m)\theta(\theta_m-\theta_s). \tag{2.2.64}$$

From Eq. (2.2.23) we find that the corresponding pdf is given by

$$f_\nu(\gamma) = \frac{2A_\nu}{1+\gamma^2}\left|\frac{\sqrt{\epsilon+(\epsilon-1)\gamma^2}+\kappa_\nu}{\sqrt{\epsilon+(\epsilon-1)\gamma^2}-\kappa_\nu}\right|^2 \theta\big(\gamma_m - |\gamma|\big), \tag{2.2.65}$$

where $\gamma_m = \tan(\theta_m/2)$. The coefficient A_ν is determined from the normalization of $f_\nu(\gamma)$.

In Fig. 2.14 we present results obtained for the case of incident light of p polarization. In Fig. 2.14(a) we present a segment of a single realization of the surface profile function of a silver surface constructed on the basis of the pdf $f_p(\gamma)$ given by Eq. (2.2.65). The wavelength of the incident light was assumed to be $\lambda = 612.7$ nm. The dielectric function of silver at this wavelength is $\epsilon(\omega) = -17.2 + i0.498$. The angle θ_m was chosen to be $\theta_m = 20°$, and the characteristic length b was $b = 20$ μm. In Fig. 2.14(b) the mean differential reflection coefficient $\langle \partial R_p/\partial\theta_s\rangle$, calculated by the method described in Section 2.2.2.1, is plotted as a function of θ_s for the scattering of p-polarized light incident normally on a silver surface defined by these parameters. Also plotted for comparison, is the result for $\langle \partial R_p/\partial\theta_s\rangle$ given by Eq. (2.2.64), which served as the input to the determination of the surface. Results for $N_p = 10\,000$ realizations of the surface profile function were averaged to obtain the plot in Fig. 2.14(b).

In Fig. 2.14(c) we plot a segment of a single realization of a photoresist surface constructed on the basis of the pdf $f_p(\gamma)$ given by (2.3.65). The wavelength of the incident light was $\lambda = 612.7$ nm, and the dielectric constant of photoresist was $\epsilon = 2.69$. The angle θ_m was chosen to be $\theta_m = 20°$, and the characteristic length b was $b = 20$ μm. The mean differential reflection coefficient $\langle \partial R_p/\partial\theta_s\rangle$ for the scattering of p-polarized light incident normally on a photoresist surface characterized by these parameters, is plotted in Fig. 2.14(d), together with the result given by Eq. (2.2.64). Results for $N_p = 10\,000$ realizations of the surface profile function were averaged to obtain the plot in Fig. 2.14(d).

In Figs. 2.15(a)–2.15(d) we present the corresponding results obtained in the case where the incident light is s polarized. The calculations are based on the pdf $f_s(\gamma)$ given by Eq. (2.2.65), and the experimental and roughness parameters are the same as those used in obtaining Figs. 2.14(a)–2.14(d).

One sees from Figs. 2.14 and 2.15 that there is not much difference between the results for a metal surface illuminated by p- and s-polarized light. However, $\langle \partial R_s/\partial\theta_s\rangle$ for a dielectric surface is slightly larger than is $\langle \partial R_p/\partial\theta_s\rangle$ for the same surface. This is consistent with the result that a dielectric surface reflects s-polarized light more strongly than it reflects p-polarized light [2.17].

The results presented in Figs. 2.14 and 2.15 show that a surface designed by the approach developed in Section 2.2.1 to act as a band-limited uniform diffuser indeed produces a scattered field with the specified angular dependence of its mean differential reflection coefficient. The use of a larger value of the characteristic length b, at the expense of increased computational effort, would sharpen the corners of the scattering patterns displayed in these figures, as it did in the results for perfectly conducting surfaces presented in Section 2.1.4.1. But even without this refinement the calculated $\langle \partial R_{p,s}/\partial\theta_s\rangle$ reproduce those used as inputs to their calculation very well. As follows from the expressions Eqs. (2.2.64)

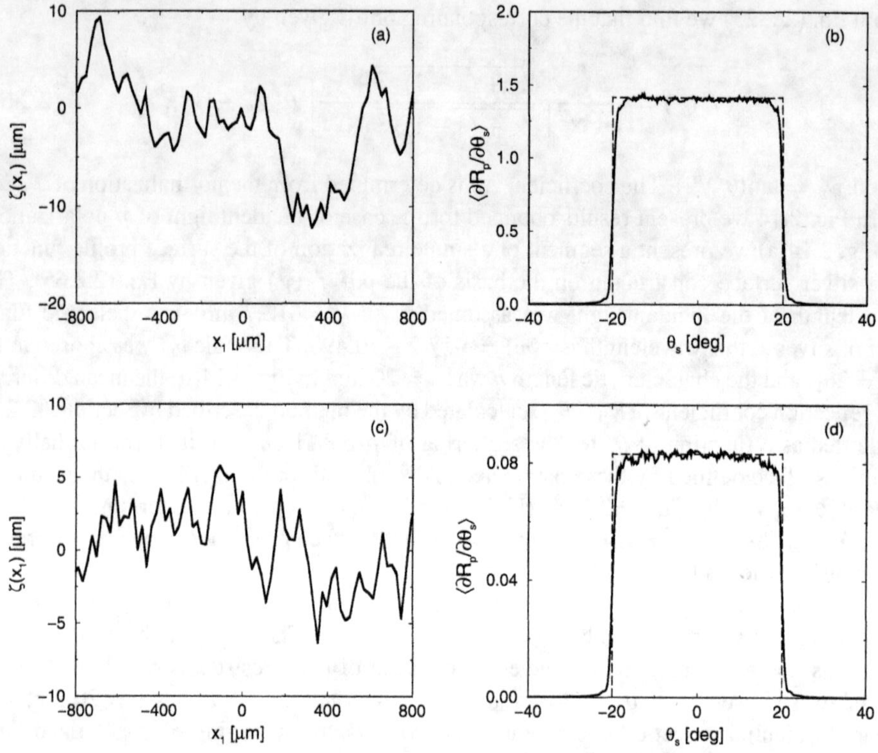

Figure 2.14. (a) A segment of a single realization of the surface profile function $\zeta(x_1)$ of a silver surface designed to act as a band-limited uniform diffuser when illuminated at normal incidence by p-polarized light, for which $f_p(\gamma)$ is given by Eq. (2.2.65). The values of the parameters employed in obtaining this surface are $\lambda = 612.7$ nm, $\epsilon(\omega) = -17.2 + i0.498$, $\theta_0 = 0°$, $\theta_m = 20°$, and $b = 20$ μm; (b) the mean differential reflection coefficient $\langle \partial R_p/\partial\theta_s \rangle$, estimated from $N_p = 10\,000$ realizations of the surface profile function, for the scattering of p-polarized light incident normally on a silver surface defined by these parameters. The calculations have been carried out by the rigorous method described in Section 2.2.2.1. The input distribution given by Eq. (2.2.64) is also plotted for comparison; (c) a segment of a single realization of the surface profile function for a photoresist surface constructed on the basis of the pdf $f_p(\gamma)$ given by Eq. (2.2.65). The value of the parameters assumed in obtaining this surface are $\lambda = 612.7$ nm, $\epsilon = 2.69$, $\theta_0 = 0°$, $\theta_m = 20°$, and $b = 20$ μm; (d) the mean differential reflection coefficient $\langle \partial R_p/\partial\theta_s \rangle$, estimated from $N_p = 10\,000$ realizations of the surface profile function, for the scattering of p-polarized light incident normally on a photoresist surface defined by these parameters. The calculations have been carried out by the rigorous method described in Section 2.2.2.1. The input distribution given by Eq. (2.2.64) is also plotted for comparison.

and (2.2.65) there should be no frequency dependence in the case where the dielectric function of the scattering medium is frequency independent. The frequency dependence of the dielectric function leads only to the change of the constant A, i.e., of the absolute value of the mean differential reflection coefficient. In Figs. 2.16(a) and 2.16(b) we present results obtained for the case of s-polarized incident light of three wavelengths, namely 457.9 nm, 533 nm, and 612.7 nm, scattered from silver (a) and photoresist (b) surfaces designed to

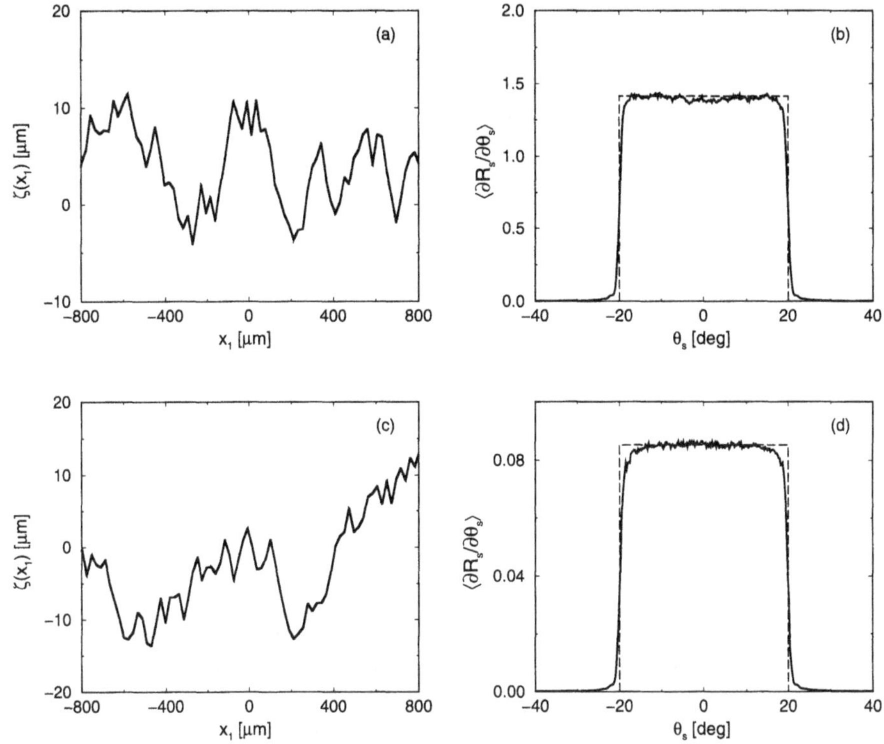

Figure 2.15. The same as Fig. 2.14, except that the incident light is s-polarized.

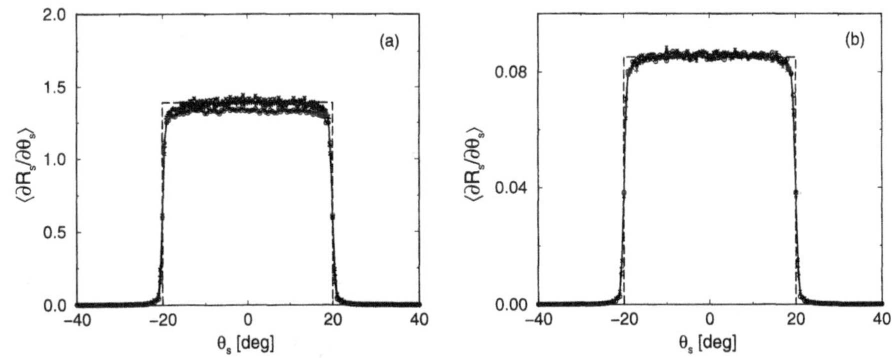

Figure 2.16. The mean differential reflection coefficient $\langle \partial R_s / \partial \theta_s \rangle$, estimated from $N_p = 20\,000$ realizations of the surface profile function, for the scattering of s-polarized light incident normally on silver (a) and photoresist (b) surfaces designed to act as band-limited uniform diffusers at the wavelength $\lambda = 633$ nm. The calculations have been carried out on the basis of a rigorous computer simulation approach to the scattering of light of three different wavelengths of the incident light: $\lambda = 458$ nm ($\circ$), $\lambda = 533$ nm ($+$), and $\lambda = 633$ nm ($\times$). The values of the parameters employed are $\theta_m = 20°$, $b = 22$ μm.

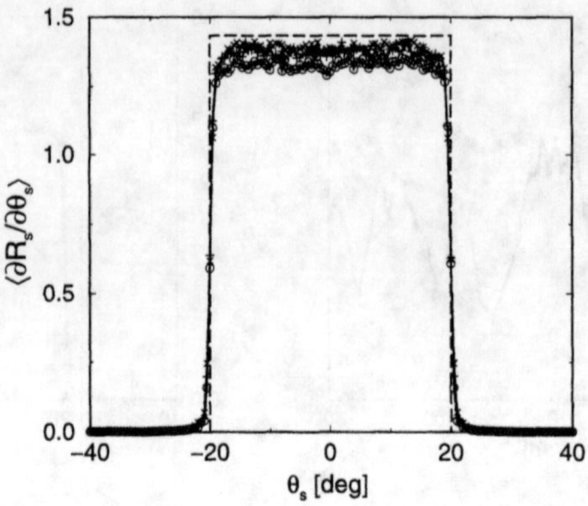

Figure 2.17. The mean differential reflection coefficient $\langle \partial R/\partial\theta_s \rangle$ estimated from $N_p = 20\,000$ realizations of the surface profile function for the case of a band-limited uniform diffuser for which $f(\gamma)$ is given by Eq. (2.1.139), but the calculations have been carried out on the basis of a rigorous computer simulation approach to the scattering of s-polarized light of three different wavelengths of the incident light: $\lambda = 458$ nm ($\circ$), $\lambda = 533$ nm ($+$), and $\lambda = 633$ nm ($\times$), from a one-dimensional silver surface. The values of the parameters employed are $\theta_m = 20^\circ$, $b = 22$ μm.

act as band-limited uniform diffusers at the wavelength $\lambda = 633$ nm. In the calculations we assumed that the dielectric function of the metal is frequency-dependent, while that of the photoresist is not. Indeed, only the absolute value of the mean differential reflection coefficient in Figs. 2.16 depends on the wavelength of the incident light.

The results presented in Figs. 2.14 and 2.15 were obtained for metal and dielectric surfaces obtained from pdfs $f(\gamma)$ (2.2.23) that explicitly take into account the penetrability of these surfaces, i.e., that depend on the dielectric constant of the scattering medium, and on the polarization of the incident light. In Section 2.1.4.1 we have obtained the pdf $f(\gamma)$ for generating a perfectly conducting surface that act as a band-limited uniform diffuser when illuminated by p- or s-polarized light. This pdf (2.1.139) is independent of the frequency and the polarization of the incident light. We conclude this section by examining the mean differential reflection coefficient that is produced when a surface designed on the assumption that the scattering medium is a perfect conductor is ruled on a finitely conducting metal surface.

In Fig. 2.17 we present results for the mean differential reflection coefficient $\langle \partial R/\partial\theta_s \rangle$ estimated from $N_p = 20\,000$ realizations of the surface profile function designed to act as a band-limited uniform diffuser in the scattering of s-polarized light from a perfectly conducting surface for which $f(\gamma)$ is given by Eq. (2.1.139). However, the calculations are carried out, on the basis of the rigorous computer simulation approach developed in Section 2.2.2.1, for the scattering of s-polarized light of three different wavelengths of the incident light from the surface profile ruled on a silver substrate.

It is seen that the fact that the scattering medium is now silver rather than the perfect conductor for which the surface was designed does not affect the region of scattering angles within which the mean differential reflection coefficient is required to be nonzero and constant. This is determined only by the value of γ_m in Eq. (2.1.139), not by the composition of the substrate. There is a weak dependence of the amplitude of the mean differential reflection coefficient on the wavelength of the incident light, which is due to the frequency dependence of the dielectric function of silver, which enters the rigorous calculations that produced Fig. 2.17. To obtain the correct dependence of the amplitude on the wavelength of the incident light one must use a random surface constructed on the basis of a pdf $f_\nu(\gamma)$ that includes the dielectric function of the scattering medium, its frequency dependence, and its polarization. Nevertheless, we see that a surface designed for a perfect conductor produces nearly the same scattering pattern when it is fabricated on a metallic substrate. Moreover, the most important features of band-limited uniform diffusers, namely the allowed band of the scattering angles and the constant amplitude within this band, are independent of the optical properties of the substrates.

2.2.3.2. A Lambertian Diffuser For a surface that is designed to act as a Lambertian diffuser when illuminated at normal incidence by ν-polarized light the mean differential reflection coefficient we wish to reproduce is

$$\left\langle \frac{\partial R_\nu}{\partial \theta_s} \right\rangle = A_\nu \cos\theta_s, \quad -\frac{\pi}{2} \leqslant \theta_s \leqslant \frac{\pi}{2}. \tag{2.2.66}$$

From Eq. (2.2.23) the corresponding pdf $f_\nu(\gamma)$ is given by

$$f_\nu(\gamma) = 2A_\nu \frac{1-\gamma^2}{(1+\gamma^2)^2} \left| \frac{\kappa_\nu + \sqrt{\epsilon + (\epsilon - 1)\gamma^2}}{\kappa_\nu - \sqrt{\epsilon + (\epsilon - 1)\gamma^2}} \right|^2 \theta\big(1 - |\gamma|\big). \tag{2.2.67}$$

The coefficient A_ν is obtained from the normalization of $f_\nu(\gamma)$.

In Fig. 2.18 we present results obtained for the case that the incident light is p polarized. A segment of a single realization of the surface profile function of a silver surface constructed on the basis of the pdf $f_p(\gamma)$ given by Eq. (2.2.67) is presented in Fig. 2.18(a). The characteristic length b was assumed to be $b = 20$ µm.

In Fig. 2.18(b) we present a plot of $\langle \partial R_p / \partial \theta_s \rangle$ calculated by the rigorous computer simulation approach described in Section 2.2.2.1, and the input result for this function given by Eq. (2.2.66). Results for $N_p = 30000$ realizations of the surface profile function were averaged to obtain the plot in this figure. It is seen that the computed result for $\langle \partial R_p / \partial \theta_s \rangle$ follows the cosine law given by Eq. (2.2.66) very closely within the noise level for all scattering angles θ_s. This is somewhat surprising, since one might have expected the geometrical optics approximation to break down for the largest scattering angles. That this is not observed in the results of the computer simulation calculations may be an indication that multiple-scattering processes are of minor importance in the scattering from the randomly rough surface, even at the largest scattering angles. The large number of realizations of the surface profile function used in obtaining this result was needed in order to reduce the noise level significantly. The reason for the slow convergence of $\langle \partial R_p / \partial \theta_s \rangle$ with increasing N_p is due to the large value of b used in the simulations, which meant that for the length L_1

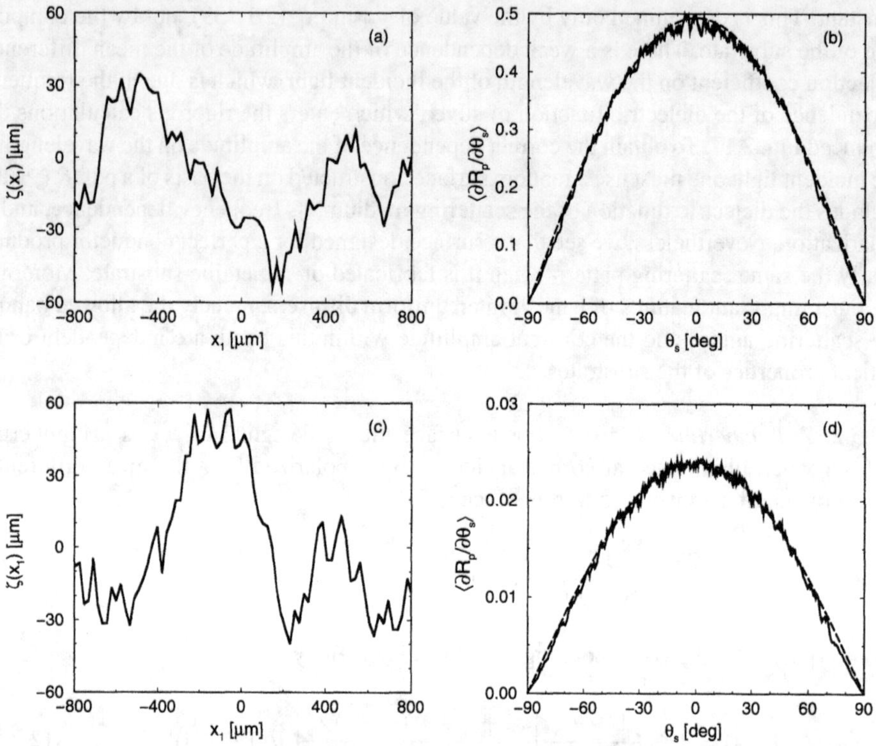

Figure 2.18. (a) A segment of a single realization of the surface profile function $\zeta(x_1)$ of a silver surface designed to act as a Lambertian diffuser when illuminated at normal incidence by p-polarized light for which $f_p(\gamma)$ is given by Eq. (2.2.67). The values of the parameters employed in obtaining this surface are $\lambda = 612.7$ nm, $\epsilon(\omega) = -17.2 + i0.498$, $\theta_0 = 0°$, and $b = 22$ μm; (b) the mean differential reflection coefficient $\langle \partial R_p/\partial\theta_s \rangle$, estimated from $N_p = 30\,000$ realizations of the surface profile function, for the scattering of p-polarized light incident normally on a silver surface defined by these parameters. The calculations have been carried out by the rigorous method described in Section 2.2.2.1. The input distribution given by Eq. (2.2.66) is also plotted for comparison; (c) a segment of a single realization of the surface profile function for a photoresist surface constructed on the basis of the pdf $f_p(\gamma)$ given by Eq. (2.2.67). The values of the parameters assumed in obtaining this surface are $\lambda = 612.7$ nm, $\epsilon = 2.69$, $\theta_0 = 0°$, and $b = 22$ μm. (d) the mean differential reflection coefficient $\langle \partial R_p/\partial\theta_s \rangle$, estimated from 30 000 realizations of the surface profile function, for the scattering of p-polarized light incident normally on a photoresist surface defined by these parameters. The calculations have been carried out by the rigorous method described in Section 2.2.2.1. The input distribution given by Eq. (2.2.66) is also plotted for comparison.

of the x_1 axis covered by the randomly rough surface assumed in these calculations only a few segments of length b could be included in each realization of the surface defined by Eq. (2.1.83). In Fig. 2.18(c) we plot a segment of a single realization of a photoresist surface constructed on the basis of the pdf $f_p(\gamma)$ given by Eq. (2.2.67). The wavelength of the incident light was $\lambda = 612.7$ nm, and $\epsilon = 2.69$. The characteristic length b was

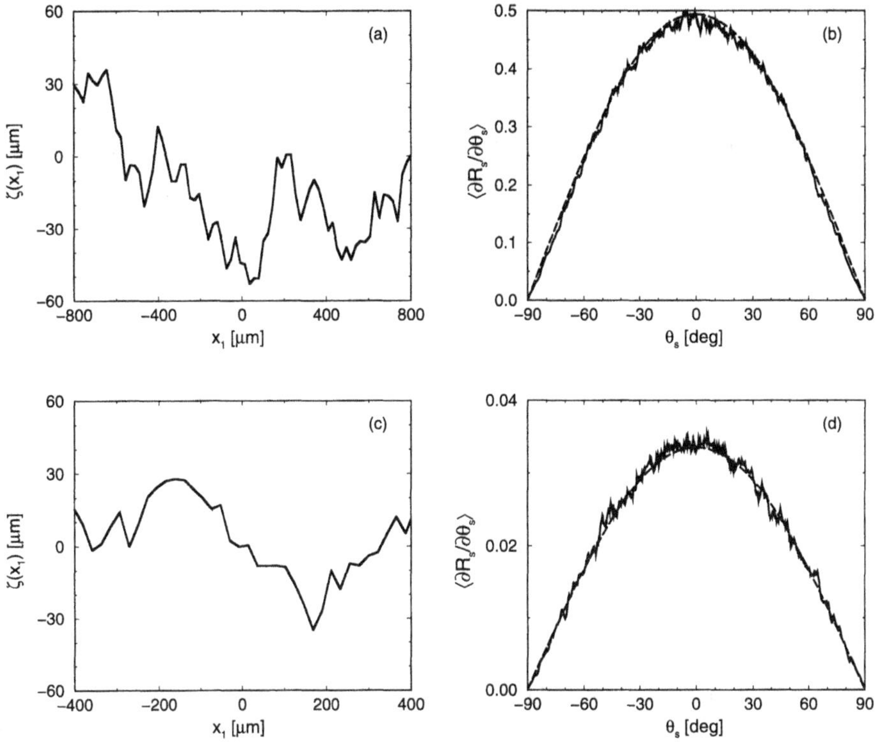

Figure 2.19. The same as Fig. 2.18, except that the incident light is s-polarized.

assumed to be $b = 20$ μm. The mean differential reflection coefficient $\langle \partial R_p/\partial\theta_s \rangle$ for the scattering of p-polarized light incident normally on a photoresist surface characterized by these parameters is presented in Fig. 2.18(d), together with the input function given by Eq. (2.2.66).

In Figs. 2.19(a)–2.19(d) we present the corresponding results in the case that the incident light is s-polarized. The computed results for $\langle \partial R_s/\partial\theta_s \rangle$ follow the cosine law (2.2.66) very closely for all scattering angles, as is the case when p-polarized light is used to illuminate the surfaces.

Again, there is not much difference between the results for the silver surface illuminated by p- and s-polarized light but, not unexpectedly, $\langle \partial R_s/\partial\theta_s \rangle$ for a dielectric surface is again larger than $\langle \partial R_p/\partial\theta_s \rangle$ for the same surface.

2.2.3.3. A Surface that Suppresses Single-Scattering Processes The mean differential reflection coefficient for a surface that suppresses single-scattering processes within a specified interval of scattering angles will be assumed to have the form given by Eq. (2.1.145), namely

$$\left\langle \frac{\partial R_\nu}{\partial \theta_s} \right\rangle = \begin{cases} 0 & |\theta_s| < \theta_m, \\ A_\nu \cos\theta_s & \theta_m < |\theta_s| < \pi/2. \end{cases} \tag{2.2.68}$$

If we assume that the angle of incidence is θ_0, then from Eqs. (2.1.146) and (2.2.22) we find that the corresponding pdf $f_\nu(\gamma)$ is given by

$$f_\nu(\gamma) = \frac{1}{|r_\nu(-\gamma|\theta_0)|^2} \frac{2A_\nu}{(1+\gamma^2)^2} \frac{\cos\theta_0}{\cos\theta_0 + \gamma\sin\theta_0} \left[(1-\gamma^2)\cos\theta_0 + 2\gamma\sin\theta_0\right]$$
$$\times \left[\theta\left(\gamma - \tan\frac{1}{2}(\theta_m + \theta_0)\right)\theta\left(\tan\frac{1}{2}\left(\frac{\pi}{2} + \theta_0\right) - \gamma\right) \right.$$
$$\left. + \theta\left(-\gamma - \tan\frac{1}{2}(\theta_m - \theta_0)\right)\theta\left(\tan\frac{1}{2}\left(\frac{1}{2} - \theta_0\right) + \gamma\right) \right]. \tag{2.2.69}$$

The coefficient A_ν is then determined from the normalization of $f_\nu(\gamma)$.

We have noted earlier that one of the main reasons for an interest in a random surface that suppresses single-scattering processes within some range of scattering angles is that it makes more readily visible multiple-scattering effects, such as enhanced backscattering, when they occur within this range of scattering angles. Enhanced backscattering is the presence of a well-defined peak in the retroreflection direction $\theta_s = -\theta_0$ in the angular dependence of the mean differential reflection coefficient. It arises in the scattering of light from a randomly rough surface due to the coherent interference of each multiply-scattered wave with its reciprocal partner, which strikes the surface at the same points as the former wave, but in the reverse order [2.21]. In order for this effect to be observable, the mean distance between consecutive hills and valleys on the surface should be of the order of, but somewhat larger than, the wavelength of the incident light [2.22,2.23]. For a sufficiently rough metal surface it occurs for both p- and s-polarized incident light [2.23]. In scattering from a sufficiently rough dielectric surface it occurs for s-polarized incident light, but not for p-polarized light unless the dielectric constant of the scattering medium is greater than approximately 3 [2.21,2.23].

To demonstrate how a surface that suppresses single-scattering processes within a certain range of scattering angles can enhance the visibility of the enhanced backscattering peak we consider the scattering of p- and s-polarized light from a one-dimensional randomly rough silver surface generated on the basis of pdf $f_s(\gamma)$ given by Eq. (2.2.69).

In Fig. 2.20(a) we present a plot of the mean differential reflection coefficient $\langle \partial R_p/\partial\theta_s \rangle$ as a function of θ_s, calculated by the rigorous computer simulation approach described in Section 2.2.2.1. The parameters employed in obtaining this figure were $\theta_0 = 2°$, $\theta_m = 40°$, $b = 1$ μm, $\lambda = 479.5$ nm, $\epsilon(\omega) = -7.5 + i0.24$, $N_s = 100$, and $N_p = 20000$. A strong suppression of $\langle \partial R_p/\partial\theta_s \rangle$ in the interval $|\theta_s| < 50°$ is clearly seen, although due to the small value of b assumed this interval is not as sharply defined as it is in Figs. 2.12(a) and 2.12(b). An enhanced backscattering peak at $\theta_s = -2°$ is also clearly seen. When the angle of incidence is increased to $\theta_0 = 5°$, with the remaining experimental and roughness

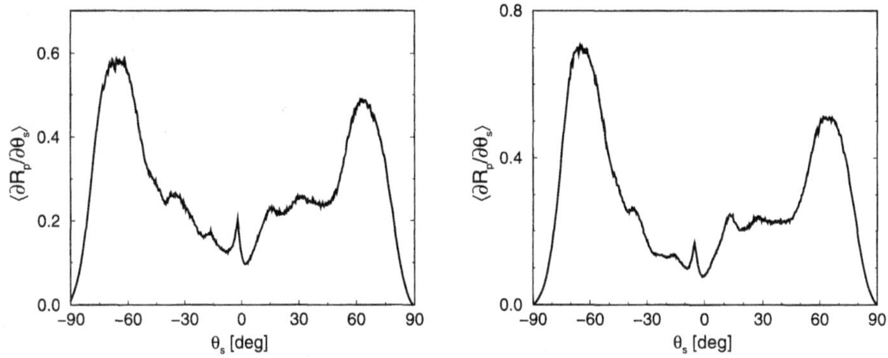

Figure 2.20. The mean differential reflection coefficient $\langle \partial R_p / \partial \theta_s \rangle$ as a function of θ_s, calculated by the rigorous computer simulation approach. The parameters employed in obtaining this result were $\theta_m = 50°$, $b = 1$ μm, $\lambda = 479.5$ nm, $\epsilon(\omega) = -7.5 + i0.24$, $N_s = 100$. (a) $\theta_0 = 2°$; (b) $\theta_0 = 5°$.

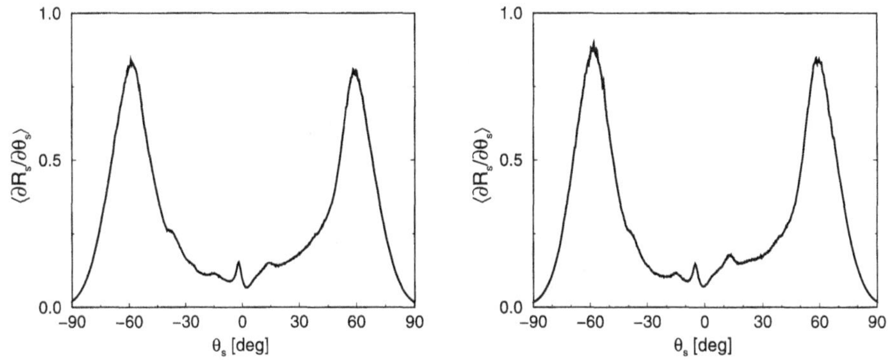

Figure 2.21. The same as Fig. 2.20, except that the incident light is s-polarized.

parameters kept unchanged, the peak in $\langle \partial R_p / \partial \theta_s \rangle$ shifts to $\theta_s = -5°$, as is shown in Fig. 2.20(b). This is what is expected of an enhanced backscattering peak.

The corresponding results for the mean differential reflection coefficient $\langle \partial R_s / \partial \theta_s \rangle$ are presented in Figs. 2.21(a) and 2.21(b), and are seen to be qualitatively and quantitatively similar to those obtained for p-polarized incident light and presented in Figs. 2.20(a) and 2.20(b), respectively.

2.3. A Random Surface that Suppresses Leakage

Not all surfaces designed to produce a specified angular dependence of the mean intensity of the scattered field are designed with the assumption that the field incident on them is a volume electromagnetic wave. In this section we consider the design of a one-dimensional randomly rough surface of finite extent, on an otherwise planar metal surface that scatters a surface electromagnetic wave—a surface plasmon polariton—incident normally on it with a specified angular dependence of the mean intensity of the field scattered into the vacuum region above the surface. Surfaces of this type can be useful in theoretical and experimental studies of the strong localization of surface plasmon polaritons by random surface roughness.

2.3.1. Surface Plasmon Polaritons

A surface plasmon polariton is an electromagnetic mode that, in its simplest form, propagates in a wavelike fashion along the planar interface between vacuum and a metal, with an amplitude that decays exponentially rapidly with increasing distance from the interface into both the vacuum and the metal [2.18]. We assume that the metal is characterized by an isotropic, frequency-dependent, complex dielectric function $\epsilon(\omega) = \epsilon_1(\omega) + i\epsilon_2(\omega)$. In this case we can assume with no loss of generality that the surface plasmon polariton propagates in the x_1 direction. Its sagittal plane, namely the plane defined by its direction of propagation and the normal to the surface, is the x_1x_3 plane. We assume initially that the surface plasmon polariton is either p-polarized or s-polarized with respect to this plane. Thus, as in Section 2.2.1, we again introduce the function $F_\nu(x_1, x_3|\omega)$, which is $H_2(x_1, x_3|\omega)$ when $\nu = p$, and is $E_2(x_1, x_3|\omega)$ when $\nu = s$. In the present case this function satisfies the Helmholtz equation

$$\left(\frac{\partial^2}{\partial x_1^2} + \frac{\partial^2}{\partial x_3^2} + \frac{\omega^2}{c^2}\right) F_\nu^>(x_1, x_3|\omega) = 0 \tag{2.3.1}$$

in the vacuum region $x_3 > 0$, and the equation

$$\left(\frac{\partial^2}{\partial x_1^2} + \frac{\partial^2}{\partial x_3^2} + \epsilon(\omega)\frac{\omega^2}{c^2}\right) F_\nu^<(x_1, x_3|\omega) = 0 \tag{2.3.2}$$

in the metal $x_3 < 0$. It satisfies the boundary conditions

$$F_\nu^>(x_1, x_3|\omega)\big|_{x_3=0} = F_\nu^<(x_1, x_3|\omega)\big|_{x_3=0}, \tag{2.3.3}$$

$$\frac{\partial}{\partial x_3} F_\nu^>(x_1, x_3|\omega)\bigg|_{x_3=0} = \frac{1}{\kappa_\nu}\frac{\partial}{\partial x_3} F_\nu^<(x_1, x_3|\omega)\bigg|_{x_3=0} \tag{2.3.4}$$

at the interface $x_3 = 0$, where $\kappa_p = \epsilon(\omega)$ and $\kappa_s = 1$. In addition, $F_\nu^>(x_1, x_3|\omega)$ $(F_\nu^<(x_1, x_3|\omega))$ tends to zero as $x_3 \to \infty$ $(x_3 \to -\infty)$.

If we assume that $F_\nu^\gtrless(x_1, x_3|\omega)$ have a wave-like dependence on x_1 of the form $\exp(ikx_1)$, the solution of Eq. (2.3.1) that satisfies the boundary condition at $x_3 = \infty$ is

$$F_\nu^>(x_1, x_3|\omega) = A^> \exp\bigl[ikx_1 - \beta_0(k, \omega)x_3\bigr], \tag{2.3.5}$$

where

$$\beta_0(k, \omega) = \left[k^2 - (\omega/c)^2\right]^{\frac{1}{2}}, \quad \operatorname{Re}\beta_0(k, \omega) > 0, \ \operatorname{Im}\beta_0(k, \omega) < 0. \tag{2.3.6}$$

Similarly, the solution of Eq. (2.3.2) that satisfies the boundary condition at $x_3 = -\infty$ is

$$F_\nu^<(x_1, x_3|\omega) = A^< \exp\left[ikx_1 + \beta(k, \omega)x_3\right], \tag{2.3.7}$$

where

$$\beta(k, \omega) = \left[k^2 - \epsilon(\omega)(\omega/c)^2\right]^{\frac{1}{2}}, \quad \operatorname{Re}\beta(k, \omega) > 0, \ \operatorname{Im}\beta(k, \omega) < 0. \tag{2.3.8}$$

The boundary conditions (2.3.3) and (2.3.4) yield the following pair of equations for the amplitudes $A^>$ and $A^<$:

$$A^> = A^<, \tag{2.3.9}$$

$$-\beta_0(k, \omega)A^> = \frac{1}{\kappa_\nu}\beta(k, \omega)A^<. \tag{2.3.10}$$

The solvability condition for this system of homogeneous equations yields the dispersion relation for these surface electromagnetic waves,

$$\frac{\beta(k, \omega)}{\beta_0(k, \omega)} = -\kappa_\nu. \tag{2.3.11}$$

In analyzing this dispersion relation it is convenient initially to simplify it by neglecting the imaginary part of the dielectric function, $\epsilon_2(\omega)$. In this case we see from Eqs. (2.3.5) and (2.3.7) that $\beta_0(k, \omega)$ and $\beta(k, \omega)$ must be real and positive in order that the boundary conditions at infinity be satisfied. In turn, it follows from Eq. (2.3.11) that κ_ν has to be negative. Since $\kappa_s = 1$ for all frequencies ω, we are led to the conclusion that surface plasmon polaritons of s polarization cannot exist at a planar vacuum–metal interface. The restoration of the imaginary part of the dielectric function $\epsilon(\omega)$ does not change this conclusion.

In the case of p polarization Eq. (2.3.11) becomes

$$\frac{\beta(k, \omega)}{\beta_0(k, \omega)} = -\epsilon(\omega). \tag{2.3.12}$$

Therefore, surface plasmon polaritons exist only in the frequency range(s) in which $\epsilon(\omega)$ is negative (for a real $\epsilon(\omega)$). In this frequency range $\beta(k, \omega)$, Eq. (2.3.8) is always positive. In order that $\beta_0(k, \omega)$ be positive, we see from Eq. (2.3.6) that we must have $k > \omega/c$.

If we square both sides of Eq. (2.3.12) and rearrange the terms in the resulting equation, we can write the dispersion relation for surface plasmon polaritons in the form

$$k(\omega) = \frac{\omega}{c}\left[\frac{\epsilon(\omega)}{\epsilon(\omega) + 1}\right]^{\frac{1}{2}} = k_1(\omega) + ik_2(\omega), \tag{2.3.13}$$

which is valid for a complex dielectric function as well as for a real one. To lowest order in $\epsilon_2(\omega)$, assumed to be small compared to unity, we find that

$$k_1(\omega) = \frac{\omega}{c}\left(\frac{|\epsilon_1(\omega)|}{|\epsilon_1(\omega)| - 1}\right)^{\frac{1}{2}} > \frac{\omega}{c}, \tag{2.3.14}$$

$$k_2(\omega) = \frac{1}{2}\frac{\omega}{c}\left(\frac{|\epsilon_1(\omega)|}{|\epsilon_1(\omega)| - 1}\right)^{\frac{1}{2}} \frac{\epsilon_2(\omega)}{|\epsilon_1(\omega)|(|\epsilon_1(\omega)| - 1)} > 0. \tag{2.3.15}$$

The single nonzero component of the magnetic field of a (p-polarized) surface plasmon polariton of frequency ω propagating along a planar vacuum–metal interface can therefore be written as

$$H_2^{>}(x_1, x_3|\omega) = \exp\bigl[ik(\omega)x_1 - \beta_0(\omega)x_3\bigr] \tag{2.3.16}$$

for $x_3 > 0$, and

$$H_2^{<}(x_1, x_3|\omega) = \exp\bigl[ik(\omega)x_1 + \beta(\omega)x_3\bigr], \tag{2.3.17}$$

where

$$\beta_0(\omega) = \beta_0\bigl(k(\omega), \omega\bigr) = \frac{\omega}{c}\left[\frac{-1}{\epsilon(\omega) + 1}\right]^{\frac{1}{2}}, \tag{2.3.18}$$

$$\beta(\omega) = \beta\bigl(k(\omega), \omega\bigr) = -\epsilon(\omega)\frac{\omega}{c}\left[\frac{-1}{\epsilon(\omega) + 1}\right]^{\frac{1}{2}}. \tag{2.3.19}$$

2.3.2. *Leakage*

A surface plasmon polariton can propagate on a weakly rough one-dimensional randomly rough vacuum–metal interface [2.19–2.21], not only on a planar interface. As it propagates across a one-dimensional randomly rough vacuum–metal interface it continuously loses energy not only through ohmic losses in the metal, but also through its roughness-induced conversion into volume electromagnetic waves in the vacuum above the random interface, that propagate away from the interface. This *leakage*, as it is called, interferes with the determination of the strong or Anderson localization length of the surface plasmon polariton by means of computer simulation calculations, or experimental measurements of its transmissivity as a function of the length of the random segment of the surface [2.22–2.27]. The use of a randomly rough surface that suppresses leakage in such studies therefore facilitates the investigation of the strong localization of surface plasmon polarizations by random surface roughness. In this section we describe a way of generating a one-dimensional randomly rough surface that suppresses leakage.

2.3.3. The Incident Field

The physical system we consider consists of vacuum in the region $x_3 > \zeta(x_1)$, and a metal, characterized by an isotropic, frequency-dependent, complex dielectric function $\epsilon(\omega) = \epsilon_1(\omega) + i\epsilon_2(\omega)$ in the region $x_3 < \zeta(x_1)$. We are interested in the frequency range in which $|\epsilon_1(\omega)| < -1$, within which, according to Eq. (2.3.14), surface plasmon polaritons exist. The surface profile function $\zeta(x_1)$ is assumed to be a single-valued function of x_1 that is nonzero only in the interval $-L/2 < x_1 < L/2$.

We write the surface profile function $\zeta(x_1)$ in the form

$$\zeta(x_1) = \Gamma(x_1)s(x_1), \tag{2.3.20}$$

where $s(x_1)$ is a single-valued function of x_1 that is differentiable and constitutes a stationary, zero-mean, Gaussian random process defined by

$$\langle s(x_1)s(x_1')\rangle = \delta^2 W(|x_1 - x_1'|), \tag{2.3.21}$$

where the angle brackets denote an average over the ensemble of realizations of $s(x_1)$, and $\delta = \langle \zeta^2(x_1)\rangle^{\frac{1}{2}}$ is the rms height of the surface defined by $s(x_1)$.

In what follows we will also need the power spectrum of the surface roughness, $g(|k|)$, that is defined by

$$W(|x_1|) = \int_{-\infty}^{\infty} \frac{dk}{2\pi} g(|k|) \exp(ikx_1), \tag{2.3.22}$$

so that

$$g(|k|) = \int_{-\infty}^{\infty} dx_1\, W(|x_1|) \exp(-ikx_1). \tag{2.3.23}$$

The function $\Gamma(x_1)$ in Eq. (2.3.20) serves to restrict the nonzero values of $\zeta(x_1)$ to the interval $-L/2 < x_1 < L/2$. One form that $\Gamma(x_1)$ can have is

$$\Gamma(x_1) = \theta\left(\frac{L}{2} + x_1\right)\theta\left(\frac{L}{2} - x_1\right). \tag{2.3.24}$$

A smoother differentiable form for $\Gamma(x_1)$ is provided by

$$\Gamma(x_1) = \frac{1 + \cosh(\kappa L/2)}{\cosh \kappa x_1 + \cosh(\kappa L/2)}, \tag{2.3.25}$$

where the parameter κ controls the range of x_1 values within which $\Gamma(x_1)$ decreases from unity to zero. In view of the factor $\Gamma(x_1)$ in Eq. (2.3.20) the surface profile function $\zeta(x_1)$ is not a stationary random process, even through $s(x_1)$ is.

We assume that a p-polarized surface plasmon polariton, whose sagittal plane is the x_1x_3 plane is incident on the rough segment of the surface from the planar region of the surface where $x_1 < -L/2$. In defining the incident field when the semi-infinite metal supporting it is lossy we have to take into account that it is attenuated as it propagates in the $+x_1$ direction. Indeed, the energy mean free path of a surface plasmon polariton at a planar vacuum–metal interface due to ohmic losses in the metal, $\ell_\epsilon(\omega)$, is

$$\ell_\epsilon(\omega) = 1/\big(2k_2(\omega)\big), \tag{2.3.26}$$

where $k_2(\omega)$ is defined by Eq. (2.3.15). But this result also means that the amplitude of this wave grows exponentially as $x_1 \to -\infty$. This causes some of the integrals that arise in the scattering theory to diverge. In order to avoid these unphysical divergences we have to create a source in the region $x_1 < -L/2$ which launches surface plasmon polaritons propagating in both the $+x_1$ direction and the $-x_1$ direction, and therefore decaying in both directions. We can do this by writing the single nonzero component of the magnetic field in the vacuum region $x_3 > \zeta(x_1)_{\max}$ in the form

$$\begin{aligned} H_2^>(x_1, x_3|\omega) = A\big\{&\theta(x_1 + L_0)\exp\big[ik(\omega)(x_1 + L_0) - \beta_0(\omega)x_3\big] \\ &+ \theta(-x_1 - L_0)\exp\big[-ik(\omega)(x_1 + L_0) - \beta_0(\omega)x_3\big]\big\} \\ &+ \int_{-\infty}^{\infty} \frac{dq}{2\pi} R^>(q, \omega)\exp\big[iqx_1 - \beta_0(q, \omega)x_3\big], \end{aligned} \tag{2.3.27}$$

and in the region of the metal $x_3 < \zeta(x_1)_{\min}$ in the form

$$\begin{aligned} H_2^<(x_1, x_3|\omega) = A\big\{&\theta(x_1 + L_0)\exp\big[ik(\omega)(x_1 + L_0) + \beta(\omega)x_3\big] \\ &+ \theta(-x_1 - L_0)\exp\big[-ik(\omega)(x_1 + L_0) + \beta(\omega)x_3\big]\big\} \\ &+ \int_{-\infty}^{\infty} \frac{dq}{2\pi} R^<(q, \omega)\exp\big[iqx_1 + \beta_0(q, \omega)x_3\big], \end{aligned} \tag{2.3.28}$$

where $L_0 > L/2$. These expressions satisfy the equations

$$\begin{aligned} &\left(\frac{\partial^2}{\partial x_1^2} + \frac{\partial^2}{\partial x_3^2} + \frac{\omega^2}{c^2}\right) H_2^>(x_1, x_3|\omega) \\ &\quad = 2iAk(\omega)\delta(x_1 + L_0)\exp\big[-\beta_0(\omega)x_3\big], \quad x_3 > \zeta(x_1)_{\max}, \end{aligned} \tag{2.3.29}$$

$$\begin{aligned} &\left(\frac{\partial^2}{\partial x_1^2} + \frac{\partial^2}{\partial x_3^2} + \epsilon(\omega)\frac{\omega^2}{c^2}\right) H_2^<(x_1, x_3|\omega) \\ &\quad = 2iAk(\omega)\delta(x_1 + L_0)\exp\big[\beta(\omega)x_3\big], \quad x_3 < \zeta(x_1)_{\min}. \end{aligned} \tag{2.3.30}$$

Thus, we have introduced a planar source perpendicular to the x_1 axis at $x_1 = -L_0$, whose strength decreases exponentially with increasing distance from the interface into the vacuum and the metal.

2.3.4. The Scattered Field

In what follows we assume that the surface roughness is sufficiently weak that the surface profile function $\zeta(x_1)$ satisfies the conditions for the validity of the Rayleigh hypothesis [2.28]. This is the assumption that the expressions for $H_2^{\gtrless}(x_2, x_3|\omega)$ given by Eqs. (2.3.27) and (2.3.28) which, strictly speaking, are valid only outside the selvedge region $\zeta(x_1)_{\min} < \zeta(x_1)_{\max}$, can be used in satisfying the boundary conditions (2.3.3) and (2.3.4) at the interface $x_3 = \zeta(x_1)$. The limits of validity of this assumption are not known for a randomly rough surface, but it is generally assumed that they are the same for a periodically corrugated surface, namely that $|\zeta'(x_1)| \ll 1$ [2.29–2.32].

When we substitute Eqs. (2.3.27) and (2.3.28) into the boundary conditions (2.3.3) and (2.3.4), we obtain a pair of coupled inhomogeneous integral equations for the scattering amplitudes $R^{>}(q, \omega)$ and $R^{<}(q, \omega)$, namely

$$
\begin{aligned}
&\int_{-\infty}^{\infty} \frac{dq}{2\pi} R^{>}(q,\omega) \exp\bigl[iqx_1 - \beta_0(q,\omega)\zeta(x_1)\bigr] \\
&\quad - \int_{-\infty}^{\infty} \frac{dq}{2\pi} R^{<}(q,\omega) \exp\bigl[iqx_1 + \beta(q,\omega)\zeta(x_1)\bigr] \\
&\quad = A\bigl\{\theta(x_1 + L_0)\exp\bigl[ik(\omega)(x_1 + L_0) + \beta(\omega)\zeta(x_1)\bigr] \\
&\qquad - \theta(x_1 + L_0)\exp\bigl[ik(\omega)(x_1 + L_0) - \beta_0(\omega)\zeta(x_1)\bigr] \\
&\qquad + \theta(-x_1 - L_0)\exp\bigl[-ik(\omega)(x_1 + L_0) + \beta(\omega)\zeta(x_1)\bigr] \\
&\qquad - \theta(-x_1 - L_0)\exp\bigl[-ik(\omega)(x_1 + L_0) - \beta_0(\omega)\zeta(x_1)\bigr]\bigr\}
\end{aligned}
\tag{2.3.31}
$$

and

$$
\begin{aligned}
&\int_{-\infty}^{\infty} \frac{dq}{2\pi} R^{>}(q,\omega)\bigl[-\zeta'(x_1)iq - \beta_0(q,\omega)\bigr]\exp\bigl[iqx_1 - \beta_0(q,\omega)\zeta(x_1)\bigr] \\
&\quad - \frac{1}{\epsilon(\omega)} \int_{-\infty}^{\infty} \frac{dq}{2\pi} R^{<}(q,\omega)\bigl[-\zeta'(x_1)iq + \beta(q,\omega)\bigr]\exp\bigl[-iqx_1 + \beta(q,\omega)\zeta(x_1)\bigr] \\
&= A\biggl\{\frac{1}{\epsilon(\omega)}\theta(x_1 + L_0)\bigl[-\zeta'(x_1)ik(\omega) + \beta(\omega)\bigr]\exp\bigl[ik(\omega)(x_1 + L_0) + \beta(\omega)\zeta(x_1)\bigr]
\end{aligned}
$$

$$+\theta(x_1+L_0)\big[\zeta'(x_1)ik(\omega)+\beta_0(\omega)\big]\exp\big[ik(\omega)(x_1+L_0)-\beta_0(\omega)\zeta(x_1)\big]$$
$$+\frac{1}{\epsilon(\omega)}\theta(-x_1-L_0)\big[\zeta'(x_1)ik(\omega)+\beta(\omega)\big]\exp\big[-ik(\omega)(x_1+L_0)+\beta(\omega)\zeta(x_1)\big]$$
$$+\theta(-x_1-L_0)\big[-\zeta'(x_1)ik(\omega)+\beta_0(\omega)\big]\exp\big[-ik(\omega)(x_1+L_0)-\beta_0\zeta(x_1)\big]\Big\}. \tag{2.3.32}$$

Since we are interested in the suppression of the light scattered into the vacuum region, it is the scattering amplitude $R^>(q,\omega)$ that is of interest to us. We can eliminate the scattering amplitude $R^<(q,\omega)$ from this pair of equations to obtain a single integral equation for $R^>(q,\omega)$ alone. We do this by multiplying Eq. (2.3.31) by $[i\zeta'(x_1)p+\beta(p,\omega)]\exp[-ipx_1+\beta(p,\omega)\zeta(x_1)]$, and Eq. (2.3.32) by $-\epsilon(\omega)\exp[-ipx_1+\beta(p,\omega)\zeta(x_1)]$; integrating the resulting equations over x_1; and adding them. In this way we obtain the reduced Rayleigh equation satisfied by $R^>(q,\omega)$:

$$\big(1-\epsilon(\omega)\big)\int_{-\infty}^{\infty}\frac{dq}{2\pi}\frac{\hat{I}(\beta(p,\omega)-\beta_0(q,\omega)|p-q)}{\beta(p,\omega)-\beta_0(q,\omega)}\big[pq-\beta(p,\omega)\beta_0(q,\omega)\big]R^>(q,\omega)$$
$$=-2iAk(\omega)\int_{-\infty}^{\infty}\frac{dq}{2\pi}\exp\big[-i(q-p)L_0\big]\frac{\hat{I}(\beta(p,\omega)+\beta(\omega)|q)}{\beta(p,\omega)+\beta(\omega)}$$
$$-2iAk(\omega)\int_{-\infty}^{\infty}\frac{dq}{2\pi}\frac{\exp[-i(q-p)L_0]}{(q-p)^2-k^2(\omega)}\frac{\hat{I}(\beta(p,\omega)-\beta_0(\omega)|q)}{\beta(p,\omega)-\beta_0(\omega)}$$
$$\times\big\{q\big[p-\epsilon(\omega)(q-p)\big]+\epsilon(\omega)k^2(\omega)-p^2+\big(1-\epsilon(\omega)\big)\beta(p,\omega)\big\}, \tag{2.3.33}$$

where

$$\hat{I}(\gamma|Q)=\int_{-\infty}^{\infty}dx_1\,\exp(-iQx_1)\exp\big(\gamma\zeta(x_1)\big). \tag{2.3.34}$$

A somewhat more convenient form for Eq. (2.3.33) is obtained if we remove the delta function from $\hat{I}(\gamma|Q)$ by writing

$$\hat{I}(\gamma|Q)=2\pi\delta(Q)+\gamma\hat{J}(\gamma|Q), \tag{2.3.35}$$

where

$$\hat{J}(\gamma|Q)=\int_{-\infty}^{\infty}dx_1\,\exp(-iQx_1)\frac{\exp(\gamma\zeta(x_1))-1}{\gamma}. \tag{2.3.36}$$

After a good deal of algebra we obtain

$$
\begin{aligned}
&R^{>}(p,\omega) \\
&\quad = \frac{\epsilon(\omega)-1}{\epsilon(\omega)} G_0(p,\omega) \int_{-\infty}^{\infty} \frac{dq}{2\pi} \hat{J}\big(\beta(p,\omega)-\beta_0(q,\omega)|p-q\big) \\
&\quad \times \big[pq-\beta(p,\omega)\beta_0(q,\omega)\big] R^{>}(q,\omega) - \frac{2iAk(\omega)}{\epsilon(\omega)} G_0(p,\omega) \\
&\quad \times \int_{-\infty}^{\infty} \frac{ds}{2\pi} \exp(isL_0) \Big\{ \hat{J}\big(\beta(p,\omega)+\beta(\omega)|p-s\big) - \epsilon(\omega)\hat{J}\big(\beta(p,\omega)-\beta_0(\omega)|p-s\big) \\
&\quad - \big(1-\epsilon(\omega)\big)\hat{J}\big(\beta(p,\omega)-\beta_0(\omega)|p-s\big) \frac{ps-\beta(p,\omega)\beta_0(\omega)}{s^2-k^2} \Big\},
\end{aligned} \tag{2.3.37}
$$

where

$$
G_0(p,\omega) = \frac{\epsilon(\omega)}{\epsilon(\omega)\beta_0(p,\omega)+\beta(p,\omega)} \tag{2.3.38}
$$

is the Green's function for a surface plasmon polariton at a planar vacuum–metal interface.

If we introduce the transition matrix $T(p|k(\omega))$ by the relation

$$
R^{>}(p,\omega) = G_0(p,\omega) T\big(p|k(\omega)\big), \tag{2.3.39}
$$

then Eq. (2.3.37) takes the form

$$
T\big(p|k(\omega)\big) = T_0(p,\omega) + \int_{-\infty}^{\infty} \frac{dq}{2\pi} V(p|q) G_0(q,\omega) T\big(q|k(\omega)\big), \tag{2.3.40}
$$

where

$$
\begin{aligned}
T_0(p,\omega) = &-\frac{2iAk(\omega)}{\epsilon(\omega)} \int_{-\infty}^{\infty} \frac{ds}{2\pi} \exp(isL_0) \\
&\times \Big\{ \hat{J}\big(\beta(p,\omega)+\beta(\omega)|p-s\big) - \epsilon(\omega)\hat{J}\big(\beta(p,\omega)-\beta_0(\omega)|p-s\big) \\
&- \big(1-\epsilon(\omega)\big)\hat{J}\big(\beta(p,\omega)-\beta_0(\omega)|p-s\big) \frac{ps-\beta(p,\omega)\beta_0(\omega)}{s^2-k^2} \Big\}
\end{aligned} \tag{2.3.41}
$$

and

$$V(p|q) = \frac{\epsilon(\omega)-1}{\epsilon(\omega)} \hat{J}(\beta(p,\omega) - \beta_0(q,\omega)|p-q)\big[pq - \beta(p,\omega)\beta_0(q,\omega)\big]. \tag{2.3.42}$$

In fact, the inhomogeneous term $T_0(p,\omega)$ in Eq. (2.3.40) is far simpler than its representation in Eq. (2.3.41) might suggest. To see this, we first consider the integral

$$I_1 = \int_{-\infty}^{\infty} \frac{ds}{2\pi} \exp(isL_0)\hat{J}(\gamma|p-s). \tag{2.3.43}$$

On introducing the expression for $\hat{J}(\gamma|p-s)$ given by Eq. (2.3.36) into Eq. (2.3.43), assuming the representation of $\zeta(x_1)$ given by Eqs. (2.3.20) and (2.3.24), and interchanging the orders of integration, we find that

$$\begin{aligned} I_1 &= \int_{-\frac{L}{2}}^{\frac{L}{2}} dx_1 \exp(-ipx_1)\frac{\exp(\gamma\zeta(x_1))-1}{\gamma} \int_{-\infty}^{\infty} \frac{ds}{2\pi} \exp\big[is(x_1+L_0)\big] \\ &= \int_{-\frac{L}{2}}^{\frac{L}{2}} dx_1 \exp(-ipx_1)\frac{\exp(\gamma\zeta(x_1))-1}{\gamma}\delta(x_1+L_0) \\ &= 0, \end{aligned} \tag{2.3.44}$$

because $x_1 = -L_0$ lies outside the interval $(-L/2, L/2)$. Thus the first two terms on the right-hand side of Eq. (2.3.41) vanish. We next consider the integral

$$\begin{aligned} I_2 &= \int_{-\infty}^{\infty} \frac{ds}{2\pi} \exp(isL_0)\hat{J}(\gamma|p-s)\frac{ps - \beta(p,\omega)\beta_0(\omega)}{s^2-k^2} \\ &= \big[pk - \beta(p,\omega)\beta_0(\omega)\big] \int_{-\infty}^{\infty} \frac{ds}{2\pi} \exp(isL_0)\frac{\hat{J}(\gamma|p-s)}{s^2-k^2} \\ &\quad + p\int_{-\infty}^{\infty} \frac{ds}{2\pi} \exp(isL_0)\frac{\hat{J}(\gamma|p-s)}{s+k} \\ &\equiv \big[pk - \beta(p,\omega)\beta_0(\omega)\big]I_{2a} + pI_{2b}. \end{aligned} \tag{2.3.45}$$

With the use of Eqs. (2.3.20), (2.3.24), and (2.3.36) we transform the integral I_{2a} into

$$
\begin{aligned}
I_{2a} &= \int_{-\frac{L}{2}}^{\frac{L}{2}} dx_1 \exp(-ipx_1) \frac{\exp(\gamma\zeta(x_1)) - 1}{\gamma} \int_{-\infty}^{\infty} \frac{ds}{2\pi} \frac{\exp[is(x_1 + L_0)]}{(s-k)(s+k)} \\
&= \frac{i}{2k} \int_{-\frac{L}{2}}^{\frac{L}{2}} dx_1 \exp(-ipx_1) \frac{\exp(\gamma\zeta(x_1)) - 1}{\gamma} \\
&\quad \times \left\{\theta(x_1 + L_0)\exp\left[ik(x_1 + L_0)\right] + \theta(-x_1 - L_0)\exp\left[-ik(x_1 + L_0)\right]\right\} \\
&= \frac{i}{2k}\exp(ikL_0) \int_{-\frac{L}{2}}^{\frac{L}{2}} dx_1 \exp\left[-i(p-k)x_1\right] \frac{\exp(\gamma\zeta(x_1)) - 1}{\gamma}.
\end{aligned}
\tag{2.3.46}
$$

In the same way we find that

$$
\begin{aligned}
I_{2b} &= \int_{-\frac{L}{2}}^{\frac{L}{2}} dx_1 \exp(-ipx_1) \frac{\exp(\gamma s(x_1)) - 1}{\gamma} \int_{-\infty}^{\infty} ds \frac{\exp[is(x_1 + L_0)]}{s+k} \\
&= -i\exp(-ikL_0) \int_{-\frac{L}{2}}^{\frac{L}{2}} dx_1 \exp\left[-i(p+k)x_1\right] \frac{\exp(\gamma\zeta(x_1)) - 1}{\gamma} \theta(-x_1 - L_0) \\
&= 0.
\end{aligned}
\tag{2.3.47}
$$

On collecting the preceding results we obtain finally

$$
\begin{aligned}
T_0(p,\omega) &= -A\frac{1-\epsilon(\omega)}{\epsilon(\omega)}\left[pk(\omega) - \beta(p,\omega)\beta_0(\omega)\right] \\
&\quad \times \exp\left(ik(\omega)L_0\right) \int_{-\frac{L}{2}}^{\frac{L}{2}} dx_1 \exp\left\{-i\left[p - k(\omega)\right]x_1\right\} \\
&\quad \times \frac{\exp\{[\beta(p,\omega) - \beta_0(\omega)]\zeta(x_1)\} - 1}{\beta(p,\omega) - \beta_0(\omega)} \\
&= A\exp\left(ik(\omega)L_0\right)V\left(p|k(\omega)\right).
\end{aligned}
\tag{2.3.48}
$$

We note that if the rounded version of the function $\Gamma(x_1)$ given by Eq. (2.3.25) is used in the definition of the surface profile function $\zeta(x_1)$, Eq. (2.3.20), in place of the version given by Eq. (2.3.24), it is still quite a good approximation to regard the integrals I_1 and I_{2b} as vanishingly small, as long as the difference $L_0 - L/2$ is greater than approximately $4.6/\kappa$.

2.3.5. *The Random Surface*

From Eq. (2.3.27) we see that the scattered field in the vacuum region is expressed in terms of $R^{>}(q,\omega)$ as

$$H_2^{>}(x_1, x_3|\omega)_{\rm sc} = \int_{-\infty}^{\infty} \frac{dq}{2\pi} R^{>}(q,\omega)\exp\bigl[iqx_1 - \beta_0(q,\omega)x_3\bigr]. \tag{2.3.49}$$

The total time-averaged power scattered from the region $-L_1/2 < x_1 < L_1/2$, $-L_2/2 < x_2 < L_2/2$ of the surface, where L_1 and L_2 are large lengths, is given by

$$P_{\rm sc} = \int_{-\frac{L_1}{2}}^{\frac{L_1}{2}} dx_1 \int_{-\frac{L_2}{2}}^{\frac{L_2}{2}} dx_2\, {\rm Re}\bigl(S_{3p}^c\bigr)_{\rm sc}, \tag{2.3.50}$$

where $(S_{3p}^c)_{\rm sc}$ is the 3-component of the complex Poynting vector of the scattered field, and is given in terms of $H_2^{>}(x_1, x_3|\omega)_{\rm sc}$ by

$$\bigl(S_{3p}^c\bigr)_{\rm sc} = -\frac{ic^2}{8\pi\omega}\frac{\partial H_2^{>}(x_1, x_3|\omega)_{\rm sc}}{\partial x_3} H_2^{>}(x_1, x_3|\omega)_{\rm sc}^{*}. \tag{2.3.51}$$

On combining Eqs. (2.3.49) and (2.3.51) with Eq. (2.3.50) the latter equation becomes

$$P_{\rm sc} = {\rm Re}\, L_2 \frac{ic^2}{8\pi\omega} \int_{-\infty}^{\infty} \frac{dq}{2\pi} \beta_0(q,\omega)\bigl|R^{>}(q,\omega)\bigr|^2 \exp\bigl[-2\,{\rm Re}\,\beta_0(q,\omega)x_3\bigr]. \tag{2.3.52}$$

If we use the results that

$${\rm Im}\,\beta_0(q,\omega) = \begin{cases} -\alpha_0(q,\omega) & |q| < (\omega/c), \\ 0 & |q| > (\omega/c), \end{cases} \tag{2.3.53}$$

which follow from Eqs. (2.1.7) and (2.3.6), we find that

$$P_{sc} = L_2 \frac{c^2}{8\pi\omega} \int_{-\frac{\omega}{c}}^{\frac{\omega}{c}} \frac{dq}{2\pi} \alpha_0(q,\omega) \left| R^{>}(q,\omega) \right|^2. \tag{2.3.54}$$

With the change of variable $q = (\omega/c)\sin\theta_s$, we obtain finally

$$P_{sc} = L_2 \frac{\omega}{16\pi^2} \int_{-\frac{\pi}{2}}^{\frac{\pi}{2}} d\theta_s \cos^2\theta_s \left| R^{>}\big((\omega/c)\sin\theta_s, \omega\big) \right|^2. \tag{2.3.55}$$

Since the integrand in Eq. (2.3.54) is non-negative, we see that the only way in which leakage can be suppressed, i.e., the only way in which P_{sc} can be made to vanish, is to design a one-dimensional randomly rough surface for which the amplitude $R^{>}(q,\omega)$ is identically zero for $-(\omega/c) < q < (\omega/c)$. To suppress the leakage we will use a surface that is characterized by the power spectrum [2.33]

$$g(|Q|) = \frac{\pi}{2\Delta k}\big[\theta(Q - k_{\min})\theta(k_{\max} - Q) + \theta(-Q - k_{\min})\theta(k_{\max} + Q)\big], \tag{2.3.56}$$

where

$$k_{\min} = 2k_1(\omega) - \Delta k, \tag{2.3.57a}$$

$$k_{\max} = 2k_1(\omega) + \Delta k, \tag{2.3.57b}$$

and Δk must satisfy the inequality

$$\Delta k \ll k_1(\omega) - \frac{\omega}{c}. \tag{2.3.58}$$

That a surface characterized by the power spectrum (2.3.56) suppresses leakage can be seen from the following argument. The incident surface plasmon polariton has a wave number whose real part is $k_1(\omega)$. After its first interaction with the surface roughness it will be scattered into waves the real parts of whose wave numbers lie in the two intervals $(3k_1(\omega) - \Delta k, 3k_1(\omega) + \Delta k)$ and $(-k_1(\omega) - \Delta k, -k_1(\omega) + \Delta k)$. This is because the wave numbers in the power spectrum of the surface roughness with which $k_1(\omega)$ can combine lie in the intervals $(2k_1(\omega) - \Delta k, 2k_1(\omega) + \Delta k)$ and $(-2k_1(\omega) - \Delta k, -2k_1(\omega) + \Delta k)$. For the same reason, after the scattered waves interact again with the surface roughness the real parts of the wave numbers of the doubly-scattered waves will lie in the three intervals $(5k_1(\omega) - 2\Delta k, 5k_1(\omega) + 2\Delta k)$, $(k_1(\omega) - 2\Delta k, k_1(\omega) + 2\Delta k)$, and $(-3k_1(\omega) - 2\Delta k, -3k_1(\omega) + 2\Delta k)$. After the third interaction with the surface roughness the real parts of the wave numbers of the scattered waves will lie in the four intervals $(7k_1(\omega) - 3\Delta k, 7k_1(\omega) + 3\Delta k)$, $(3k_1(\omega) - 3\Delta k, 3k_1(\omega) + 3\Delta k)$, $(-k_1(\omega) - 3\Delta k, -k_1(\omega) + 3\Delta k)$, and $(-5k_1(\omega) - 3\Delta k, -5k_1(\omega) + 3\Delta k)$, and so on. Thus, for example, if we wish that $-k_1(\omega) + 3\Delta k$ not fall into the radiative region $(-(\omega/c), (\omega/c))$, we must require that $-k_1(\omega) + 3\Delta k < -(\omega/c)$, so that $\Delta k < (1/3)(k_1(\omega) - (\omega/c))$. When this condition is satisfied, after triple scattering the surface polaritons will not be converted into volume electromagnetic waves. In general, if we wish the surface plasmon polariton to scatter n times from the surface roughness without being converted into volume electromagnetic waves, we must require that $\Delta k < (1/n)(k_1(\omega) - (\omega/c))$.

A randomly rough surface characterized by a power spectrum in the form of two Gaussian peaks centered at $q = \pm k_1(\omega)$ was assumed in Ref. [2.34] in a search for a localization-induced enhancement of the surface plasmon polariton field. However, such a power spectrum is nonzero in the interval $(-(\omega/c), (\omega/c))$ and, therefore, such a surface does not suppress leakage.

2.3.6. *Generation of a One-Dimensional Random Surface*

In order to be able to solve the integral equation (2.3.40) numerically, either rigorously or approximately, it is necessary to be able to generate the surface profile function $\zeta(x_1)$ numerically. In view of Eq. (2.3.20) this means that we have to be able to generate a function $s(x_1)$ that is a single-valued function of x_1 that is differentiable and constitutes a stationary, zero-mean, Gaussian random process defined by Eq. (2.3.21). There are several ways of doing this [2.35–2.38]. However, for our purposes the following approach [2.39] is particularly convenient.

The characteristic functional describing a random function $f(x)$ is defined in general by

$$\Phi[u] \equiv \left\langle \exp\left[i \int dx\, u(x) f(x) \right] \right\rangle, \tag{2.3.59}$$

where the angle brackets denote an average over the ensemble of realizations of $f(x)$. Equation (2.3.59) can be rewritten in the form

$$\Phi[u] \equiv \exp\left\langle \exp\left[i \int dx\, u(x) f(x) \right] - 1 \right\rangle_c, \tag{2.3.60}$$

where $\langle\ \rangle_c$ denotes the cumulant average [2.40]. Since for a zero-mean Gaussian random function all multipoint cumulant averages of higher than second order vanish [2.40], it follows that if $f(x)$ is a zero-mean Gaussian random process its characteristic functional is given by [2.41]

$$\Phi[u] = \exp\left\{ -\frac{1}{2} \int dx \int dx'\, u(x) C(x, x') u(x') \right\}, \tag{2.3.61}$$

where

$$C(x,x') = \langle f(x)f(x')\rangle. \tag{2.3.62}$$

We now consider the random function $s(x)$ defined by

$$s(x) = \sum_n g_n\phi_n(x), \tag{2.3.63}$$

where the $\{g_n\}$ are independent Gaussian numbers with zero mean $\langle g_n\rangle = 0$ and arbitrary variances, while the $\{\phi_n(x)\}$ constitute a complete orthonormal set of functions

$$\int dx\, \phi_m(x)\phi_n(x) = \delta_{mn}. \tag{2.3.64}$$

Then $s(x)$ is a zero-mean Gaussian random function with a two-point correlation function

$$\langle s(x)s(x')\rangle = B(x,x') = \sum_n \langle g_n^2\rangle\phi_n(x)\phi_n(x'). \tag{2.3.65}$$

To see this, we consider the characteristic functional for the function $s(x)$. On substituting Eq. (2.3.63) into Eq. (2.3.59) we obtain

$$\Phi[u] = \left\langle \exp\left[i\sum_n g_n \int u(x)\phi_n(x)\,dx\right]\right\rangle_{\{g_n\}}, \tag{2.3.66}$$

where the averaging is performed over all g_n. With the definition

$$u_n \equiv \int u(x)\phi_n(x)\,dx, \tag{2.3.67}$$

and because all g_n are statistically independent, it follows from Eq. (2.3.66) that

$$\begin{aligned}\Phi[u] &= \left\langle \exp i\sum_n g_n u_n\right\rangle_{\{g_n\}} = \left\langle \prod_n \exp(ig_n u_n)\right\rangle_{\{g_n\}}\\ &= \prod_n \langle \exp(ig_n u_n)\rangle_{g_n}.\end{aligned} \tag{2.3.68}$$

For a Gaussian g_n we have

$$\langle \exp(ig_n u_n)\rangle_{g_n} = \exp\left(-\langle g_n^2\rangle u_n^2/2\right), \tag{2.3.69}$$

and Eq. (2.3.68) becomes

$$\begin{aligned}\Phi[u] &= \exp\left\{-\frac{1}{2}\sum_n \langle g_n^2\rangle u_n^2\right\}\\ &= \exp\left\{-\frac{1}{2}\sum_n \langle g_n^2\rangle \int dx \int dx'\, u(x)u(x')\phi_n(x)\phi_n(x')\right\}\\ &= \exp\left\{-\frac{1}{2}\int dx \int dx'\, u(x)u(x')\left[\sum_n \langle g_n^2\rangle \phi_n(x)\phi_n(x')\right]\right\}\\ &= \exp\left\{-\frac{1}{2}\int dx \int dx'\, u(x)B(x,x')u(x')\right\},\end{aligned} \tag{2.3.70}$$

where we have used Eqs. (2.3.65) and (2.3.67). On comparing this result with Eq. (2.3.61) we find that $s(x)$ is a zero-mean Gaussian random function defined by the two-point correlation function (2.3.65).

In computer simulation studies of scattering from one-dimensional randomly rough surfaces we require a finite segment of such a surface that covers the x axis from $x = -L/2$ to $x = L/2$. This is because computers cannot deal with infinitely long surfaces. In practice, for a reason that will be clear below, what is often done is to generate such a surface by an approach that produces a function $s(x_1)$ in a larger interval $(-\mathcal{L}/2, \mathcal{L}/2)$, and then to use the central position of this interval, namely the interval $(-L/2, L/2)$ with $L = \frac{1}{2}\mathcal{L}$, say, as the segment of rough surface used in the computer simulations.

To apply the general method just presented for generating the profile function for such a surface to this case we need a complete set of functions $\{\phi_n(x)\}$ orthonormal over the interval $-\mathcal{L}/2 < x < \mathcal{L}/2$. A convenient set is given by [2.39]

$$\phi_n(x) = \begin{cases} \frac{1}{\sqrt{\mathcal{L}}} & \text{for } n = 0,\\ \sqrt{\frac{2}{\mathcal{L}}}\sin(\frac{2\pi kx}{\mathcal{L}}) & \text{for } n = 2k-1,\\ \sqrt{\frac{2}{\mathcal{L}}}\cos(\frac{2\pi kx}{\mathcal{L}}) & \text{for } n = 2k.\end{cases} \tag{2.3.71}$$

When these functions are substituted into Eq. (2.3.65), we obtain

$$\begin{aligned}B(x,x') &= \frac{\langle g_0^2\rangle}{\mathcal{L}} + \sum_{k=1}^{\infty}\frac{2\langle g_{2k-1}^2\rangle}{\mathcal{L}}\sin\frac{2\pi kx}{\mathcal{L}}\sin\frac{2\pi kx'}{\mathcal{L}}\\ &\quad + \sum_{k=1}^{\infty}\frac{2\langle g_{2k}^2\rangle}{\mathcal{L}}\cos\frac{2\pi kx}{\mathcal{L}}\cos\frac{2\pi kx'}{\mathcal{L}}\\ &= \frac{\langle g_0^2\rangle}{\mathcal{L}} + \frac{1}{\mathcal{L}}\sum_{k=1}^{\infty}\left[\langle g_{2k}^2\rangle + \langle g_{2k-1}^2\rangle\right]\cos\frac{2\pi k(x-x')}{\mathcal{L}}\end{aligned}$$

$$+\frac{1}{\mathcal{L}}\sum_{k=1}^{\infty}\left[\langle g_{2k}^2\rangle-\langle g_{2k-1}^2\rangle\right]\cos\frac{2\pi k(x+x')}{\mathcal{L}}. \tag{2.3.72}$$

If we wish $s(x)$ to be a stationary random process, we must require that

$$\langle g_{2k}^2\rangle=\langle g_{2k-1}^2\rangle. \tag{2.3.73}$$

In addition, if we require that $B(x,x')$ vanish as $|x-x'|\to\infty$, we must set $\langle g_0^2\rangle=0$. We denote the resulting correlation function by $\delta^2 W(|x-x'|)$, and obtain

$$\delta^2 W\big(|x-x'|\big)=\frac{2}{\mathcal{L}}\sum_{k=1}^{\infty}\langle g_{2k}^2\rangle\cos\frac{2\pi k(x-x')}{\mathcal{L}}. \tag{2.3.74}$$

By the use of Eq. (2.3.22) we can write that

$$\delta^2 W\big(|x-x'|\big)=\frac{\delta^2}{\pi}\int_0^{\infty}dk\,g(k)\cos\big[k(x-x')\big]. \tag{2.3.75}$$

With the use of the result that

$$\int_0^{\infty}dk\,f(k)\cong\frac{2\pi}{\mathcal{L}}\sum_{k=1}^{\infty}f\left(\frac{2\pi k}{\mathcal{L}}\right), \tag{2.3.76}$$

we can rewrite Eq. (2.3.75) as

$$\delta^2 W\big(|x-x'|\big)=\frac{\delta^2}{\pi}\frac{2\pi}{\mathcal{L}}\sum_{k=1}^{\infty}g\left(\frac{2\pi k}{\mathcal{L}}\right)\cos\frac{2\pi k(x-x')}{\mathcal{L}}. \tag{2.3.77}$$

From a comparison of Eqs. (2.3.74) and (2.3.77) we find that

$$\langle g_{2k}^2\rangle=\delta^2 g\left(\frac{2\pi k}{\mathcal{L}}\right). \tag{2.3.78}$$

We now return to Eq. (2.3.63) which we rewrite as

$$s(x)=\sum_{k=1}^{\infty}\left[g_{2k-1}\sqrt{\frac{2}{\mathcal{L}}}\sin\left(\frac{2\pi kx}{\mathcal{L}}\right)+g_{2k}\sqrt{\frac{2}{\mathcal{L}}}\cos\left(\frac{2\pi kx}{\mathcal{L}}\right)\right]. \tag{2.3.79}$$

In view of Eqs. (2.3.73) and (2.3.78), and the fact that g_{2k-1} and g_{2k} are independent random numbers, we can set

$$g_{2k-1} = \delta\sqrt{g(2\pi k/\mathcal{L})}\alpha_{2k-1}, \qquad g_{2k} = \delta\sqrt{g(2\pi k/\mathcal{L})}\alpha_{2k}, \tag{2.3.80}$$

where the $\{\alpha_m\}$ are independent Gaussian random numbers with zero mean and unit variance,

$$\langle \alpha_m \rangle = 0, \qquad \langle \alpha_m^2 \rangle = 1. \tag{2.3.81}$$

As a result, we obtain the algorithm

$$s(x) = \delta\sqrt{\frac{2}{\mathcal{L}}}\sum_{k=1}^{\infty}\sqrt{g\left(\frac{2\pi k}{\mathcal{L}}\right)}\left[\alpha_{2k-1}\sin\left(\frac{2\pi kx}{\mathcal{L}}\right) + \alpha_{2k}\cos\left(\frac{2\pi kx}{\mathcal{L}}\right)\right] \tag{2.3.82}$$

for generating a stationary, zero-mean Gaussian random process with a prescribed power spectrum.

The $\{\alpha_m\}$ can be generated by the Marsaglia and Bray modification of the Box–Muller transformation of a pair of uniform deviates between zero and one obtained from a linear congruential generator [2.42].

The function $s(x)$ generated by Eq. (2.3.82) is a periodic function of x with period $\mathcal{L}$. It is to minimize end effects associated with this periodicity of $s(x)$ that only that portion of this function which corresponds to values of x in the interval $(-L/2, L/2)$, where $L = \frac{1}{2}\mathcal{L}$ is used in computer simulations of scattering and transmission problems.

An attractive feature of the result given by Eq. (2.3.82) is that it allows calculating $s(x)$ at arbitrary values of x without any interpolation being required, in contrast with other approaches [2.35–2.38] that generate $s(x)$ only at equally spaced points along the x axis.

In Fig. 2.22 we present a plot of a segment of one realization of the function $s(x)$ obtained from Eq. (2.3.82) for a silver surface. The surface roughness was characterized by the parameters $\Delta k = 0.3(k_1(\omega) - (\omega/c))$ and $\delta = 3$ nm. The vacuum wavelength of the surface plasmon polaritons propagating across this surface was assumed to be $\lambda = (2\pi c/\omega) = 457.9$ nm. The dielectric function of silver at this wavelength is $\epsilon(\omega) = -7.5 + i0.24$ [2.43]. The corresponding wave number of the surface plasmon polariton is $k(\omega) = k_1(\omega) + ik_2(\omega) = (1.0741 + i0.0026)(\omega/c)$. The surface designed with these parameters should suppress leakage due to scattering processes of up to, and including, third order.

2.3.7. *Solution of the Scattering Problem*

The integral equation (2.3.40) for the transition matrix $T(p|k(\omega))$ can be solved either purely numerically or in an approximate analytical fashion. We consider both approaches in this section.

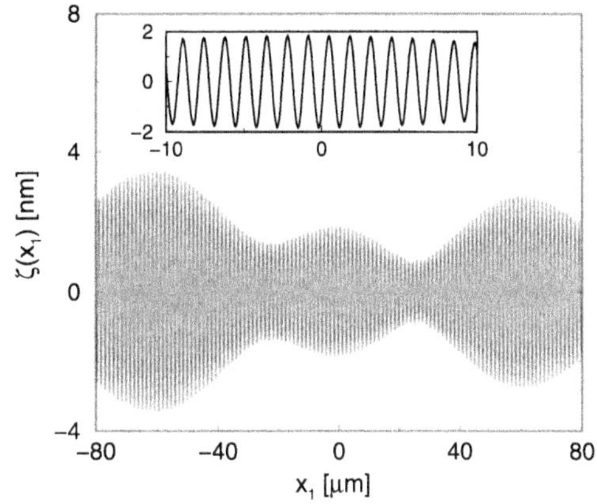

Figure 2.22. A segment of a single realization of the random surface profile function $s(x_1)$ generated numerically by the approach described in Section 2.3.6. The values of the parameters used in generating this realization were $\Delta k = 0.3[k_1(\omega) - (\omega/c)]$ where $k_1(\omega) = 1.0741(\omega/c)$, and $\delta = 3$ nm.

2.3.7.1. An Approximate Analytical Solution In deriving Eq. (2.3.40) we have assumed that the conditions for the validity of the Rayleigh hypothesis are satisfied. The scattering potential $V(p|q)$ given by Eq. (2.3.42) does not have any poles in the complex planes of the variables p and q, and the inequality [2.44]

$$\{\exp[(\beta(k,\omega) - \beta_0(k,\omega))\zeta(x_1)] - 1\} \ll 1 \tag{2.3.83}$$

is satisfied. It should be noted that the scattering potential $V(p|q)$ can diverge as $|p| \to \infty$ and $|q| \to \infty$. These singularities do not contribute to the integral term of Eq. (2.3.40). Nevertheless special care should be taken when solving Eq. (2.3.40) numerically. In view of the power spectrum of the surface roughness assumed, Eq. (2.3.56), the main contribution to the integral term in Eq. (2.3.40) comes from the poles of the Green's function $G_0(q,\omega)$. In the vicinity of its poles $G_0(q,\omega)$ can be written as

$$G_0(q,\omega) \cong \theta\left(q - \frac{\omega}{c}\right)\frac{C(\omega)}{q - k(\omega)} - \theta\left(-q - \frac{\omega}{c}\right)\frac{C(\omega)}{q + k(\omega)}, \tag{2.3.84}$$

where $k(\omega)$ is given by Eq. (2.3.13), and the residue $C(\omega)$ is

$$C(\omega) = \frac{|\epsilon_1(\omega)|^{3/2}}{|\epsilon_1(\omega)|^2 - 1}. \tag{2.3.85}$$

Then, by restricting the integration over q in Eq. (2.3.40) to the segments $(-\infty, -(\omega/c))$ and $(\omega/c, \infty)$, and generating closed contours by going into the complex q plane (see

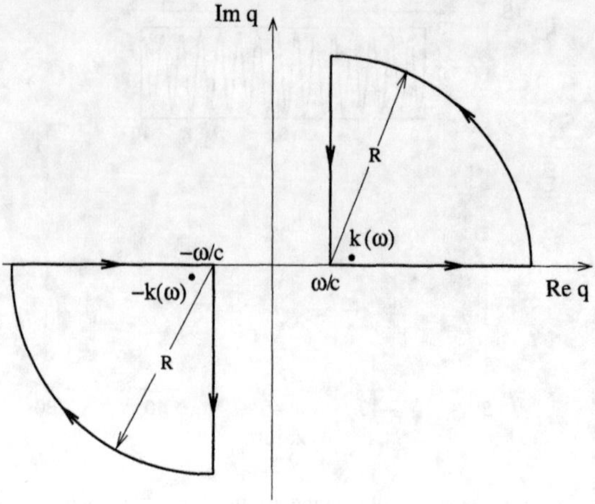

Figure 2.23. The contour in the complex q plane used in obtaining Eq. (2.3.86) from Eq. (2.3.40), with the aid of Eq. (2.3.84).

Fig. 2.23) as this was done in Ref. [2.45], we can rewrite Eq. (2.3.40) as

$$T(p|k) = T_0(p) + iC(\omega)V(p|k)T(k|k) + iC(\omega)V(p|-k)T(-k|k), \tag{2.3.86}$$

where the wave number k here denotes $k(\omega)$. To solve this equation we set $p = k$ and $p = -k$ in turn, to obtain

$$T(k|k) = iC(\omega)V(k|-k)T(-k|k), \tag{2.3.87a}$$

$$T(-k|k) = T_0(-k) + iC(\omega)V(-k|k)T(k|k). \tag{2.3.87b}$$

In writing these equations we have used the results that

$$V(k|k) = V(-k|-k) = 0, \qquad T_0(k) = 0, \tag{2.3.88}$$

which follow from Eq. (2.3.42) and Eqs. (2.3.13), (2.3.18), and (2.3.19), which show that $\beta(k,\omega)\beta_0(k,\omega) \equiv k^2$. As a result we obtain

$$T(k|k) = \frac{iC(\omega)V(k|-k)T_0(-k)}{1 + C^2(\omega)V(k|-k)V(-k|k)}, \tag{2.3.89a}$$

$$T(-k|k) = \frac{T_0(-k)}{1 + C^2(\omega)V(k|-k)V(-k|k)}, \tag{2.3.89b}$$

so that

$$T(q|k) = T_0(q) + \frac{iC(\omega)V(q|-|k) - C^2(\omega)V(q|k)V(k|-k)}{1 + C^2(\omega)V(k|-k)V(-k|k)}T_0(-k). \tag{2.3.90}$$

The scattering amplitude $R^{>}(q,\omega)$ is obtained from this solution with the use of Eqs. (2.3.38) and (2.3.39). The result is then used to calculate the mean value $(\omega/c)^2\langle|R^{>}(q,\omega)|^2\rangle = (\omega/c)^2|G_0(q,\omega)|^2\langle|T(q|k(\omega))|^2\rangle$. The averaging of $|T(q|k)|^2$ is carried out in the following way. An ensemble of N_p realizations of the function $s(x_1)$ is generated by the method described in Section 2.3.6. From each realization a surface profile function $\zeta(x_1)$ is calculated by the use of Eq. (2.3.20), with the function $\Gamma(x_1)$ given by Eq. (2.3.24), and with its use the functions entering the expression (2.3.90) for $T(q|k)$ are calculated. In particular, the function $\hat{J}(\gamma|Q)$ is calculated by means of the expansion

$$\hat{J}(\gamma|Q) = \sum_{n=1}^{\infty} \frac{\gamma^{n-1}}{n!} \int_{-\infty}^{\infty} dx_1\, \zeta^n(x_1) \exp(-iQx_1). \tag{2.3.91}$$

The presence of the factor $\Gamma(x_1)$ in the definition of $\zeta(x_1)$, Eq. (2.3.20), effectively limits the domain of integration in Eq. (2.3.91) to the interval $|x_1| < L/2$. The function $|T(q|k)|^2$ is then calculated for each realization of $\zeta(x_1)$, and an arithmetic average of the N_p results for this function yields the average $\langle|T(q|k)|^2\rangle$.

In using the pole approximation, thereby reducing the integral equation (2.3.40) to a system of linear algebraic equations, we lose the contribution from the nonsingular part of the integrand. This might be significant if the transition matrix has strong peaks. These peaks can arise in the vicinities of $q = \pm(2n+1)k_1(\omega)$ due to the narrow windows in the power spectrum of the surface roughness. The heights of the peaks can be expected to be of the order of $1/\Delta k$, and since $\Delta k \gg k_2(\omega)$, the contributions of these possible peaks to the integral term in Eq. (2.3.40) are much smaller than the contribution from the poles of the Green's function $G_0(q)$ and can be neglected.

2.3.7.2. Numerical Simulations To obtain a purely numerical solution to the problem of surface plasmon polariton scattering by a randomly rough surface one can use the rigorous numerical simulation method presented in Section 2.2.2.1. In the case that the incident field is a surface plasmon polariton Eqs. (2.2.29) and (2.2.30) have slightly different forms. We recall that the single nonzero component of the field $H_2^{>}(x_1,x_3|\omega)$ in the vacuum satisfies Eq. (2.3.29), while $H_2^{<}(x_1,x_3|\omega)$ in the metal satisfies Eq. (2.3.30). If we now apply the results given by Eqs. (2.3.29) and (2.2.24) and Green's second integral identity in the plane, Eq. (2.1.8), to the region $x_3 > \zeta(x_1)$ we obtain the following equation for $H_2^{>}(x_1,x_3|\omega)$:

$$\begin{aligned}
&\theta\big(x_3-\zeta(x_1)\big)H_2^{>}(x_1,x_3|\omega)\\
&\quad = \tilde{H}_2^{>}(x_1,x_3|\omega)_{\mathrm{inc}} + \frac{1}{4\pi}\int_S ds'\bigg[\bigg(\frac{\partial}{\partial n'}G_0(x_1,x_3|x_1',x_3')\bigg)H_2^{>}(x_1',x_3'|\omega)\\
&\qquad - G_0(x_1,x_3|x_1',x_3')\frac{\partial}{\partial n'}H_2^{>}(x_1',x_3'|\omega)\bigg],
\end{aligned} \tag{2.3.92}$$

where $\tilde{H}_2^>(x_1, x_3|\omega)_{\rm inc}$ is given by

$$\begin{aligned}
&\tilde{H}_2^>(x_1, x_3|\omega)_{\rm inc} \\
&= \frac{ik(\omega)}{2\pi} \int_0^\infty dx_3' \, G_0(x_1, x_3|-L_0, x_3') \exp(-\beta_0(\omega)x_3) \\
&= A\{\theta(x_1 + L_0)\exp[ik(\omega)(x_1 + L_0) - \beta_0(\omega)x_3] \\
&\quad + \theta(-x_1 - L_0)\exp[-ik(\omega)(x_1 + L_0) - \beta_0(\omega)x_3]\} \\
&\quad + k(\omega) \int_{-\infty}^{\infty} \frac{dq}{2\pi} \frac{1}{\alpha_0(q)(i\alpha_0(q) + \beta_0(\omega))} \exp[-iq(x_1 + L_0) + i\alpha_0(q)x_3].
\end{aligned} \tag{2.3.93}$$

We next apply the results given by Eqs. (2.2.39) and (2.2.24) and Green's second integral identity in the plane, Eq. (2.1.8), to the region $x_3 < \zeta(x_1)$ and obtain the following equation for $H_2^<(x_1, x_3|\omega)$:

$$\begin{aligned}
\theta(\zeta(x_1) - x_3) H_2^<(x_1, x_3|\omega) = \tilde{H}_2^<(x_1, x_3|\omega)_{\rm inc} & \\
- \frac{1}{4\pi} \int_S ds' \bigg[\left(\frac{\partial}{\partial n'} G_\epsilon(x_1, x_3|x_1', x_3') \right) & H_2^<(x_1', x_3'|\omega) \\
- G_\epsilon(x_1, x_3|x_1', x_3') \frac{\partial}{\partial n'} & H_2^<(x_1', x_3'|\omega) \bigg],
\end{aligned} \tag{2.3.94}$$

where $\tilde{H}_2^<(x_1, x_3|\omega)_{\rm inc}$ is given by

$$\begin{aligned}
\tilde{H}_2^<(x_1, x_3|\omega)_{\rm inc} &= \frac{ik(\omega)}{2\pi} \int_0^\infty dx_3' \, G_\epsilon(x_1, x_3|-L_0, x_3') \exp(\beta(\omega)x_3) \\
&= A\{\theta(x_1 + L_0)\exp[ik(\omega)(x_1 + L_0) + \beta(\omega)x_3] \\
&\quad + \theta(-x_1 - L_0)\exp[-ik(\omega)(x_1 + L_0) + \beta(\omega)x_3]\} \\
&\quad + k(\omega) \int_{-\infty}^{\infty} \frac{dq}{2\pi} \frac{1}{\alpha(q)(i\alpha(q) + \beta(\omega))} \exp[-iq(x_1 + L_0) - i\alpha(q)x_3].
\end{aligned} \tag{2.3.95}$$

Proceeding in the manner described in Section 2.2.2.1 one obtains a pair of matrix equations that can be solved numerically.

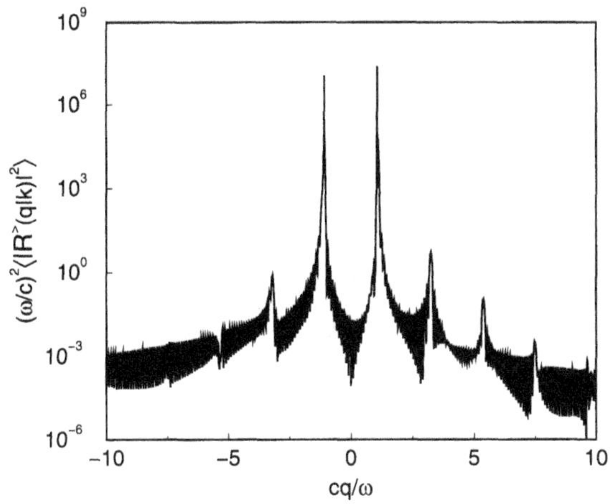

Figure 2.24. Plot of $(\omega/c)^2\langle|R^>(q,\omega)|^2\rangle$ as a function of cq/ω, calculated by averaging numerically the result for this function obtained from Eqs. (2.3.38), (2.3.39), and (2.3.90), for a silver surface characterized by the parameters $\Delta k = 0.3[k_1(\omega) - (\omega/c)]$ and $\delta = 3$ nm. The rough portion of the surface has the length $L = 20\lambda$. The wave number of the surface plasmon polariton, $k(\omega) = k_1(\omega) + ik_2(\omega) = (1.0741 + i0.0026)(\omega/c)$, corresponds to a vacuum wavelength of $\lambda = 457.9$ nm, and the dielectric function of silver at this frequency is $\epsilon(\omega) = -7.5 + i0.24$.

2.3.8. Results

To show explicitly that surfaces generated in the manner described in this section suppress leakage, i.e., that the scattering amplitude $R^>(q,\omega)$ vanishes for q in the radiative region $(-(\omega/c), (\omega/c))$, in Fig. 2.24, we present a plot of $(\omega/c)^2\langle|R^>(q,\omega)|^2\rangle$ as a function of cq/ω for a silver surface whose randomly rough portion has a length $L = 20\lambda$. The result plotted in Fig. 2.24 was calculated from the approximate analytic expression for $T(q|k)$, Eq. (2.3.90), in the manner described in Section 2.3.7.1. The experimental and roughness parameters assumed in these calculations were the same ones used in calculating the segment of a single realization of the surface profile function plotted in Fig. 2.22. It is clear from Fig. 2.24 that $\langle|R^>(q,\omega)|^2\rangle$ is indeed suppressed in the radiative region.

The analogous result obtained by means of a purely numerical solution of Eqs. (2.3.92) and (2.3.94) in the manner described in Section 2.2.2.1 is presented in Fig. 2.25 for the case in which the rms height of the surface roughness was taken to be $\delta = 10$ nm and the smooth differentiable form for $\Gamma(x_1)$ given by Eq. (2.3.25) was assumed, with the remaining experimental and roughness parameters retaining the values used in obtaining the results presented in Figs. 2.22 and 2.24. It is seen from this figure that although $(\omega/c)^2\langle|R^>(q,\omega)|^2\rangle$ is strongly suppressed in the region of small values of $|q| \ll (\omega/c)$, it is far from zero for almost grazing directions of radiation $|q| \lesssim (\omega/c)$. This is due to the strong higher order surface plasmon polariton scattering processes that are possible for such a strongly rough surface. Eight peaks, corresponding to q equal to $\pm k_1(\omega)$, $\pm 3k_1(\omega)$, $\pm 5k_1(\omega)$, and $\pm 7k_1(\omega)$

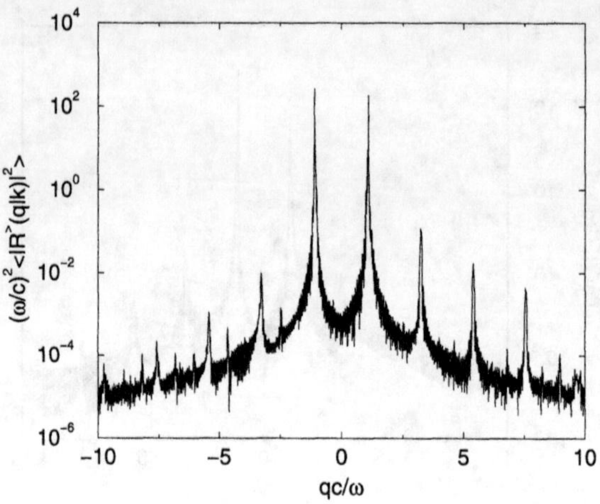

Figure 2.25. Plot of $(\omega/c)^2\langle|R^>(q,\omega)|^2\rangle$ as a function of cq/ω, calculated by means of a numerical solution of Eqs. (2.3.92) and (2.3.94) for a silver surface characterized by the parameters $\Delta k = 0.3[k_1(\omega) - (\omega/c)]$ and $\delta = 10$ nm. The rough portion of the surface has length $L = 20\lambda$. The wave number of the surface plasmon polariton, $k(\omega) = k_1(\omega) + ik_2(\omega) = (1.0741 + i0.0026)(\omega/c)$, corresponding to a vacuum wavelength $\lambda = 457.9$ nm, and the dielectric function of silver at this frequency is $\epsilon(\omega) = -7.5 + i0.24$.

are clearly seen in this figure. It is also seen that when $|q| \geqslant 7(\omega/c)$, $(\omega/c)^2\langle|R^>(q,\omega)|^2\rangle$ becomes flatter. This flattening is due to absorption occurring at all such scattering wave numbers, rather than just at well-defined wave numbers, as is the case for $|q| < 7(\omega/c)$.

The results obtained in this section show that it is possible to generate a one-dimensional randomly rough metal surface that suppresses leakage when a surface plasmon polariton propagates across it. The choice of an incident surface plasmon polariton field of the form present in Eqs. (2.3.27) and (2.3.28) may also be useful in other problems where a surface plasmon polariton interacts with a surface defect on an otherwise planar lossy metal surface [2.46–2.52].

2.4. Surfaces that Display Enhanced Backscattering for Only a Single Specified Angle of Incidence

The enhanced backscattering of light from a randomly rough surface is the presence of a well-defined peak in the retroreflection direction in the dependence of the mean differential reflection coefficient on the scattering angle. It is a remarkable property of this effect that it occurs for any angle of incidence, although its strength decreases as the angle of incidence increases.

In some optical experiments it is required to position a source and the detector at the same specified direction with respect to the site at which the scattering surface is located. Thus, the detector is positioned in the retroreflection direction. Moreover, it is also required that the light be scattered in the retroreflection direction for only a single, specified, angle of incidence, and to have the highest possible efficiency for scattering into that direction.

It is well known [2.53] that efficient backscattering can be achieved in single-scattering from a deterministic, rather than a random, rough surface when that surface is a classical diffraction grating in the Littrow mount under the so-called perfect blazing conditions [2.54–2.56]. This is the case where the diffraction grating produces only two propagating diffracted orders, namely the (0) and (-1) orders, and the amplitude of the grating corrugations is chosen so that the (0) order is completely suppressed. The physical system to describe the Littrow mount consists of a vacuum in the region $x_3 > \zeta(x_1)$ and a metal characterized by the dielectric function $\epsilon(\omega)$ in the region $x_3 < \zeta(x_1)$. The surface profile function $\zeta(x_1)$ represents a periodic grating $\zeta(x_1) = \zeta(x_1 + a)$, where a is the grating period. Such a grating scatters light of wavelength λ incident on it at an angle θ_0 into discrete diffracted orders at the scattering angles $\theta_s^{(n)}$ determined by

$$\sin\theta_s^{(n)} = \sin\theta_0 + \frac{\lambda}{a}n, \tag{2.4.1}$$

where n is an integer. If the period of the grating is chosen to be $a = \lambda/(2\sin\theta_c)$, where θ_c is some specified angle, the (-1) diffracted order will occur at the scattering angle $\sin\theta_s^{(-1)} = \sin\theta_0 - 2\sin\theta_c$. Therefore, if the angle of incidence θ_0 coincides with this specified angle θ_c, $\theta_0 = \theta_c$, the (-1) diffracted order will occur in the backscattering direction $\theta_s^{(-1)} = -\theta_c$. No higher order diffraction peaks will be present in the diffraction pattern if the angle θ_c is large enough that $\sin\theta_c > 1/3$ ($\theta_c > 19.5°$), so that the $(+1)$ order and higher order diffraction peaks fall into the nonradiative region. The (0) order diffraction peak, i.e., the specular peak, is present in the angular distribution of the intensity of the scattered light. As was shown in [2.57], the efficiency of the diffraction depends quasi-periodically on the amplitude of the grating. Therefore the amplitude of the grating could be chosen such that the intensity of the specular peak reaches its minimum, while the intensity of the (-1) order peak is at its maximum at this particular angle of incidence θ_c. In the absence of damping, e.g., in the case of diffraction from a perfectly conducting or a dielectric grating, perfect blazing can be achieved for both p- and s-polarized light. However, in the case of a dielectric grating the intensity of the backscattering peak is weak, because the majority of the incident light is transmitted rather than reflected. In Fig. 2.26(a) as an example, we present a gray level plot of the intensity of the s-polarized light scattered from a perfectly conducting diffraction grating in the Littrow mount whose amplitude is chosen to produce perfect blazing at $\theta_c = 25°$. In Fig. 2.26(b) the intensities of the (0) (dashed line) and (-1) order (solid line) diffraction peaks are plotted as functions of the angle of scattering in the vicinity of $\theta_0 = \theta_c$. Both the (0) and (-1) order peak intensities are smooth functions of the angle.

In Ref. [2.58] it was shown how to design a one-dimensional randomly rough metal surface that produces an enhanced backscattering peak for only a single, specified, angle of incidence, when it is illuminated by p-polarized light of frequency ω (wavelength λ),

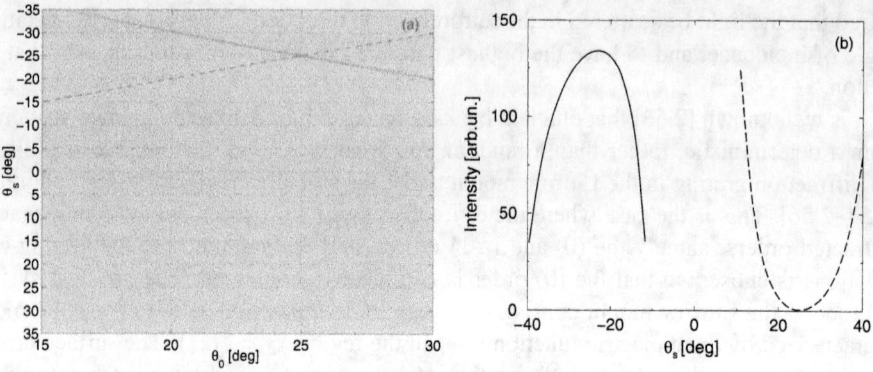

Figure 2.26. A gray-level plot of the angular dependence of the intensity of s-polarized light scattered from a classical diffraction grating. The grating was designed to produce an enhanced backscattering peak and no specular peak when the angle of incidence $\theta_c = 25°$. The dashed line shows the retroreflection direction.

whose plane of incidence is perpendicular to the generators of the surface. However, the surface designed in Ref. [2.58] also produced a peak in the specular direction whose magnitude was comparable to that of the enhanced backscattering peak.

The surface in Ref. [2.58] was designed on the basis of the roughness induced excitation of surface plasmon polaritons by the incident light and their subsequent multiple scattering by the surface roughness. As a result, the height of the enhanced backscattering peak was only twice as high as the background at its position, when the contribution to the background from single-scattering processes was subtracted off.

Here we present a different approach to the design of a one-dimensional rough metal surface that produces an enhanced backscattering peak for only a single, specified, angle of incidence, but does not give rise to a specular peak. It is based on representing the surface profile function as a sum of two periodic profiles with different periods $\zeta(x_1) = s_1(x_1) + s_2(x_1)$, where $s_1(x_1 + a_1) = s_1(x_1)$ and $s_2(x_1 + a_2) = s_2(x_1)$. The periods of the two gratings are chosen to produce a resonant excitation of forward and backward propagating surface plasmon polaritons supported by the metal–vacuum interface by the light incident at the specified angle of incidence θ_c. Thus, the period of the first grating is $a_1 = \lambda/(\sin\theta_c + n_{sp}(\lambda))$, where $n_{sp}(\lambda) = \mathrm{Re}\sqrt{\epsilon(\omega)/(\epsilon(\omega)+1)}$ is the refractive index of the surface plasmon polariton of frequency ω, and $\epsilon(\omega)$ is the dielectric function of the metal. The grating of this period efficiently excites forward propagating surface waves. The period of the second grating is $a_2 = \lambda/(\sin\theta_c - n_{sp}(\lambda))$, so that this grating efficiently excites backward propagating surface plasmon polaritons. Furthermore, the forward propagating surface plasmon polaritons interact with the grating of period a_2 and are converted into volume electromagnetic waves in the vacuum that are radiated in the retroreflection direction. Analogously, the backward propagating surface plasmon polaritons interact with the grating of the period a_1 and are converted into volume waves in the vacuum that are also radiated in the retroreflection direction. These waves interfere constructively to produce the enhanced backscattering peak. A strong specular peak is also present in the intensity of the scattered light. However, we show that a careful choice of the amplitudes of the two gratings leads to an additional enhancement of the peak in the retroreflection direction and a complete elimination of the specular peak, and thus leads to perfect blazing. Apart from weak ohmic losses all the energy of the incident light is retroreflected.

The physical system we consider in this section consists of a vacuum in the region $x_3 > \zeta(x_1)$ and a metal characterized by a complex, frequency dependent dielectric function $\epsilon(\omega)$ in the region $x_3 < \zeta(x_1)$. The surface profile function $\zeta(x_1)$ is assumed to represent a quasi-grating

$$\zeta(x_1) = s_1(x_1) + s_2(x_2), \tag{2.4.2a}$$

$$s_{1,2}(x_1) = A_{1,2}\cos(G_{1,2}x_1), \tag{2.4.2b}$$

where $G_{1,2}$ are the wavevectors of the two components of the grating and $A_{1,2}$ are their amplitudes. It is intuitively clear that such a quasi-grating scatters light incident on it at an angle θ_0 into discrete diffracted orders at the scattering angles θ_s determined by

$$\sin\theta_s = \sin\theta_0 + m\frac{G_1 c}{\omega} + n\frac{G_2 c}{\omega}, \tag{2.4.3}$$

where n and m are integers. Generally, if the wavevectors of the two gratings are related by $G_{1,2} = K \mp k_c$, where K is an arbitrary wave number, $k_c = (\omega/c)\sin\theta_c$, and θ_c is a prescribed angle, the diffracted order $(1,-1)$ will be radiated into the direction determined by the condition $\sin\theta_s = \sin\theta_0 - 2\sin\theta_c$, and, therefore into the retroreflection direction if $\theta_0 = \theta_c$. In the case where $G_{1,2} = k_1(\omega) \mp k_c$, where $k_1(\omega) = \mathrm{Re}\,k(\omega)$ is the real part of the wavevector of surface plasmon polaritons supported by the vacuum-metal interface, the $(+1,0)$ and $(0,-1)$ diffraction orders are surface plasmon polaritons rather than radiative or evanescent fields. Then the intensity of the $(1,-1)$ diffracted order is enhanced due to the constructive interference of the diffraction processes.

The surface $x_3 = \zeta(x_1)$ is illuminated from the vacuum by a p-polarized plane wave of frequency ω, whose plane of incidence is the x_1x_3 plane. In the vacuum region $x_3 > \zeta(x_1)_{\max}$ the single nonzero component of the magnetic field is the sum of an incoming incident field and a superposition of outgoing scattered waves,

$$H_2^{>}(x_1, x_3|\omega) = \exp\bigl[ikx_1 - i\alpha_0(k)x_3\bigr] + \int_{-\infty}^{\infty} \frac{dq}{2\pi} R(q|k)\exp\bigl[iqx_1 + i\alpha_0(q)x_3\bigr], \tag{2.4.4}$$

where $\alpha_0(q) = [(\omega/c)^2 - q^2]^{\frac{1}{2}}$, with $\mathrm{Re}\,\alpha_0(q) > 0$, $\mathrm{Im}\,\alpha_0(q) > 0$.

We assume that the quasi-grating is sufficiently weak so that the conditions for the validity of the Rayleigh hypothesis [2.28] are fulfilled. Then the reduced Rayleigh equation for the scattering amplitude $R(q|k)$ can be derived in exactly the same manner as described in Section 2.3.4 and has the form

$$\int_{-\infty}^{\infty} \frac{dq}{2\pi} \frac{pq + i\beta(p)\alpha_0(q)}{i\beta(p) - \alpha_0(q)} \hat{I}\big(\beta(q) + i\alpha_0(q)|p - q\big)R(q|k)$$
$$= \frac{pq - i\beta(p)\alpha_0(k)}{i\beta(p) + \alpha_0(k)} \hat{I}\big(\beta(q) - i\alpha_0(k)|p - k\big), \tag{2.4.5}$$

where $\beta(q) = [q^2 - \epsilon(\omega)(\omega/c)^2]^{\frac{1}{2}}$, with $\operatorname{Re}\beta(q) > 0$, $\operatorname{Im}\beta(q) < 0$, and $\hat{I}(\gamma|Q)$ is given by Eq. (2.3.34)

For a sinusoidal surface profile function $s_j(x_1) = A_j \cos(G_j x_1)$ we can introduce the function

$$\hat{I}^{(j)}(\gamma|Q) = \int_{-\infty}^{\infty} dx_1\, e^{-iQx_1} e^{\gamma s_j(x_1)} = \sum_{m=-\infty}^{\infty} \mathcal{I}_m^{(j)}(\gamma A_j) 2\pi\delta(Q - nG_j), \tag{2.4.6}$$

where

$$\mathcal{I}_m^{(j)}(\gamma A_j) = \frac{1}{a_j} \int_{-a_j/2}^{a_j/2} dx_1\, e^{-i\frac{2pi}{a_j}x_1} e^{\gamma A_j}. \tag{2.4.7}$$

Using Eqs. (2.4.6) and (2.4.7) we can rewrite the expression for $\hat{I}(\gamma|Q)$ (2.3.34) as

$$\hat{I}(\gamma|Q) = \int_{-\infty}^{\infty} \frac{dp}{2\pi} I^{(1)}(\gamma|Q - p) I^{(2)}(\gamma|p)$$
$$= \sum_{m=-\infty}^{\infty} \sum_{n=-\infty}^{\infty} \mathcal{I}_m^{(1)}(\gamma A_1) \mathcal{I}_n^{(2)}(\gamma A_2) 2\pi\delta(Q - mG_1 - nG_2). \tag{2.4.8}$$

With the aid of representation (2.4.8) the reduced Rayleigh equation (2.4.5) takes the matrix form

$$\sum_{m=-\infty}^{\infty} \sum_{n=-\infty}^{\infty} \frac{p(p - mG_1 - nG_2) + i\beta(p)\alpha_0(p - mG_1 - mG_2)}{i\beta(p) - \alpha_0(p - mG_1 - nG_2)}$$
$$\times \mathcal{I}_m^{(1)}\big((\beta(p) - i\alpha_0(q - mG_1 - nG_2))A_1\big)$$
$$\times \mathcal{I}_n^{(2)}\big((\beta(p) - i\alpha_0(p - mG_1 - nG_2))A_2\big) R(p - mG_1 - nG_2|k)$$

$$= \sum_{m=-\infty}^{\infty} \sum_{n=-\infty}^{\infty} 2\pi\delta(p - k_{m,n}) \frac{k_{m,n}k - i\beta(k_{m,n})\alpha_0(k)}{i\beta(k_{m,n}) + \alpha_0(k)}$$
$$\times \mathcal{I}_m^{(1)}\big((\beta(k_{m,n}) + i\alpha_0(k))A_1\big)\mathcal{I}_n^{(2)}\big((\beta(k_{m,n}) + i\alpha_0(k))A_2\big), \tag{2.4.9}$$

where we have introduced the notation $k_{m,n} = k + mG_1 + nG_2$. In order to satisfy this equation we represent the scattering amplitude $R(q|k)$ in the form

$$R(q|k) = \sum_{m=-\infty}^{\infty} \sum_{n=-\infty}^{\infty} R_{m,n}(k) 2\pi\delta(q - k - mG_1 - nG_2). \tag{2.4.10}$$

Thus the amplitudes $R_{m,n}$ are the scattering amplitudes into the (m, n) order of diffraction, where the first index shows the order of the diffraction by the grating with the period a_1 and the second index shows the order of the diffraction by the grating with the period a_2. When this representation is substituted into Eq. (2.4.9) we obtain

$$\sum_{l=-\infty}^{\infty} \sum_{l'=-\infty}^{\infty} \frac{k_{m,l}k_{n,l'} + i\beta(k_{m,l})\alpha_0(k_{n,l'})}{i\beta(k_{m,l}) - \alpha_0(k_{n,l'})}$$
$$\times \mathcal{I}_{m-l}^{(1)}\big((\beta(k_{m,l}) - i\alpha_0(k_{n,l'}))A_1\big)\mathcal{I}_{n-l'}^{(2)}\big((\beta(k_{m,l}) - i\alpha_0(k_{n,l'}))A_2\big) R_{l,l'}(k)$$
$$= \frac{k_{m,n}k - i\beta(k_{m,n})\alpha_0(k)}{i\beta(k_{m,n}) + \alpha_0(k)}$$
$$\times \mathcal{I}_m^{(1)}\big((\beta(k_{m,n}) + i\alpha_0(k))A_1\big)\mathcal{I}_n^{(2)}\big((\beta(k_{m,n}) + i\alpha_0(k))A_2\big). \tag{2.4.11}$$

To determine the optimal amplitudes of the gratings we truncate the system of algebraic equations (2.4.11) and equate to zero the specular amplitude of diffraction $R_{0,0}$. The number of diffraction orders to be taken into account in this system depends on the angle θ_c at which the blazing effect is desired. In the case where $\theta_c > 19.5°$ (as in the Littrow mount case) it is sufficient to keep only four orders, namely $(0, 0)$, $(1, 0)$, $(0, -1)$, $(1, -1)$, when calculating the amplitude $R_{0,0}$. The smaller the angle θ_c the larger the number of the diffraction orders that fall into the radiative region and, in principle, should be taken into account. However, for weakly rough gratings the diffraction amplitudes of higher orders (possible cross-talks) are weak and can be neglected. In most cases it is sufficient to keep only seven orders to calculate the amplitude $R_{0,0}$. The intensities of the cross-talk peaks that fall into the radiative region cannot be set to zero simultaneously, but they are weak being due to higher order scattering processes. Note, that the same method can be used to determine the optimal amplitude of the grating in the Littrow mount configuration, where only two orders of diffraction can be kept in the matrix equation. However, in this case the reduced Rayleigh equation should be used with care since the optimal amplitude of the grating is sufficiently large to violate the applicability of the Rayleigh hypothesis [2.28].

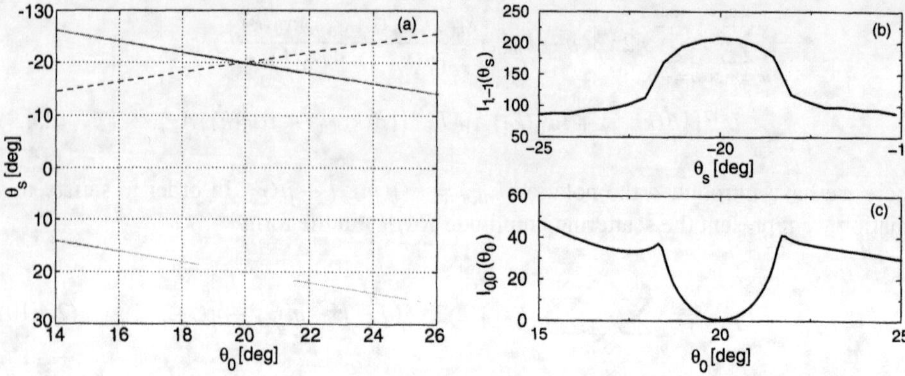

Figure 2.27. (a) A gray-level plot of the angular dependence of the scattered intensity. The gratings were designed to produce an enhanced backscattering peak and no specular peak when the angle of incidence $\theta_c = 20°$. The dashed line shows the retroreflection direction. The intensities of the $(1, -1)$ diffracted order (b) and $(0, 0)$ (specular) diffracted order (c) as functions of the angle of scattering in the vicinity of the retroreflection and specular direction, respectively.

To obtain a purely numerical solution to the problem of the scattering of the incident p-polarized light by a quasi-grating we use the rigorous numerical simulation method presented in Section 2.2.2.1. The numerical calculations were carried out for the case where the quasi-grating defined by Eq. (2.4.2) is ruled on a silver surface. The wavelength of the incident light was chosen to be $\lambda = 457.9$ nm. The dielectric function of the silver at this frequency is $\epsilon(\omega) = -7.25 + i0.24$. The specified angle of incidence θ_c was chosen to be $\theta_c = 20°$. The optimal amplitudes of the quasi-grating required to eliminate the specular peak for p-polarized incident light were determined by the solution of the system of 4 algebraic equations (2.4.11) and are $A_1 = 0.14$ μm, $A_2 = 0.031$ μm. The length of the grating was taken to be $L = 280$ μm. In Fig. 2.27(a) we present a gray-level plot of the intensity of the scattered light, when the angle of incidence is in the vicinity of the specified direction defined by θ_c. In Figs. 2.27(b) and 2.27(c) the intensities of the $(1, -1)$ order (b) ($k - 2(\omega/c)\sin\theta_c$ peak) and $(0, 0)$ order (c) (specular peak) are plotted as functions of the angle of scattering in the vicinity of the angle θ_c. The enhancement of the intensity of the $(1, -1)$ order peak in the vicinity of $\theta_s = -\theta_c$ is due to the coherent interference of the double scattering processes mediated by surface plasmon polaritons exactly as in the case of the enhanced backscattering phenomenon in light scattering from randomly rough surfaces [2.44]. In the case where the optimal amplitudes of the gratings are chosen the enhancement reaches a factor of 2. For arbitrary amplitudes of the gratings the enhancement of the $(1, -1)$ order is smaller, but still exists. This enhancement was demonstrated experimentally by O'Donnell et al. [2.59].

2.5. Surfaces that Synthesize Infrared Absorption Spectra

Correlation spectroscopy [2.60–2.62] is a diagnostic tool in which the degree of correlation between the transmission or reflection of an unknown sample and that of a reference cell containing a known compound is determined over a fixed spectral range, as a mean of identifying the unknown sample. In the case that the known compound in the reference cell is toxic and/or corrosive, or is short-lived, it is useful to have a diffractive optical element that synthesizes the infrared spectrum of that compound for use in a correlation spectrometer.

The production of a synthetic spectrum requires the design of a diffractive element, a one-dimensional rough surface, with the property that at a fixed diffraction angle the spectrum of the light diffracted from the surface accurately reproduces a desired spectrum. In the existing theoretical approaches to the solution of this design problem [2.63–2.66] the Fraunhofer approximation [2.67] is used to express the diffracted field as a function of the wavelength of the incident infrared field, which is a plane wave that illuminates the surface at normal incidence. In [2.63,2.64] it is assumed that the surface imposes only a phase profile $\phi_\lambda(x_1)$ on the incident optical field, where x_1 is the distance measured along the surface. The subscript λ indicates that the phase profile is expected to depend on the wavelength of the incident radiation. A modified Gerchberg–Saxton algorithm [2.68–2.70] is used to obtain a phase profile $\phi(x_1)$ that produces a close approximation to the desired spectrum. The phase profile determined in this way is then converted into a surface-relief profile $d(x_1)$, where $d(x_1) = \phi(x_1)\lambda_1/4\pi$, and λ_1 is a specific wavelength within the spectral range of interest (usually the wavelength that corresponds to the largest feature in the spectrum). The resulting function $d(x_1)$ is represented in the form of a grating-like structure of N lines each of width Δ, whose depths relative to a base level are drawn from a set of discrete values. This approach produces theoretical spectra in good agreement with experimental spectra. However, the phase profile of the resulting diffractive element matches the one obtained by the use of the Gerchberg–Saxton algorithm, only at the wavelength λ_1. At other wavelengths design errors are present, and a simulated annealing approach [2.71] to eliminating them is suggested.

In [2.65] the diffractive optical element is represented by a grating-like structure consisting of N micromirror elements each of width Δ. The vertical positions of these micromirror elements are adjusted by the Davidon–Fletcher–Powell iterative optimization procedure [2.72] to produce a wavelength dependence of the intensity of the diffracted field at the prescribed diffraction angle that matches the infrared spectrum of a given compound throughout the entire spectral range of interest.

In related work Belikov and Solgaard [2.66] presented an analytic approach to the generation of a two-dimensional rough surface, defined by $x_3 = \zeta(\mathbf{x}_\parallel)$, where $\mathbf{x}_\parallel = (x_1, x_2, 0)$, that produces a scattered field in a specified direction whose complex angular spectrum has a prescribed dependence on the wavelength of the incident light. In the Fraunhofer approximation the angular spectrum of the scattered field is expressed as the one-dimensional Fourier transform of the fractional level density $s(h)$ of $\zeta(\mathbf{x}_\parallel)$, where $s(h)\,dh$ is the fraction of the area S of the surface within which $h < \zeta(\mathbf{x}_\parallel) < h + dh$. The inversion of this transform to obtain $s(h)$ is straightforward, and from knowledge of this function a realization of the surface profile function can be generated. The authors suggest that such an optical element could be fabricated as a two-dimensional array of dual-state tiltable mirrors, similar to the digital micromirror device made by Texas Instruments.

In this section we present two approaches to the design of a one-dimensional rough surface that synthesizes a specified experimental spectrum consisting of N lines at the N frequencies $\omega_1, \omega_2, \ldots, \omega_N$ within the spectral range $0 < \omega_{\text{min}} < \omega < \omega_{\text{max}}$. We assume that $\omega_1 < \omega_2 < \cdots < \omega_N$. The first is a probabilistic approach. It leads to a surface in the form of a random lamellar grating, the depths of whose grooves relative to a base level are independent, identically distributed random deviates, obtained from a probability density function that is the solution of the problem of reconstructing a function from a knowledge of the modulus of its Fourier transform. The latter function is expressed in terms of the spectrum to be reproduced. The second approach is a deterministic one, in which the surface profile function is represented as a superposition of cosinusoidal gratings, whose number equals the number of lines in the spectrum to be reproduced. The periods and amplitudes of these gratings are obtained from the frequencies and intensities of the spectral lines, respectively. The widths of the lines are reproduced by modulating the cosines by Gaussian functions, whose $1/e$ half widths are obtained from the experimental line widths. The results obtained by these two approaches are validated by computer simulation calculations.

2.5.1. Random Surfaces

The physical system we consider initially consists of vacuum in the region $x_3 > \zeta(x_1)$ and a perfect conductor in the region $x_3 < \zeta(x_1)$. A perfect conductor is a good approximation to a metal in the infrared range of the optical spectrum. The surface profile function $\zeta(x_1)$ is assumed to be a single-valued function of x_1 that constitutes a random process, but not necessarily a stationary one. The surface $x_3 = \zeta(x_1)$ is illuminated by an s-polarized plane wave of frequency ω, whose plane of incidence is the x_1x_3 plane. The angle of incidence is θ_0. The intensity of the scattered field is measured as a function of ω in the far field at a fixed scattering angle θ_s.

2.5.1.1. The Scattered Field Our starting point is the expression for the single nonzero component of the scattered electric field given by Eq. (2.1.21),

$$E_2(x_1, x_3|\omega)_{\text{sc}} = -\frac{1}{4\pi} \int_{-\infty}^{\infty} dx_1' \left[G_0(x_1, x_3|x_1', x_3')\right]\big|_{x_3'=\zeta(x_1')} F(x_1'|\omega), \tag{2.5.1}$$

the representation of the Green's function $G_0(x_1, x_3|x_1', x_3')$ given by Eq. (2.1.6a), and the expression for the source function $F(x_1|\omega)$ in the Kirchhoff approximation. According to the discussion following Eq. (2.1.58) the latter is given by the first term on the right-hand side of Eq. (2.1.41) together with Eqs. (2.1.42) and (2.1.11), which results in

$$F(x_1|\omega) = -2i\big[k\zeta'(x_1) + \alpha_0(k)\big]\exp\big[ikx_1 - i\alpha_0(k)\zeta(x_1)\big]. \tag{2.5.2}$$

We assume that the observation point is in the far zone, so that $r = (x_1^2 + x_3^2)^{\frac{1}{2}}$ is much larger than x_1' and $\zeta(x_1)_{\max}$. In this limit we have

$$\begin{aligned}\left[(x_1 - x_1')^2 + (x_3 - \zeta(x_1')^2\right]^{\frac{1}{2}} &= r\left[1 + \frac{x_1'^2 - 2x_1'x_1 - 2x_3\zeta(x_1') + \zeta^2(x_1')}{r^2}\right]^{\frac{1}{2}} \\ &\cong r - x_1'\frac{x_1}{r} - \zeta(x_1')\frac{x_3}{r} + \cdots \end{aligned} \tag{2.5.3}$$

to the lowest nonzero order in x_1' and $\zeta(x_1')$. The use of this result together with the large argument expression for the Hankel function $H_0^{(1)}(z)$ [2.73],

$$H_0^{(1)}(z) \cong \sqrt{\frac{2}{\pi z}}\exp\left[iz - i(\pi/4)\right], \tag{2.5.4}$$

yields the Green's function $G_0(x_1, x_3|x_1', \zeta(x_1'))$ in the form

$$\begin{aligned}&G_0\left(x_1, x_3|x_1', \zeta(x_1')\right) \\ &\quad\cong i\left(\frac{2\pi c}{\omega r}\right)^{\frac{1}{2}}\exp\left\{i(\omega/c)\left[r - (x_1/r)x_1' - (x_3/r)\zeta(x_1')\right] - i(\pi/4)\right\}.\end{aligned} \tag{2.5.5}$$

On combining this result with the expression for the source function $F(x_1'|\omega)$ given by Eq. (2.5.2), we find that the scattered field is given by

$$\begin{aligned}E_2(x_1, x_3|\omega)_{\mathrm{sc}} = &-\left(\frac{\omega}{2\pi cr}\right)^{\frac{1}{2}}\exp\left[i(\omega/c)r - i(\pi/4)\right] \\ &\times\int_{-\infty}^{\infty} dx_1' \exp\left[-i(\omega/c)x_1'(\sin\theta_s - \sin\theta_0)\right] \\ &\times\exp\left[-i(\omega/c)\zeta(x_1')(\cos\theta_s + \cos\theta_0)\right]\left[\cos\theta_0 + \zeta'(x_1')\sin\theta_0\right].\end{aligned} \tag{2.5.6}$$

In writing Eq. (2.5.6) we have used the results that $(x_1/r) = \sin\theta_s$, where θ_s is the scattering angle, measured clockwise from the x_3 axis, while $(x_3/r) = \cos\theta_s$.

We can transform Eq. (2.5.6) into a form in which no derivative of the surface profile function is present. To do this we introduce the function $I(\gamma|Q)$ through the relation

$$\exp\left[-i\gamma\zeta(x_1)\right] = \int_{-\infty}^{\infty}\frac{dQ}{2\pi}I(\gamma|Q)\exp(iQx_1). \tag{2.5.7}$$

It follows that

$$-i\gamma\zeta'(x_1)\exp[-i\gamma\zeta(x_1)] = \int_{-\infty}^{\infty} \frac{dQ}{2\pi} iQI(\gamma|Q)\exp(iQx_1). \tag{2.5.8}$$

With these results Eq. (2.5.6) can be rewritten as

$$\begin{aligned} E_2(x_1, x_3|\omega)_{sc} &= -\left(\frac{\omega}{2\pi cr}\right)^{\frac{1}{2}} \exp\bigl[i(\omega/c)r - i(\pi/4)\bigr] \\ &\quad \times \int_{-\infty}^{\infty} dx_1' \exp\bigl[-i(\omega/c)x_1'(\sin\theta_s - \sin\theta_0)\bigr] \\ &\quad \times \int_{-\infty}^{\infty} \frac{dQ}{2\pi} \exp(iQx_1')I\bigl((\omega/c)(\cos\theta_s + \cos\theta_0)|Q\bigr) \\ &\quad \times \left[\cos\theta_0 - \sin\theta_0 \frac{Q}{(\omega/c)(\cos\theta_s + \cos\theta_0)}\right] \\ &= -\left(\frac{\omega}{2\pi cr}\right)^{\frac{1}{2}} \exp\bigl[i(\omega/c)r - i(\pi/4)\bigr] \frac{1 + \cos(\theta_s + \theta_0)}{\cos\theta_s + \cos\theta_0} \\ &\quad \times I\bigl((\omega/c)(\cos\theta_s + \cos\theta_0)|(\omega/c)(\sin\theta_s + \sin\theta_0)\bigr). \end{aligned} \tag{2.5.9}$$

If we now use the inverse of Eq. (2.5.7),

$$I(\gamma|Q) = \int_{-\infty}^{\infty} dx_1 \exp(-iQx_1)\exp\bigl(-i\gamma\zeta(x_1)\bigr), \tag{2.5.10}$$

we obtain finally

$$\begin{aligned} &E_2(x_1, x_3|\omega)_{sc} \\ &\quad = -\left(\frac{\omega}{2\pi cr}\right)^{\frac{1}{2}} \frac{1 + \cos(\theta_s + \theta_0)}{\cos\theta_s + \cos\theta_0} \exp\bigl[i(\omega/c)r - i(\pi/4)\bigr] \\ &\quad \times \int_{-\infty}^{\infty} dx_1 \exp\bigl[-i(\omega/c)(\sin\theta_s - \sin\theta_0)x_1 - i(\omega/c)(\cos\theta_s + \cos\theta_0)\zeta(x_1)\bigr]. \end{aligned} \tag{2.5.11}$$

2.5.1.2. The Mean Scattered Intensity The squared modulus of the scattered field, averaged over the ensemble of realizations of the surface profile function, is

$$\begin{aligned}\langle|E_2(x_1,x_3|\omega)_{sc}|^2\rangle &= \left(\frac{\omega}{2\pi cr}\right)\left[\frac{1+\cos(\theta_s+\theta_0)}{\cos\theta_s+\cos\theta_0}\right]^2 \\ &\quad\times\int_{-\infty}^{\infty}dx_1'\int_{-\infty}^{\infty}dx_1''\exp\bigl[-i(\omega/c)(\sin\theta_s-\sin\theta_0)(x_1'-x_1'')\bigr] \\ &\quad\times\bigl\langle\exp\bigl[-i(\omega/c)(\cos\theta_s+\cos\theta_0)\bigl(\zeta(x_1')-\zeta(x_1'')\bigr)\bigr]\bigr\rangle. \end{aligned}\tag{2.5.12}$$

Our goal is to find a surface profile function $\zeta(x_1)$ for which the right-hand side of this equation reproduces the frequency dependence of the expression on the left-hand side.

To evaluate the double integral on the right-hand side of Eq. (2.5.12) we begin by assuming that the surface profile function has the form

$$\zeta(x_1)=\alpha x_1+b_1d_n,\quad nb<x_1<(n+1)b,\ n=-N_s,-N_s+1,\ldots,N_s,\tag{2.5.13}$$

where N_s is a large integer, the $\{d_n\}$ are independent identically distributed random deviates, b and b_1 are characteristic lengths, and α is a characteristic slope of the surface that will be determined below. Because the $\{d_n\}$ are independent identically distributed random deviates, the probability density function of d_n,

$$f(\gamma)=\bigl\langle\delta(\gamma-d_n)\bigr\rangle,\tag{2.5.14}$$

where the angle brackets now denote an average over the ensemble of realizations of d_n, is independent of n. The assumption that the surface profile function has the form given by Eq. (2.5.13) violates our initial assumption that $\zeta(x_1)$ is a single-valued function of x_1. However, we will see in what follows that the discontinuities of the surface profile function at $x_1=nb$, where $n=0,\pm1,\pm2,\ldots$, do not compromise our determination of $f(\gamma)$ or of the frequency dependence of the intensity of the scattered field.

The double integral on the right-hand side of Eq. (2.5.12) is now rewritten as

$$\begin{aligned}P &= \int_{-\infty}^{\infty}dx_1'\int_{-\infty}^{\infty}dx_1''\exp\bigl[-ig(x_1'-x_1'')\bigr]\bigl\langle\exp\bigl[-ih\bigl(\zeta(x_1')-\zeta(x_1'')\bigr)\bigr]\bigr\rangle \\ &= \sum_{m=-N_s}^{N_s-1}\sum_{n=-N_s}^{N_s-1}\int_{mb}^{(m+1)b}dx_1'\int_{nb}^{(n+1)b}dx_1''\exp\bigl[-ig(x_1'-x_1'')\bigr] \\ &\quad\times\bigl\langle\exp\bigl[-ih(\alpha x_1'+b_1d_m)\bigr]\exp\bigl[ih(\alpha x_1''+b_1d_n)\bigr]\bigr\rangle,\end{aligned}\tag{2.5.15}$$

where, to simplify the notation, we have introduced the functions

$$g(\theta_s, \theta_0) = (\omega/c)(\sin\theta_s - \sin\theta_0), \tag{2.5.16a}$$

$$h(\theta_s, \theta_0) = (\omega/c)(\cos\theta_s + \cos\theta_0). \tag{2.5.16b}$$

Note that the function $h(\theta_s, \theta_0)$ was called $a(\theta_s, \theta_0)$ in Eq. (2.1.61). In the double sum in Eq. (2.5.15) we separate the terms with $n = m$ from those with $n \neq m$, and use the fact that the $\{d_n\}$ are independent random deviates, to obtain

$$\begin{aligned} P = &\sum_{m=-N_s}^{N_s-1} \int_{mb}^{(m+1)b} dx_1' \int_{mb}^{(m+1)b} dx_1'' \exp[-i(g+\alpha h)(x_1' - x_1'')] \\ &+ \sum_{m=-N_s}^{N_s-1} {\sum_{n=-N_s}^{N_s-1}}' \int_{mb}^{(m+1)b} dx_1' \int_{nb}^{(n+1)b} dx_1'' \exp[-i(g+\alpha h)(x_1' - x_1'')] \\ &\times \langle \exp(-ihb_1 d_m)\rangle \langle \exp(ihb_1 d_n)\rangle, \end{aligned} \tag{2.5.17}$$

where the prime on the double sum indicates that the terms with $n = m$ are omitted. If we define the function $F(\nu)$ by

$$F(\nu) = \int_{-\infty}^{\infty} d\gamma\, f(\gamma) \exp(-i\nu\gamma), \tag{2.5.18}$$

we can rewrite Eq. (2.5.17) as

$$\begin{aligned} P = &\sum_{m=-N_s}^{N_s-1} \int_{mb}^{(m+1)b} dx' \exp[-i(g+\alpha h)x_1'] \\ &\times \int_{mb}^{(m+1)b} dx_1'' \exp[i(g+\alpha h)x_1''][1 - |F(hb_1)|^2] \\ &+ \sum_{m=-N_s}^{N_s-1} \sum_{m=-N_s}^{N_s-1} \int_{mb}^{(m+1)b} dx_1' \exp[-i(g+\alpha h)x_1'] \\ &\times \int_{nb}^{(n+1)b} dx_1'' \exp[i(g+\alpha h)x_1''] |F(hb_1)|^2. \end{aligned} \tag{2.5.19}$$

The integrals and the resulting sums are readily carried out, with the result that

$$P = 2N_s b^2 \operatorname{sinc}^2\big[(g+\alpha h)(b/2)\big]\big[1 - |F(hb_1)|^2\big] + (2N_s b)^2 \operatorname{sinc}^2\big[(g+\alpha h)N_s b\big]|F(hb_1)|^2. \tag{2.5.20}$$

The use of the normalization condition for $f(\gamma)$ enables showing that $|F(hb_1)|^2 \leqslant 1$, so that the right-hand side of Eq. (2.5.20) is non-negative, as it must be.

For a given angle of incidence θ_0 we choose the characteristic slope α by requiring that the argument of each sinc function in Eq. (2.5.20) vanishes,

$$\alpha = -\frac{g(\theta_s, \theta_0)}{h(\theta_s, \theta_0)} = -\frac{\sin\theta_s - \sin\theta_0}{\cos\theta_s + \cos\theta_0} = -\tan\left(\frac{\theta_s - \theta_0}{2}\right), \tag{2.5.21}$$

where we have used Eqs. (2.5.16) and (2.1.90). This result means that the direction of observation is the direction of specular reflection from each segment of the surface.

Since sinc(0) equals unity, we have obtained the result that

$$\langle |E_2(x_1, x_3|\omega)_{\mathrm{sc}}|^2 \rangle = \left(\frac{\omega}{2\pi c r}\right)\left[\frac{1+\cos(\theta_s+\theta_0)}{\cos\theta_s + \cos\theta_0}\right]^2 \times 2N_s(2N_s - 1)b^2\left[\big|F\big((\omega/c)(\cos\theta_s + \cos\theta_1)b_1\big)\big|^2 + \frac{1}{2N_s - 1}\right]. \tag{2.5.22}$$

Let us define the experimental intensity $I(\omega)$ by

$$\langle |E_2(x_1, x_3|\omega)_{\mathrm{sc}}|^2 \rangle = 2N_s(2N_s - 1)b^2\left(\frac{\omega}{2\pi c r}\right)\left[\frac{1+\cos(\theta_s+\theta_0)}{\cos\theta_s + \cos\theta_0}\right]^2 I(\omega), \tag{2.5.23}$$

so that

$$I(\omega) = \left[\left|F\left(\frac{\omega b_1}{c}(\cos\theta_s + \cos\theta_0)\right)\right|^2 + \frac{1}{2N_s - 1}\right]. \tag{2.5.24}$$

It follows from this result that

$$\left|F\left(\frac{\omega b_1}{c}(\cos\theta_s + \cos\theta_0)\right)\right| = \left[I(\omega) - \frac{1}{2N_s - 1}\right]^{\frac{1}{2}}. \tag{2.5.25}$$

The problem of determining the surface profile function $\zeta(x_1)$ thus reduces to the problem of obtaining the pdf $f(\gamma)$ from a knowledge of the modulus of its Fourier transform. This is a classical inverse problem, and we now turn to a description of its solution.

2.5.1.3. Determination of $f(\gamma)$ from $|F(\nu)|$ In this section we describe the approach used to obtain the pdf of d_n, $f(\gamma)$, from a knowledge of the modulus of its Fourier transform $|F(\nu)|$. We begin by rewriting Eq. (2.5.18) in the form

$$\int_{-\infty}^{\infty} d\gamma\, f(\gamma)\exp(-i\nu\gamma) = F(\nu) \equiv \left|F(\nu)\right| \exp\left[i\chi(\nu)\right]. \tag{2.5.26}$$

If $F(\nu)$ were known, $f(\gamma)$ would be given by

$$f(\gamma) = \int_{-\infty}^{\infty} \frac{d\nu}{2\pi} F(\nu)\exp(i\gamma\nu). \tag{2.5.27}$$

However, we know only $|F(\nu)|$, which is given by Eq. (2.5.25),

$$\left|F(\nu)\right| = \left[I\left(\frac{c\nu}{b_1(\cos\theta_s + \cos\theta_0)}\right) - \frac{1}{2N_s - 1}\right]^{\frac{1}{2}}. \tag{2.5.28}$$

We therefore adopt an iterative approach [2.68–2.70] to obtain $f(\gamma)$ from $|F(\nu)|$.

The experimental spectrum is known in a frequency range that encompasses the frequencies $\omega_1, \omega_2, \ldots, \omega_N$ at which its lines are centered. We will denote this range by

$$\omega_{\min} \leqslant \omega \leqslant \omega_{\max}, \tag{2.5.29}$$

where

$$\omega_{\min} < \omega_1, \omega_N < \omega_{\max}. \tag{2.5.30}$$

This means that $|F(\nu)|$ is known only for positive values of ν, in a range that we denote by

$$\nu_{\min} < \nu < \nu_{\max}, \tag{2.5.31}$$

where

$$\nu_{\min} = \frac{\omega_{\min}}{c}(\cos\theta_s + \cos\theta_0)b_1, \tag{2.5.32a}$$

$$\nu_{\max} = \frac{\omega_{\max}}{c}(\cos\theta_s + \cos\theta_0)b_1. \tag{2.5.32b}$$

However, it is convenient for what follows to assume that $|F(\nu)|$ exists in the interval

$$-\nu_{\max} < \nu < \nu_{\max}, \tag{2.5.33}$$

and is an even function of ν,

$$\left|F(-\nu)\right| = \left|F(\nu)\right|. \tag{2.5.34}$$

We can do this with no loss of generality, because the form of $|F(\nu)|$ for ν outside the interval (2.5.31) is of no interest to us.

Moreover, instead of working with Fourier integral transforms we will use discrete Fourier transforms. Thus, we introduce the definitions

$$T_0 = 2\nu_{\max}, \tag{2.5.35}$$

$$\nu_k = \frac{T_0}{2M}k, \quad k = -M, -M+1, \ldots, M, \tag{2.5.36}$$

$$\gamma_n = \frac{2\pi}{T_0}n, \quad n = -M, -M+1, \ldots, M, \tag{2.5.37}$$

and

$$F(\nu_k) \equiv F_k, \qquad f(\gamma_n) \equiv f_n. \tag{2.5.38}$$

It follows from Eqs. (2.5.35) and (2.5.37) that the domain of existence of $f(\gamma)$ is the interval

$$-\frac{\pi}{\nu_{\max}}M < \gamma < \frac{\pi}{\nu_{\max}}M. \tag{2.5.39}$$

The discrete analogues of Eqs. (2.5.26) and (2.5.27) are

$$F_k = \sum_{n=-M}^{M} f_n e^{-i\frac{\pi kn}{M}}, \quad k = -M, -M+1, \ldots, M, \tag{2.5.40}$$

$$f_n = \frac{1}{2M}\sum_{k=-M}^{M} c_k F_k e^{i\frac{\pi nk}{M}}, \quad n = -M, -M+1, \ldots, M, \tag{2.5.41}$$

where

$$c_k = \begin{cases} \frac{1}{2} & k = -M, M, \\ 1 & k \neq -M, M. \end{cases} \tag{2.5.42}$$

Because it is a probability density function, $f(\gamma)$ must be real, non-negative, and normalized to unity. The first of these conditions has the consequence that the phase $\chi(\nu)$ of the function $F(\nu) = |F(\nu)|\exp[i\chi(\nu)]$ must be an odd function of ν,

$$\chi(-\nu) = -\chi(\nu). \tag{2.5.43}$$

The last of these conditions requires that

$$\chi(0) = 0. \tag{2.5.44}$$

We begin the iterative determination of $f(\gamma)$ by generating, for positive values of k, a random sequence of real numbers $\{\chi_0(\nu_k)\}$, drawn from a uniform probability density function in the interval $(-\pi, \pi)$, and defining $\chi_0(\nu_{-k}) = -\chi_0(\nu_k)$, with $\chi(0) = 0$. We then construct the complex functions $F_k^{(0)} = |F(\nu_k)| \exp[i\chi_0(\nu_k)]$. From the $\{F^{(0)}\}$ we evaluate the sum

$$f_n^{(0)} = \frac{1}{2M} \sum_{k=-M}^{M} c_k F_k^{(0)} e^{i\frac{\pi nk}{M}}. \tag{2.5.45}$$

We implement the constraint $f(\gamma) \geqslant 0$ in the following way. Each coefficient $f_n^{(0)}$ is checked to determine whether it is negative or positive. If it is negative it is set equal to zero. If $f_n^{(0)}$ is positive it is kept as it is. The resulting set of values of $f_n^{(0)}$ is denoted by $\{\tilde{f}_n^{(0)}\}$ after it has been scaled to satisfy the normalization condition

$$\frac{2\pi}{T_0} \sum_{n=-M}^{M} \tilde{f}_n^{(0)} = 1. \tag{2.5.46}$$

The $\{\tilde{f}_n^{(0)}\}$ are then used to generate a new set of values $\{F_k^{(1)}\}$,

$$F_k^{(1)} = \sum_{n=-M}^{N} \tilde{f}_n^{(0)} e^{i\frac{\pi kn}{M}} = \left|F_k^{(1)}\right| e^{i\chi_1(\nu_k)}. \tag{2.5.47}$$

From the set $\{F_k^{(1)}\}$ we generate a new set $\{\tilde{F}_k^{(1)}\}$ according to

$$\tilde{F}_k^{(1)} = \left|F(\nu_k)\right| e^{i\chi_1\nu_k}. \tag{2.5.48}$$

From this set the next iterate for f_n is obtained according to

$$f_n^{(1)} = \frac{1}{2M} \sum_{k=-M}^{M} c_k \tilde{F}_k^{(1)} e^{i\frac{\pi nk}{M}}. \tag{2.5.49}$$

Each coefficient $f_n^{(1)}$ is then checked to determine whether it is negative or positive. Depending on the outcome of this check, this coefficient is set equal to zero of left unchanged, according to the criteria stated above. After the resulting set of coefficients is scaled so that it satisfies the normalization conditions imposed on $f(\gamma)$, it is denoted by $\{\tilde{f}_n^{(1)}\}$.

The set $\{\tilde{f}_n^{(1)}\}$ is used to generate a new set of values $\{F_k^{(2)}\}$,

$$F_k^{(2)} = \sum_{n=-M}^{M} \tilde{f}_n^{(1)} e^{-i\frac{\pi kn}{M}} = \left|F_k^{(2)}\right| e^{i\chi_2(\nu_k)}, \tag{2.5.50}$$

and from it we construct a new set $\{\tilde{F}_k^{(2)}\}$ according to

$$\tilde{F}_k^{(2)} = \left|F(\nu_k)\right|e^{i\chi_2(\nu_k)}, \tag{2.5.51}$$

and the iteration scheme proceeds as before. In each iteration the resulting function $|F_k^{(n)}|$ is compared with the input function $F(\nu_k)$. When the inequality $(|F(\nu_k)| - |F_k^{(n)}|)/(|F(\nu_k)| + |F_k^{(n)}|) \ll 1$ is satisfied for each value of k over the entire interval $(\nu_{\min}, \nu_{\max})$ the iteration procedure is considered to have converged.

Five to eight iterations usually suffice to produce a good result for $f(\gamma)$.

Once the function $f(\gamma)$ has been determined, a long sequence of $\{d_n\}$ is obtained from it by, e.g., the rejection method [2.6], and a realization of the surface profile function is constructed on the basis of Eq. (2.5.13).

2.5.1.4. The Solution of the Scattering Problem The surface profile defined by Eq. (2.5.13) is a multi-valued function of x_1 because of the jump discontinuities it displays at $x_1 = nb$, $n = 0, \pm 1, \pm 2, \ldots$. This means that in a rigorous numerical solution of the problem of the scattering of an electromagnetic wave from it we can no longer replace integration along the surface profile by integration along the x_1 axis, as was done in going from Eq. (2.1.14) to Eq. (2.1.16), for example. It is now necessary to integrate along the profile. In this section we indicate how this can be done, following the approach of Mendoza-Suárez and Méndez [2.73], as modified by Simonsen [2.74]. The presentation of this approach will be in the context of the problem studied in this section, namely the scattering of s-polarized light from a perfectly conducting surface defined by Eq. (2.5.13). However, as reference to the work of Mendoza-Suárez and Méndez [2.73] shows, it is applicable to scattering from much more general multi-valued surface profiles. The results for the wavelength dependence of the intensity of the scattered field obtained by this approach are then used to judge how well the surface generated in the manner described in Sections 2.5.1.2–2.5.1.3 reproduces the specified infrared absorption spectrum.

We begin by representing the surface profile by the continuous vector-valued function $\mathbf{R} = (\xi, \eta)$, where (ξ, η) are the coordinates of a point on the profile. In the case of a single-valued surface profile these coordinates would be $(x_1, \zeta(x_1))$. In the case of a multi-valued profile a more general notation is needed.

From Eq. (2.1.13) we can represent the scattered field in the form

$$E_2(\mathbf{r}|\omega)_{\rm sc} = \frac{1}{4\pi}\int_s ds'\, G_0(\mathbf{r}|\mathbf{R}')F(\mathbf{R}'|\omega), \tag{2.5.52}$$

where ds is the element of path length along the curve s, $\mathbf{r} = (x_1, x_3)$, and the source function $F(\mathbf{R}|\omega)$ is defined as

$$F(\mathbf{R}|\omega) = \hat{\mathbf{n}}(\mathbf{R}) \cdot \left(\nabla E_2(\mathbf{r}|\omega)\right)\Big|_{\mathbf{r}=\mathbf{R}}. \tag{2.5.53}$$

In Eq. (2.5.53) $\hat{\mathbf{n}}(\mathbf{R})$ is the unit vector normal to the surface at each point, directed from the scattering medium into the vacuum, and the operator ∇ denotes the gradient, expressed as

$$\nabla = \left(\frac{\partial}{\partial x_1}, \frac{\partial}{\partial x_3}\right). \tag{2.5.54}$$

The Green's function $G_0(\mathbf{r}|\mathbf{r}')$ is given by

$$G_0(\mathbf{r}|\mathbf{r}') = i\pi H_0^{(1)}\big((\omega/c)|\mathbf{r} - \mathbf{r}'|\big). \tag{2.5.55}$$

If we use the large argument representation of the Hankel function $H_0^{(1)}(z)$ [2.75],

$$H_0^{(1)}(z) \sim \left(\frac{2}{\pi z}\right)^{\frac{1}{2}} \exp\big[i\big(z - (\pi/4)\big)\big], \quad |z| \to \infty, \tag{2.5.56}$$

Eq. (2.5.52) yields the scattered field in the far zone in the form

$$E_2(\mathbf{r}|\omega)_{\mathrm{sc}} = -\frac{i}{4}\left(\frac{2c}{\pi r\omega}\right)^{\frac{1}{2}} \exp\big[i\big[(\omega/c)r - (\pi/4)\big]\big] r_s(\theta_s), \tag{2.5.57}$$

with

$$r_s(\theta_s) = \int_{-\infty}^{\infty} ds\, \exp\big[-i(\omega/c)\hat{\mathbf{r}} \cdot \mathbf{R}\big] F(\mathbf{R}|\omega). \tag{2.5.58}$$

In Eq. (2.5.58) the unit vector $\hat{\mathbf{r}}$ is given by $(\sin\theta_s, \cos\theta_s)$.

The source function $F(\mathbf{R}|\omega)$ can be determined from the integral equation (the analogue of Eq. (2.1.38))

$$E_2(\mathbf{R}'|\omega)_{\mathrm{inc}} = \frac{1}{4\pi} \int_{-\infty}^{\infty} ds\, G_0\big(\mathbf{R}' + \varepsilon\hat{\mathbf{n}}(\mathbf{R}')|\mathbf{R}\big) F(\mathbf{R}|\omega), \tag{2.5.59}$$

where ε is a positive infinitesimal. For the incident field we assume the plane wave

$$E_2(\mathbf{r}|\omega)_{\mathrm{inc}} = \exp\big[i(\omega/c)\hat{\mathbf{r}}_0 \cdot \mathbf{r}\big], \tag{2.5.60}$$

where the unit vector $\hat{\mathbf{r}}_0$ is expressed in terms of the angle of incidence θ_0 by $\hat{\mathbf{r}}_0 = (\sin\theta_0, -\cos\theta_0)$.

In solving the integral equation (2.5.59) we begin by replacing the infinitely long surface by a surface of finite length, which is represented by the curve Γ. We seek to describe the surface profile parametrically in terms of the arc length along the curve Γ, so that we can

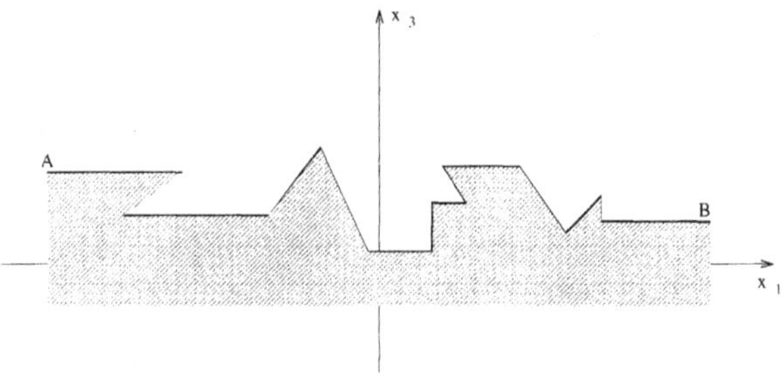

Figure 2.28. The geometry of the scattering problem studied in this section.

express the equation for the scattering of an s-polarized wave from this surface in terms of this parameter.

We assume that the surface profile Γ is a rectifiable Jordan arc [2.76] with a finite number of singular points [2.77], such that Γ is a directed path from point A to point B (see Fig. 2.28). With these conditions it is possible to find a parameter t, defined in a closed interval, such that the curve Γ can be described by a continuous single- and vector-valued function $\mathbf{R}$ of this parameter. It is convenient to choose the parameter t equal to the length of the curve Γ measured from A, the initial point of the curve. The profile of the surface can then be represented by the vector function $\mathbf{R}(s)$, such that the interval within which s is defined, I_T, is $[0, L_T]$, where L_T represents the total length of the curve Γ, viz. $L_T = \int_\Gamma ds$.

We assume that the first and second derivatives of the function $\mathbf{R}(s)$ exist at each regular point of Γ. On the other hand, Eq. (2.5.59) is not well defined at the singular points of the curve Γ, and to deal with this problem we proceed as follows. We assume that the curve Γ possesses M singular points (where M can be zero), given by the vectors $\mathbf{R}(s = L_1)$, $\mathbf{R}(s = L_2), \ldots, \mathbf{R}(s = L_M)$. Knowing the M values of the parameters at which the singularities occur, we can divide the interval $[0, L_T]$ into $M + 1$ subintervals: $I_0 = [0, L_1)$, $I_1 = (L_1, L_2), \ldots, I_M = (L_M, L_T]$, so that their union is a new parametric interval I'_T that has no singular points. Although the function $\mathbf{R}(s)$ for the interval I'_T is not continuous, this is not a problem, at least in principle, as it can be shown that the contribution from the singular points is zero, and the source function $F(\mathbf{R}|\omega)$ at the regular points is not affected by the existence of the singular points [2.78]. However, it should be noted that, because of the possibility of having strong variations of the field in the vicinity of the singular points, some difficulties in the representation of the source function in this vicinity may arise, constituting a possible source of cumulative errors.

It is now convenient to express the two components (ξ, η) of the vector-valued function $\mathbf{R}(s)$ as functions of s. Then, at each regular point of the interval I'_T we can write

$$\mathbf{R}(s) = \big(\xi(s), \eta(s)\big). \tag{2.5.61}$$

The two parametric functions $\xi(s)$ and $\eta(s)$ are of central importance to the present formulation of the scattering problem. Given a general profile that satisfies our assumptions, it is always possible to find, analytically or numerically, the functions $\xi(s)$ and $\eta(s)$, as well as their first and second derivative at each point of the internal I'_T, starting from the relation

$$(ds)^2 = (d\xi)^2 + (d\eta)^2. \tag{2.5.62}$$

At least formally, in each subinterval I_j the functions $\xi(s)$ and $\eta(s)$ can be found by inversion of the functions $s(\xi)$ and $s(\eta)$ given by

$$s(\xi) = s(\xi_j) + \int_{\xi_j}^{\xi} \left[1 + \left(\frac{d\eta(\xi)}{d\xi}\right)^2\right]^{\frac{1}{2}} d\xi, \tag{2.5.63a}$$

$$s(\eta) = s(\eta_j) + \int_{\eta_j}^{\eta} \left[1 + \left(\frac{d\xi(\eta)}{d\eta}\right)^2\right]^{\frac{1}{2}} d\eta. \tag{2.5.63b}$$

In Eqs. (2.5.63) (ξ_j, η_j) are the coordinates of the point $\mathbf{R}(s = L_j)$, and $\eta(\xi)$ and $\xi(\eta)$ are functions that describe the surface profile in the interval explored by the variables of the line integrals. This provides a formal procedure for calculating the parametric functions $\xi(s)$ and $\eta(s)$ for each subinterval.

We now introduce the preceding results in the solution of the integral equation (2.5.59) for the source function $F(\mathbf{R}|\omega)$, by converting it into a matrix equation, which can then be solved numerically. This is done in the following way.

For each subinterval I_j we introduce the set $\{R_i^{(j)}\}$ of N_j equally spaced points $R_i^{(j)} = (\xi(s_i^{(j)}), \eta(s_i^{(j)}))$ with $i = 1, 2, \ldots, N_j$. The sampling points $s_i^{(j)}$ are given by $s_i^{(j)} = L_j + (i - 1/2)\Delta s_j$, with $j = 0, 1, 2, \ldots, M$. We take $L_0 = 0$. To define the sampling interval Δs_j we proceed as follows. We first specify the desired sampling interval to be used along the surface profile, say Δ. Then, in terms of the length of the surface between two consecutive singular points, $L_{j+1} - L_j$ for the subinterval I_j, we define Δs_j by [2.74]

$$\Delta s_j = \frac{L_{j+1} - L_j}{[(L_{j+1} - L_j)/\Delta]}, \tag{2.5.64}$$

where $[\cdots]$ denotes the integer part rounded upwards (so that its smallest possible value is one). The value of N_j is then $[(L_{j+1} - L_j)/\Delta]$.

The integral on the right-hand side of Eq. (2.5.59) can now be written in the form

$$\begin{aligned} \text{RHS} &= \frac{1}{4\pi} \int_{\Gamma} ds\, G_0\big(\mathbf{R}' + \varepsilon\hat{\mathbf{n}}(\mathbf{R})|\mathbf{R}\big) F(\mathbf{R}|\omega) \\ &= \frac{1}{4\pi} \sum_{j=0}^{M} \int_{I_j} ds\, G_0\big(\mathbf{R}'|\mathbf{R}(s)\big) F\big(\mathbf{R}(s)|\omega\big), \end{aligned} \tag{2.5.65}$$

where we have passed to the limit $\varepsilon \to 0$, because $G_0(\mathbf{r}|\mathbf{r}')$ has an integrable singularity as $\mathbf{r} \to \mathbf{r}'$. We can write Eq. (2.5.65) as

$$\text{RHS} = \frac{1}{4\pi} \sum_{j=0}^{M} \sum_{i=1}^{N_j} \int\limits_{s_i^{(j)}-\frac{1}{2}\Delta s_j}^{s_i^{(j)}+\frac{1}{2}\Delta s_j} ds\, G\big(\mathbf{R}'|\mathbf{R}(s)\big) F\big(\mathbf{R}(s)|\omega\big). \tag{2.5.66}$$

On the assumption that $F(\mathbf{R}(s)|\omega)$ is a slowly varying function of s in each interval $(s_i^{(j)} - \frac{1}{2}\Delta s_j, s_i^{(j)} + \frac{1}{2}\Delta s_j)$, we evaluate it at the midpoint of each of these intervals, and remove it from the integral. In this way we obtain

$$\text{RHS} \cong \frac{1}{4\pi} \sum_{j=0}^{M} \sum_{i=1}^{N_j} F\big(\mathbf{R}(s_i^{(j)})|\omega\big) \int\limits_{s_i^{(j)}-\frac{1}{2}\Delta s_j}^{s_i^{(j)}+\frac{1}{2}\Delta s_j} ds\, G\big(\mathbf{R}'|\mathbf{R}(s)\big). \tag{2.5.67}$$

We next make the change of variable $s = s_i^{(j)} + \zeta$ to obtain

$$\text{RHS} = \frac{1}{4\pi} \sum_{j=0}^{M} \sum_{i=1}^{N_j} \left\{ \int\limits_{-\frac{1}{2}\Delta s_j}^{\frac{1}{2}\Delta s_j} d\zeta\, G_0\big(\mathbf{R}'|\mathbf{R}(s_i^{(j)} + \zeta)\big) \right\} F\big(\mathbf{R}(s_i^{(j)})|\omega\big). \tag{2.5.68}$$

We now return to the integral equation (2.5.59) and replace the vector $\mathbf{R}'$ by $\mathbf{R}(s_{i'}^{(j')})$, where $i' = 1, 2, \dots, N_{j'}$ and $j' = 0, 1, 2, \dots, M$, and obtain the matrix equation

$$E_2\big(\mathbf{R}(s_{i'}^{(j')})|\omega\big)_{\text{inc}} = \sum_{j=0}^{M} \sum_{i=1}^{N_j} L_{i'\ \ i}^{(j')(j)} F\big(\mathbf{R}(s_i^{(j)})|\omega\big), \tag{2.5.69}$$

where

$$\begin{aligned} L_{i'\ \ i}^{(j')(j)} &= \frac{1}{4\pi} \int\limits_{-\frac{1}{2}\Delta s_j}^{\frac{1}{2}\Delta s_j} d\zeta\, G_0\big(\mathbf{R}(s_{i'}^{(j')})|\mathbf{R}(s_i^{(j)} + \zeta)\big) \\ &= \frac{i}{4} \int\limits_{-\frac{1}{2}\Delta s_j}^{\frac{1}{2}\Delta s_j} d\zeta\, H_0^{(1)}\big((\omega/c)\big|\mathbf{R}(s_{i'}^{(j')}) - \mathbf{R}(s_i^{(j)} + \zeta)\big|\big) \end{aligned}$$

$$= \frac{i}{4} \int_{-\frac{1}{2}\Delta s_j}^{\frac{1}{2}\Delta s_j} d\zeta\, H_0^{(1)}\big((\omega/c)\big[\big(\xi(s_{i'}^{(j')}) - \xi(s_i^{(j)} + \zeta)\big)^2$$

$$+ \big(\eta(s_{i'}^{(j')}) - \eta(s_i^{(j)} + \zeta)\big)^2\big]^{\frac{1}{2}}\big). \tag{2.5.70}$$

In the case where $(i', j') \neq (i, j)$ the argument of the Hankel function never vanishes, and to lowest nonzero order in Δs_j we find

$$L_{i'\ i}^{(j')(j)} = \frac{i\Delta s_j}{4} H_0^{(1)}\big((\omega/c)\big[\big(\xi(s_{i'}^{(j')}) - \xi(s_i^{(j)})\big)^2 + \big(\eta(s_{i'}^{(j')}) - \eta(s_i^{(j)})\big)^2\big]^{\frac{1}{2}}\big),$$

$$(i', j') \neq (i, j). \tag{2.5.71}$$

The evaluation of the diagonal elements requires a bit more care. We have

$$L_{i\ i}^{(j)(j)} = \frac{i}{4} \int_{-\frac{1}{2}\Delta s_j}^{\frac{1}{2}\Delta s_j} d\zeta\, H_0^{(1)}\big((\omega/c)\big[\big(\xi(s_i^{(j)}) - \xi(s_i^{(j)} + \zeta)\big)^2 + \big(\eta(s_i^{(j)}) - \eta(s_i^{(j)} + \zeta)\big)^2\big]^{\frac{1}{2}}\big)$$

$$\cong \frac{i}{4} \int_{-\frac{1}{2}\Delta s_j}^{\frac{1}{2}\Delta s_j} d\zeta\, H_0^{(1)}\big((\omega/c)|\zeta|\big[\big(\xi'(s_i^{(j)})\big)^2 + \big(\eta'(s_i^{(j)})\big)^2\big]^{\frac{1}{2}}\big), \tag{2.5.72}$$

where the prime denotes differentiation with respect to argument. From Eq. (2.5.62) we find that

$$\big[\big(\xi'(s_i^{(j)})\big)^2 + \big(\eta'(s_i^{(j)})\big)^2\big]^{\frac{1}{2}} = 1. \tag{2.5.73}$$

Consequently, Eq. (2.5.72) becomes

$$L_{i\ i}^{(j)\ (j)} = \frac{i}{2} \int_0^{\frac{1}{2}\Delta s_j} d\zeta\, H_0^{(1)}\big((\omega/c)\zeta\big). \tag{2.5.74}$$

With the use of the small argument expansion of $H_0^{(1)}(z)$, Eq. (2.2.51), we find that

$$L_{i\ i}^{(j)(j)} = \frac{i\Delta s_j}{4}\left\{\frac{2i}{\pi}\ell n\left(\frac{1}{2}\frac{\omega}{c}\frac{\Delta s_j}{2e}\right) + i\gamma + 1\right\}. \tag{2.5.75}$$

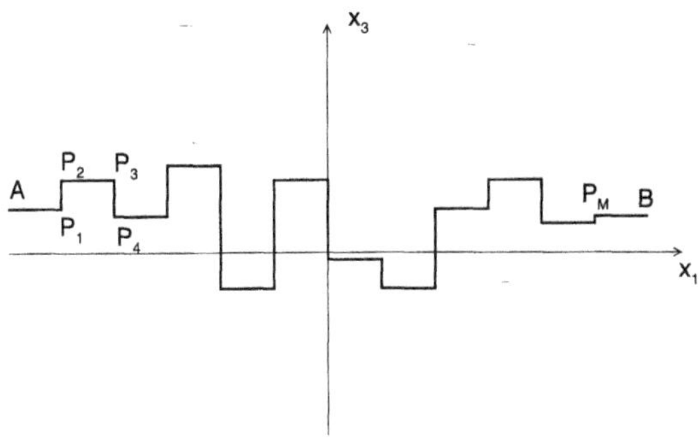

Figure 2.29. A realization of a surface profile defined by Eq. (2.5.13).

A second use of Eqs. (2.2.51) yields

$$L_{i\ i}^{(j)(j)} = i\frac{\Delta s_j}{4} H_0^{(1)}\left(\frac{\omega}{c}\frac{\Delta s_j}{2e}\right). \tag{2.5.76}$$

When the source function $F(\mathbf{R}(s_i^{(j)})|\omega)$ has been determined by solving Eq. (2.5.69), the scattering amplitude $r_s(\theta_s)$ can be calculated from

$$r_s(\theta_s) = \sum_{j=0}^{M}\sum_{i=1}^{N_j} \Delta s_j\, F\big(\xi\big(s_i^{(j)}\big), \eta\big(s_i^{(j)}\big)|\omega\big)$$
$$\times \exp\big\{-i(\omega/c)\big[\xi\big(s_i^{(j)}\big)\sin\theta_s + \eta\big(s_i^{(j)}\big)\cos\theta_s\big]\big\}. \tag{2.5.77}$$

It is then convenient to define the experimental mean intensity by

$$I(\omega) = \frac{1}{N_T}\big\langle\big|r_s(\theta_s)\big|^2\big\rangle, \tag{2.5.78}$$

where N_T is the total number of sampling points

$$N_T = \sum_{j=0}^{M} N_j. \tag{2.5.79}$$

We now apply the preceding results to the problem of the scattering of s-polarized light from a perfectly conducting surface defined by Eq. (2.5.13) in the case that $\alpha = 0$. A realization of this surface is depicted in Fig. 2.29, for the case that $M = 20$. The initial point of the profile is $A = P_0$, and the path of integration is directed to the point $B = P_{M+1}$. Between these two points we have $M = 4K$ singular points $P_1, P_2, \ldots, P_M$. The horizontal segments of the surface are each of length b. The height of the surface in the interval I_{2j} is $d_j b_1$, where $j = 0, 1, 2, \ldots, 2K$.

For the initial and end points the arc lengths measured from A are $s_A = 0$ and $s_B = L_T = (2K+1)\,b + \sum_{j=1}^{2K} |d_j - d_{j-1}| b_1$.

In the horizontal subinterval I_{2j}, between the points P_{2j} and P_{2j+1}, where $j = 0, 1, \ldots, 2K$, we have

$$\mathbf{R}(s) = \mathbf{R}(s_{2j}) + (s - s_{2j}, 0), \tag{2.5.80}$$

where the vector $\mathbf{R}(s_{2j})$ indicates the coordinates of the point P_{2j},

$$\mathbf{R}(s_{2j}) = (jb, d_j b_1). \tag{2.5.81}$$

In the vertical subinterval I_{2j+1}, between the points P_{2j+1} and P_{2j+2}, where $j = 0, 1, \ldots, 2K-1$, we have

$$\mathbf{R}(s) = \mathbf{R}(s_{2j+1}) + \big(0, (s - s_{2j+1})\,\mathrm{sgn}(d_{j+1} - d_j)\big), \tag{2.5.82}$$

where the vector $\mathbf{R}(s_{2j+1})$ indicates the coordinates of the point P_{2j+1},

$$\mathbf{R}(s_{2j+1}) = \big((j+1)b, d_j b_1\big). \tag{2.5.83}$$

From the preceding results and the definition (2.5.61) we obtain finally that in the interval I_{2j}

$$\xi\big(s_i^{(j)}\big) = jb + \left(i - \frac{1}{2}\right)\Delta s_{2j}, \tag{2.5.84a}$$

$$\eta\big(s_i^{(j)}\big) = d_j b_1, \tag{2.5.84b}$$

where $j = 0, 1, 2, \ldots, 2K$ and $i = 1, 2, \ldots, N_{2j}$, while in the interval I_{2j+1}

$$\xi\big(s_i^{(j)}\big) = (j+1)b, \tag{2.5.85a}$$

$$\eta\big(s_i^{(j)}\big) = d_j b_1 + \left(i - \frac{1}{2}\right)\mathrm{sgn}(d_{j+1} - d_j)\Delta s_{2j+1}, \tag{2.5.85b}$$

where $j = 0, 1, 2, \ldots, 2K-1$ and $i = 1, 2, \ldots, N_{2j+1}$.

2.5.1.5. Example To illustrate the approach developed here to the design of a one-dimensional randomly rough surface that produces a scattered field with a specified dependence of its intensity on the wavelength, we consider the synthesis of the IR spectrum of HF in the region 3600–4300 cm^{-1} [2.64]. In this region it consists of fifteen sharp rotational lines superimposed on a broad vibrational background [2.79]. The experimental spectrum is modeled by a sum of narrow Gaussian lines with a weak background,

$$I(\omega) = \sum_{j=1}^{15} A_j \{\exp[-(\omega - \omega_j)^2/\Delta^2] + \exp[-(\omega + \omega_j)^2/\Delta^2]\} + I_0, \qquad (2.5.86)$$

where the amplitude of the strongest peak, A_{10}, is chosen to be $A_{10} = 0.9$, to ensure that $|F(\omega b_1/c)(\cos\theta_s + \cos\theta_0)| \leqslant 1$ over the entire spectral range. To ensure the satisfaction of the condition $|F(0)| = 1$, which follows from the normalization of $f(\gamma)$, and to simplify the iteration procedure, we add to the intensity a Gaussian peak at $\omega = 0$ with an amplitude A_0 that forces $|F(0)|$ to equal unity. Therefore, $I(\omega)$ takes the form

$$I(\omega) = A_0 \exp(-\omega^2/\Delta^2) + \sum_{j=1}^{15} A_j \{\exp[-(\omega - \omega_j)^2/\Delta^2] + \exp[-(\omega + \omega_j)^2/\Delta^2]\} + I_0, \qquad (2.5.87)$$

where

$$A_0 = 1 + 1/(2N - 1) - I_0 - 2\sum_{j=1}^{15} A_j \exp(-\omega_j^2/\Delta^2). \qquad (2.5.88)$$

To simplify the iteration procedure still further, we assume that $I_0 = 1/(2N_s - 1)$. Then by using the procedure described in Sections 2.5.1.2–2.5.1.3 we designed a surface that synthesizes this spectrum for an angle of incidence $\theta_0 = 0°$, and for an angle of scattering $\theta_s = 15°$. Hence the slope α has the value $\alpha = -\tan 7.5° = -0.1317$. Since only the relative intensities of the peaks in the spectrum are of interest, the iteration procedure was stopped as soon as the relative amplitudes of the target spectrum were correct.

The designed surface consisted of $2N_s = 5000$ segments, each of length $b = 10$ μm, for a total length of the optical element of 5 cm. The characteristic depth b_1 was $b_1 = 0.1$ μm. A segment of one realization of the resulting surface profile is presented in Fig. 2.30. The pdf of d_n, $f(\gamma)$, used together with the rejection method and Eq. (2.5.13) to generate this surface profile is presented in Fig. 2.31.

The experimental IR spectrum of HF is plotted in Fig. 2.32(a) [2.79]. The intensity $I(\omega)$, calculated by means of the Kirchhoff approximation for scattering from a single realization of the surface profile function, is plotted in Fig. 2.32(b). The Kirchhoff approximation was used because a surface as long as 5 cm was needed to reproduce such a complicated spectrum. It displays peaks at the correct wavelengths, whose relative amplitudes are very close to those of the peaks in the initial spectrum $I(\omega)$. The absolute amplitudes of the peaks in the reconstructed spectrum are about 40% smaller than those of the corresponding peaks in the initial spectrum. We note that the function we are calculating, $I(\omega)$, is to within a factor $1 - (b/L)$ the ratio of the intensity of the light scattered from the random surface into the direction defined by the scattering angle $\theta_0 = 15°$ and the intensity of the light scattered from a planar surface in the same direction. Thus, a kind of universal attenuation limit on the power efficiency in our case is $1 - (b/L)$. The iteration procedure,

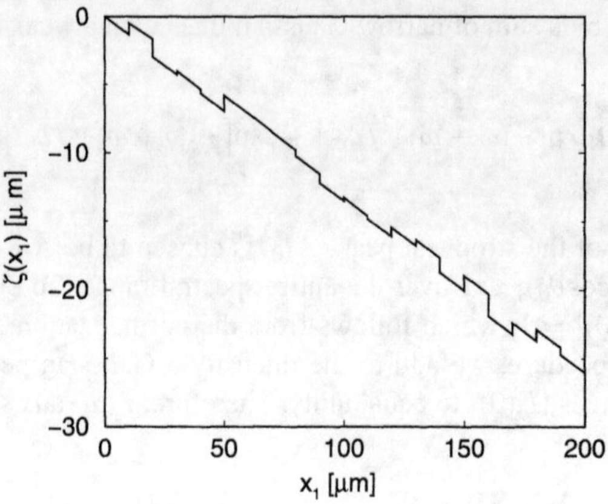

Figure 2.30. A segment of a single realization of a one-dimensional random surface that has been designed to synthesize the infrared absorption spectrum of HF in the region 3600–4300 cm^{-1}.

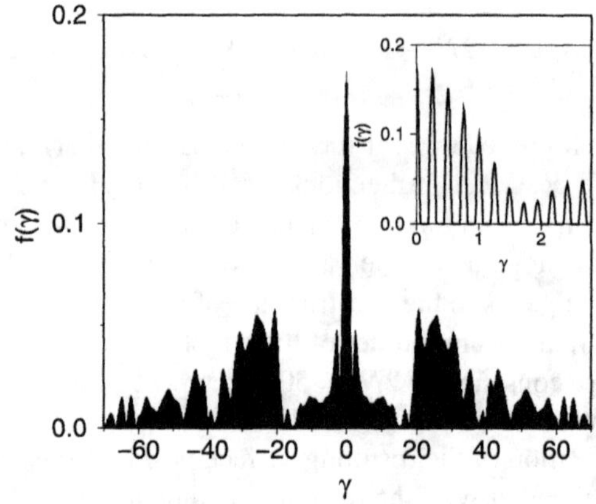

Figure 2.31. The probability density function $f(\gamma)$ used to generate the surface profile function presented in Fig. 2.30.

although it provides the correct relative amplitudes of the peaks in the spectrum, can lead to considerable power losses. This can be improved by using more iterations in determining $f(\gamma)$. However, if the amplitudes of the peaks in the experimental spectrum satisfy the condition $\sum_{j=1}^{N} A_j < 1$, the correct relative and absolute amplitudes of the peaks can be reproduced by using very few iterations.

We stress that the wavelength dependence of the intensity of light scattered from any realization of the surface obtained by the use of the pdf $f(\gamma)$ depicted in Fig. 2.31 reproduces the required spectrum with a very good accuracy. It may seem surprising that scattering from a single realization of a random surface that was constructed by requiring

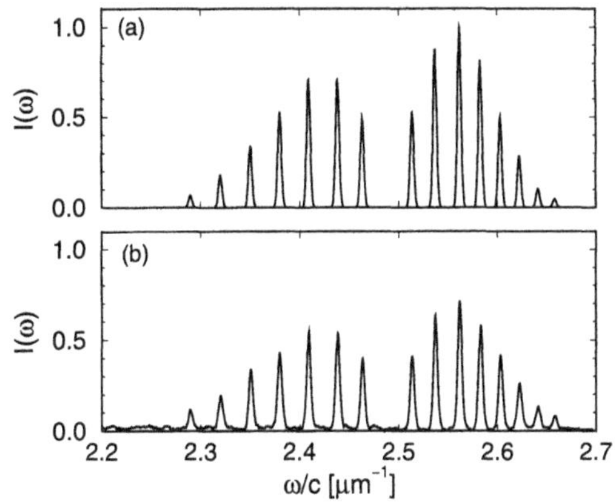

Figure 2.32. (a) The infrared absorption spectrum of HF from 3600 cm^{-1} to 4300 cm^{-1}. (b) The infrared absorption spectrum of HF in the same frequency interval synthesized by the approach presented in Section 2.5.1.

that the *mean* intensity of the scattered field, viz. the intensity of the scattered field averaged over the ensemble of realizations of the surface profile function, have a specified dependence on the wavelength of the incident field, yields a theoretical spectrum in such close agreement with the input experimental spectrum. The reason that it is not necessary to generate an ensemble of N_p realizations of the surface profile function, and to calculate the arithmetic mean of the intensities of the fields scattered from these realizations, is that the intensity is calculated as a function of wavelength at a fixed scattering angle. In this case no speckles arise along the line of sight that have to be averaged over, as is the case if the intensity is calculated as a function of scattering angle at a fixed wavelength. However, the random surface has to be long enough in order that scattering from a single realization of it produces a spectrum in good agreement with the experimental spectrum. The more complicated the spectrum that we seek to reproduce is, the more complicated is the resulting $f(\gamma)$ and, as a result, the longer the surface should be to represent well the statistics of the surface that are required. In the calculations whose results are depicted in Figs. 2.30–2.32, a value of $2N_s = 5000$ was found to be sufficient.

It should be noted that the interval of values that d_n can assume, Eq. (2.5.39), is quite large. This can result in the occurrence of quite a deep structure of the surface generated. Such a surface can be discarded in favor of a smoother one obtained from the same pdf $f(\gamma)$.

In Fig. 2.33(a) we present the experimental IR spectrum of carbonyl fluoride (COF_2) (gray line) [2.79] together with the intensity $I(\omega)$, calculated by means of the rigorous numerical approach described in Section 2.5.1.4, for scattering from a single realization of the surface profile function. A value of the sampling interval Δ in Eq. (2.5.64) given by $\Delta = 0.2$ μm was used in obtaining this result. The peaks in this spectrum are considerably wider than in the case of HF and, as a result, the pdf $f(\gamma)$ presented in Fig. 2.33(b) is confined to smaller values of γ. This allows using considerably shorter surfaces. The

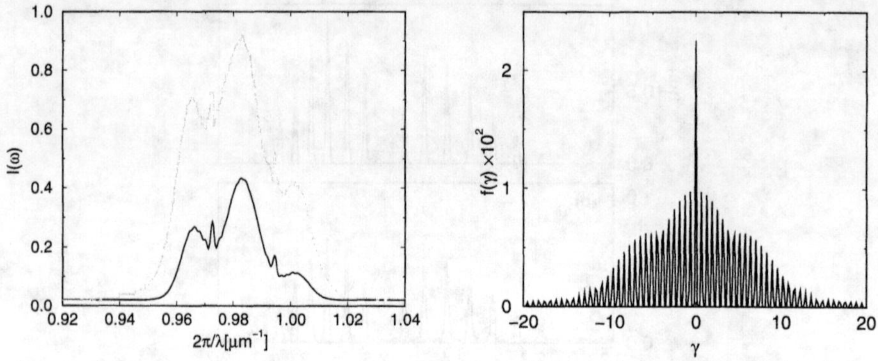

Figure 2.33. (a) The infrared absorption spectrum of COF_2 from 950 cm^{-1} to 1100 cm^{-1} (gray line) and the reconstructed spectrum (black line). (b) The probability density function $f(\gamma)$ used to generate the surface profile function.

length of the surfaces in the calculations was taken to be 1000 μm. The shape of the reconstructed spectrum is very close to the initial spectrum $I(\omega)$, but the absolute intensity of the reconstructed spectrum is about 50% smaller.

Thus, a method has been proposed in this section for designing a one-dimensional randomly rough perfectly conducting surface that scatters light at a fixed scattering angle with a mean intensity that depends in a specified way on the wavelength of an s-polarized plane wave incident normally on it. This method has been validated by the results of a calculation of the wavelength dependence of the intensity of light scattered from a single realization of such a surface carried out by a rigorous numerical approach.

2.5.2. *Deterministic Surfaces*

Although the probabilistic approach to the design of a one-dimensional randomly rough surface that synthesizes a specified infrared spectrum presented in Section 2.5.1 is an effective one, that can be used in the design of optical elements for other applications, a simpler approach to the solution of this design problem exists, which produces a deterministic surface that can be fabricated by well-known techniques. We describe it in this section.

The physical system we consider is the same one assumed in the discussion in Section 2.5.1. It consists of vacuum in the region $x_3 > \zeta(x_1)$ and a perfect conductor in the region $x_3 < \zeta(x_1)$. The surface $x_3 = \zeta(x_1)$ is illuminated from the vacuum by an s-polarized plane wave of frequency ω, whose plane of incidence is the x_1x_3 plane. In the vacuum region $x_3 > \zeta(x_1)_{\max}$ the single nonzero component of the electric field is the sum of an incoming incident field and a superposition of outgoing scattered waves,

$$E_2(x_1, x_3|\omega) = \exp\bigl[ikx_1 - i\alpha_0(k)x_3\bigr] + \int_{-\infty}^{\infty} \frac{dq}{2\pi} R(q|k) \exp\bigl[iqx_1 + i\alpha_0(q)x_3\bigr]. \quad (2.5.89)$$

The scattered field is given by the second term on the right-hand side of this equation,

$$E_2(x_1, x_3|\omega)_{sc} = \int_{-\infty}^{\infty} \frac{dq}{2\pi} R(q|k) \exp\bigl[iqx_1 + i\alpha_0(q)x_3\bigr]. \tag{2.5.90}$$

In the far zone, where $x_1 \gg \lambda$ and $x_3 \gg \lambda$, where $\lambda = 2\pi c/\omega$ is the wavelength of the incident field, the scattered field can be determined by the method of stationary phase [2.80]. This method requires expanding the function $qx_1 + \alpha_0(q)x_3$ in the exponent in the integrand on the right-hand side of Eq. (2.5.90) about the stationary point $q = q_0$. The point q_0 is the solution of the equation

$$\frac{\partial}{\partial q}\bigl[qx_1 + \alpha_0(q)x_3\bigr] = x_1 - \frac{q}{\alpha_0(q)}x_3 = 0, \tag{2.5.91}$$

from which we find that

$$q_0 = \frac{\omega}{c}\frac{x_1}{r}, \tag{2.5.92}$$

where $r = (x_1^2 + x_3^2)^{\frac{1}{2}}$. Thus, on expanding $[qx_1 + \alpha_0(q)x_3]$ about $q = q_0$, we obtain

$$\bigl[qx_1 + \alpha_0(q)x_3\bigr] = (\omega/c)r - (1/2)\bigl(cr^3/\omega x_3^2\bigr)(q - q_0)^2 + \cdots \tag{2.5.93}$$

through terms of second order in $(q - q_0)$. The scattered field is then given by

$$\begin{aligned} E_2^{>}(x_1, x_3|\omega)_{sc} &\cong \exp\bigl[i(\omega/c)r\bigr]R(q_0|k)\int_{-\infty}^{\infty}\frac{dq}{2\pi}\exp\bigl[-i(1/2)\bigl(cr^3/\omega x_3^2\bigr)(q-q_0)^2\bigr] \\ &= \left(\frac{\omega}{2\pi cr}\right)^{\frac{1}{2}}\cos\theta_s \exp\bigl\{i\bigl[(\omega/c)r - \pi/4\bigr]\bigr\}R(q_0|k), \end{aligned} \tag{2.5.94}$$

where θ_s is the scattering angle measured clockwise from the x_3 axis. It follows that $x_1/r = \sin\theta_s$, $x_3/r = \cos\theta_s$, so that

$$q_0 = (\omega/c)\sin\theta_s. \tag{2.5.95}$$

To obtain the scattering amplitude $R(q|k)$, we invoke the Rayleigh hypothesis [2.28], namely that we can use the expression (2.5.89) for the field in the vacuum region, which is valid only in the region $x_3 > \zeta(x_1)_{\max}$, in satisfying the boundary condition

$$E_2^{>}(x_1, x_3|\omega)\big|_{x_3=\zeta(x_1)} = 0 \tag{2.5.96}$$

on the surface $x_3 = \zeta(x_1)$. With this assumption, $R(q|k)$ becomes the solution of the integral equation

$$\exp\big[ikx_1 - i\alpha_0(k)\zeta(x_1)\big] + \int_{-\infty}^{\infty} \frac{dq}{2\pi} R(q|k) \exp\big[iqx_1 + i\alpha_0(q)\zeta(x_1)\big] = 0. \tag{2.5.97}$$

We assume that the surface roughness is sufficiently weak that small-amplitude perturbation theory can be used in solving Eq. (2.5.97). In this approach the scattering amplitude $R(q|k)$ is expanded in powers of the surface profile function according to

$$R(q|k) = \sum_{n=0}^{\infty} \frac{i^n}{n!} R_n(q|k), \tag{2.5.98}$$

where the subscript denotes the order of the corresponding term in $\zeta(x_1)$. When Eq. (2.5.98) is substituted into Eq. (2.5.97) and terms of the same order in $\zeta(x_1)$ on both sides of the resulting equation are equated, we obtain the following equation:

$$\sum_{m=0}^{n} \binom{n}{m} \int_{-\infty}^{\infty} \frac{dq}{2\pi} \alpha_0^{n-m}(q) \hat{\zeta}^{(n-m)}(p-q) R_m(q|k) = (-1)^{n+1} \alpha_0^n(k) \hat{\zeta}^{(n)}(p-k), \tag{2.5.99}$$

where

$$\hat{\zeta}^{(0)}(Q) = 2\pi\delta(Q), \tag{2.5.100a}$$

$$\hat{\zeta}^{(n)}(Q) = \int_{-\infty}^{\infty} dx_1\, \zeta^n(x_1) \exp(-iQx_1), \quad n \geqslant 1. \tag{2.5.100b}$$

Equation (2.5.99) can be rewritten as a recurrence relation for determining the $\{R_n(q|k)\}$:

$$R_0(q|k) = -2\pi\delta(q-k), \tag{2.5.101a}$$

$$R_n(q|k) = \big[1 + (-1)^{n+1}\big] \alpha_0^n(k) \hat{\zeta}^{(n)}(q-k) - \sum_{m=1}^{n-1} \binom{n}{m} \int_{-\infty}^{\infty} \frac{dp}{2\pi} \hat{\zeta}^{(n-m)}(q-p) \alpha_0^{n-m}(p) R_m(p|k), \quad n \geqslant 1. \tag{2.5.101b}$$

With the use of Eqs. (2.5.98) and (2.5.101) we find that through the first nonzero term in the surface profile function $\zeta(x_1)$,

$$R(q|k) = -2\pi\delta(q-k) + 2i\alpha_0(k)\int_{-\infty}^{\infty} dx_1\,\zeta(x_1)\exp\bigl[-i(q-k)x_1\bigr] + O(\zeta^2). \quad (2.5.102)$$

To simplify the following discussion we will assume that the surface $x_3 = \zeta(x_1)$ is illuminated at normal incidence ($k = 0$), while the scattering angle is assumed to be nonzero. It follows from Eqs. (2.5.94), (2.5.95) and (2.5.102) that the leading nonspecular contribution to the intensity of the scattered field is

$$\bigl|E_2^>(x_1,x_3)_{\mathrm{sc}}\bigr|^2 = \frac{2}{\pi r}\left(\frac{\omega}{c}\right)^3\cos^2\theta_s\left|\int_{-\infty}^{\infty} dx_1\,\zeta(x_1)\exp\bigl[-i(\omega/c)\sin\theta_s x_1\bigr]\right|^2. \quad (2.5.103)$$

We now represent the surface profile function $\zeta(x_1)$ as a superposition of N cosinusoidal gratings, one for each line in the spectrum that we seek to reproduce. The period of the jth grating a_j is chosen so that the grating produces the $(+1)$ order Bragg diffraction peak in the direction defined by θ_s at a given frequency ω_j, i.e., it is determined from the condition $(\omega_j/c)\sin\theta_s = 2\pi/a_j$. If the spectral range of interest is sufficiently narrow, namely $2\omega_1 > \omega_N$, then the $(+2)$ order Bragg peak produced by the grating with the longest period does not fall into this spectral range because $4\pi/a_1 = 2(\omega_1/c)\sin\theta_s > (\omega_N/c)\sin\theta_s$. For the same reason the cross-talk peaks whose positions are determined by the condition $(\omega/c)\sin\theta_s = \pm[(2\pi/a_i) - (2\pi/a_j)] = \pm[(\omega_i - \omega_j)/c]\sin\theta_s$, are also outside the spectral range we are interested in if $2\omega_1 > \omega_N$. The widths and the shapes of the spectral lines can be reproduced by modulating the gratings by slowly varying envelope functions.

Thus, we assume that the surface profile function $\zeta(x_1)$ is given by

$$\zeta(x_1) = \sum_{j=1}^{N} A_j\exp\bigl(-x_1^2/L_j^2\bigr)\cos(2\pi x_1/a_j). \quad (2.5.104)$$

The Gaussian function $\exp(-x_1^2/L_j^2)$ in this expression is an envelope function that limits the length of the jth grating and shapes the spectral lines to be Gaussian. A finite $1/e$ half width L_j for the jth grating imparts a finite width to each of the lines in the spectrum. With this choice of $\zeta(x_1)$ the integral in Eq. (2.5.103) becomes

$$\int_{-\infty}^{\infty} dx_1\,\zeta(x_1)\exp\bigl[-i(\omega/c)\sin\theta_s x_1\bigr]$$
$$= \frac{\sqrt{\pi}}{2}\sum_{j=1}^{N} A_j L_j\left\{\exp\left[-\left(\frac{\omega}{c}\sin\theta_s - \frac{2\pi}{a_j}\right)^2\frac{L_j^2}{4}\right] + \exp\left[-\left(\frac{\omega}{c}\sin\theta_s + \frac{2\pi}{a_j}\right)^2\frac{L_j^2}{4}\right]\right\}. \quad (2.5.105)$$

The second term in braces on the right-hand side of this expression produces no narrow peak for positive values of ω, because the argument of the Gaussian function does not vanish in this case. We therefore omit it in what follows. We also assume that the widths of the peaks corresponding to $(\omega/c)\sin\theta_s - (2\pi/a_j) = 0$ in the first term in braces on the right-hand side of Eq. (2.5.105) are sufficiently narrow that different peaks do not overlap in the squared modulus of this term, which can therefore be approximated by

$$\left|\frac{\sqrt{\pi}}{2}\sum_{j=1}^{N} A_j L_j \exp\left[-\left(\frac{\omega}{c}\sin\theta_s - \frac{2\pi}{a_j}\right)^2 \frac{L_j^2}{4}\right]\right|^2$$

$$\cong \frac{\pi}{4}\sum_{j=1}^{N}(A_j L_j)^2 \exp\left[-\left(\frac{\omega}{c}\sin\theta_s - \frac{2\pi}{a_j}\right)^2 \frac{L_j^2}{2}\right]. \tag{2.5.106}$$

From the relation $(2\pi c)/(a_j \sin\theta_s) = \omega_j$, the period a_j is given by

$$a_j = \frac{2\pi c}{\omega_j \sin\theta_s} \equiv \frac{\lambda_j}{\sin\theta_s}, \tag{2.5.107}$$

where $\lambda_j = 2\pi c/\omega_j$ is the wavelength of the jth line in the spectrum. If the $1/e$ halfwidth of the jth line in the spectrum we seek to synthesize is Δ_j, the $1/e$ half width L_j of the jth grating in the representation of $\zeta(x_1)$ given by Eq. (2.5.104) is obtained from

$$L_j = \frac{\sqrt{2}c}{\Delta_j \sin\theta_s}. \tag{2.5.108}$$

With the preceding results the scattered intensity (2.5.103) can be written in the form

$$\left|E_2^{>}(x_1, x_3|\omega)_{\mathrm{sc}}\right|^2 = \frac{1}{2r}\left(\frac{\omega}{c}\right)^3 \cos^2\theta_s \sum_{j=1}^{N}(A_j L_j)^2 \exp\left[-\left(\frac{\omega-\omega_j}{\Delta_j}\right)^2\right]. \tag{2.5.109}$$

We define the experimental intensity $I(\omega)$ by

$$I(\omega) = r\left|E_2^{>}(x_1, x_3|\omega)_{\mathrm{sc}}\right|^2, \tag{2.5.110}$$

so that

$$I(\omega) = \frac{1}{2}\left(\frac{\omega}{c}\right)^3 \cos^2\theta_s \sum_{j=1}^{N}(A_j L_j)^2 \exp\left[-\left(\frac{\omega-\omega_j}{\Delta_j}\right)^2\right]. \tag{2.5.111}$$

The relative values of the amplitudes $\{A_j\}$ are obtained by noting that the intensities of the lines in the experimental spectrum at the frequencies $\{\omega_j\}$ are determined by $\omega_j^3(A_j L_j)^2$.

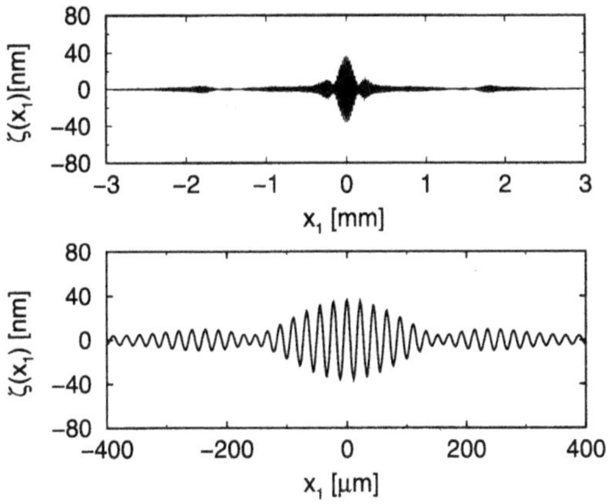

Figure 2.34. A segment of a numerically generated one-dimensional deterministic perfectly conducting rough surface designed to synthesize the infrared absorption spectrum of HF in the region 3600–4300 cm^{-1}.

The amplitude of the grating producing the strongest peak in the spectrum is then chosen to be small enough that our use of small-amplitude perturbation theory is valid. This is accomplished on the basis of the following considerations. If the frequency range of interest is narrow, $2\omega_1 > \omega_N$, so that no peaks of second order fall into it, the contribution to $I(\omega_j)$ of the second order terms in the small-amplitude perturbation theory for $R(q|k)$ gives rise to an exponentially small background. However, the contribution to $I(\omega)$ from the third order term in the surface profile function in $R(q|k)$ gives rise to peaks at the frequencies $\{\omega_j\}$, as well as to cross-talk peaks, which can fall into the frequency range $\omega_1 < \omega < \omega_N$. Therefore, to ensure that the relative heights of the peaks in the synthesized spectrum are obtained correctly, this contribution must be made negligibly small compared to the result given by Eq. (2.5.111). This requires that the amplitude of the gratings should be small compared to the wavelengths of the corresponding peaks, $(A_j/\lambda_j)^2 \ll 1$, for all j. In this way the parameters defining the surface profile function $\zeta(x_1)$, Eq. (2.5.104) are determined.

To illustrate the approach presented in this section to the design of a one-dimensional perfectly conducting, rough surface that produces a scattered field with a prescribed spectrum at a specified scattering angle, we apply it to the design of a surface that synthesizes the infrared spectrum of HF in the region 3600–4300 cm^{-1}, at a scattering angle $\theta_s = 15°$. A segment of this surface is presented in Fig. 2.34. The intensity $I(\omega)$ calculated by solving the problem of the scattering of an s-polarized plane wave incident normally on this surface in the Kirchhoff approximation, Eqs. (2.1.22) and (2.1.58a), is plotted in Fig. 2.35(a). For comparison, in Fig. 2.35(b) we have plotted the experimental spectrum of HF [2.79] that served as the input for the calculation of $I(\omega)$. It is seen that the synthesized spectrum displays peaks at the correct frequencies, whose relative amplitudes reproduce those of the peaks in the experimental spectrum.

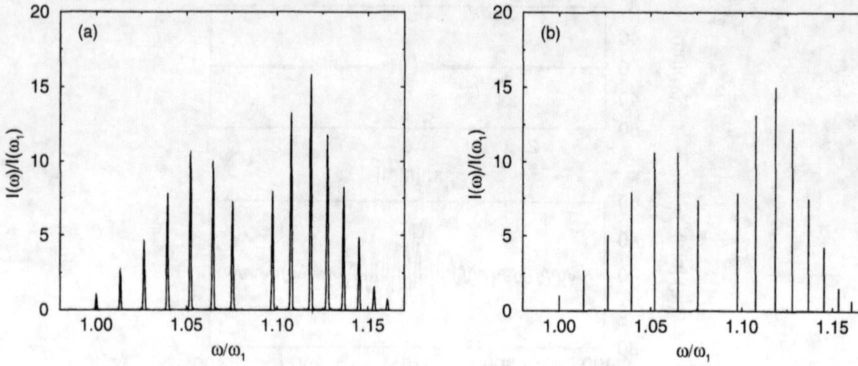

Figure 2.35. (a) The infrared absorption spectrum of HF from 3600 cm^{-1} to 4300 cm^{-1} synthesized by the approach presented in Section 2.5.2. (b) The experimental absorption spectrum of HF in the same spectral range.

The approach presented in this section can be generalized in two ways. First, each cosine in the expression (2.5.104) for the surface profile function can be replaced by a cosine with an arbitrary phase in its argument, namely $\cos[(2\pi x_1/a_j)+\phi_j]$, without altering the intensity of the scattered light under the assumptions made here. Second, each Gaussian in the expression (2.5.104) for the surface profile function can be shifted arbitrarily along the x_1 axis, that is replaced by $\exp[-(x_1-X_j)^2/L_j^2]$, without altering the intensity of the scattered light.

2.6. Surfaces that Produce Specified Thermal Emissivities

The emissivity of a surface is denoted by $e_\nu(\theta,\phi)$, where the subscript ν defines the polarization of the emitted radiation, while θ is the polar observation angle, and ϕ is the azimuthal observation angle. Through the principles of energy conservation and reciprocity, Kirchhoff's law relates the emissivity of an opaque surface to the reflectivity of the surface [2.81],

$$e_\nu(\theta,\phi)=1-r_\nu(\theta,\phi). \tag{2.6.1}$$

In Eq. (2.6.1) $r_\nu(\theta,\phi)$ is the reflectivity of the surface when it is illuminated by a $\nu=p,s$ polarized wave whose polar and azimuthal angles of incidence are θ and ϕ, respectively.

It follows from Eq. (2.6.1) that the emissivity of a surface is affected by the roughness of the surface since it is known that the reflectivity of a surface is affected by its roughness [2.82,2.83]. Indeed there have been several calculations of the emissivity of periodically corrugated surfaces [2.84,2.85] and of randomly rough surfaces [2.86,2.87]. In these calculations the surface profile was specified, in the case of a periodically corrugated surface, or its statistical properties were specified in the case of a randomly rough surface. What was sought in these calculations was the angular or wavelength dependence of the emissivity. It was shown, for example, that the existence of surface phonon polaritons at a periodically corrugated SiC surface, which produces a dip in the wavelength dependence of the reflectivity at a fixed angle of incidence at the wavelength at which these surface electromagnetic waves are most efficiently excited through the surface periodicity by the incident light, produces an enhancement of the emissivity in a limited spectral range [2.88]. In a subsequent study the reflectivity and emissivity of singly and doubly periodic metallic surfaces were calculated and measured [2.89]. The existence of thermally excited surface plasmon polaritons supported by these surfaces leads to an increase of their emissivity in a restricted spectral range. However, in all of this work the surface was known in advance and the emissivity it produced was determined.

In this section we consider an inverse version of this problem, namely we wish to specify the wavelength dependence of the emissivity at a given angle of emission, and to design a random surface that produces that emissivity. Such a surface could serve as a frequency-stable source of thermal emission if the radiation it produces is a constant within some desired wavelength region and vanishes outside this region.

2.6.1. Scattering Theory

The surface we consider is a one-dimensional randomly rough surface defined by $x_3 = \zeta(x_1)$. The region $x_3 > \zeta(x_1)$ is vacuum, while the region $x_3 < \zeta(x_1)$ is an opaque dielectric medium, characterized by a dielectric constant ϵ. The surface profile function $\zeta(x_1)$ is initially assumed to be a continuous single-valued function of x_1 that is differentiable and constitutes a random process. Since the surface of the emitter is randomly rough, its reflectivity differs from the value of unity it takes when the surface is planar, so that the emissivity is nonzero.

The surface $x_3 = \zeta(x_1)$ is illuminated from the vacuum by a ν-polarized plane wave of frequency ω, whose plane of incidence is the x_1x_3 plane. In terms of the function $F_\nu(x_1, x_3|\omega)$ introduced in Section 2.1.1, the single nonzero component of the electric (s polarization) or magnetic (p polarization) field in the vacuum region $x_3 > \zeta(x_1)_{\max}$ that satisfies the boundary conditions at infinity is the sum of an incoming incident field and of outgoing scattered waves,

$$F_\nu^>(x_1, x_3|\omega) = \exp\bigl[ikx_1 - i\alpha_0(k)x_3\bigr] + \int_{-\infty}^{\infty} \frac{dq}{2\pi} R_\nu(q|k) \exp\bigl[iqx_1 + i\alpha_0(q)x_3\bigr], \quad (2.6.2)$$

where $\alpha_0(q) = [(\omega/c)^2 - q^2]^{\frac{1}{2}}$, with $\mathrm{Re}\,\alpha_0(q) > 0$, and $\mathrm{Im}\,\alpha_0(q) > 0$.

We have seen earlier (Eq. (2.1.37)) that the mean differential reflection coefficient $\langle \partial R_\nu / \partial \theta_s \rangle$ is given by

$$\left\langle \frac{\partial R_\nu}{\partial \theta_s} \right\rangle = \frac{1}{L_1} \left(\frac{\omega}{2\pi c} \right) \frac{\cos^2\theta_s}{\cos\theta_0} \bigl\langle \bigl|R_\nu(q|k)\bigr|^2 \bigr\rangle. \quad (2.6.3)$$

If we decompose $R_\nu(q|k)$ into the sum of its mean value, $\langle R_\nu(q|k)\rangle$, and its fluctuation about this mean value, $R_\nu(q|k) - \langle R_\nu(q|k)\rangle$,

$$R_\nu(q|k) = \langle R_\nu(q|k)\rangle + \big[R_\nu(q|k) - \langle R_\nu(q|k)\rangle\big], \tag{2.6.4}$$

we find that each of these terms contributes separately to the mean differential reflection coefficient,

$$\left\langle \frac{\partial R_\nu}{\partial \theta_s} \right\rangle = \frac{1}{L_1}\left(\frac{\omega}{2\pi c}\right)\frac{\cos^2\theta_s}{\cos\theta_0}\big|\langle R_\nu(q|k)\rangle\big|^2 + \frac{1}{L_1}\left(\frac{\omega}{2\pi c}\right)\frac{\cos^2\theta_s}{\cos\theta_0}\Big[\big|R_\nu(q|k) - \langle R_\nu(q|k)\rangle\big|^2\Big]. \tag{2.6.5}$$

The first term on the right-hand side of this equation gives the contribution to the mean differential reflection coefficient from the light that has been scattered coherently (specularly), while the second gives the contribution from the light that has been scattered incoherently (diffusely).

Strictly speaking, the preceding descriptions of the two contributions to the mean differential reflection coefficient given by the two terms on the right-hand side of Eq. (2.6.5) are valid only when the surface profile function is a stationary random process. For such a surface profile function the mean value of the scattering amplitude $R_\nu(q|k)$ is diagonal in q and k, $\langle R_\nu(q|k)\rangle = 2\pi\delta(q-k)R_\nu(k)$. Since in one dimension

$$\big[2\pi\delta(q-k)\big]^2 = 2\pi\delta(q-k)2\pi\delta(0) = L_1 2\pi\delta(q-k), \tag{2.6.6}$$

we see that the first term on the right-hand side of Eq. (2.6.5) indeed describes the angular and frequency dependence of the specular component of the scattered light. The second term then describes the angular and frequency dependence of the diffuse component of the scattered light. In what follows we will see that the use of the expression

$$\left\langle \frac{\partial R}{\partial \theta_s} \right\rangle_{\text{coh}} = \frac{1}{L_1}\left(\frac{\omega}{2\pi c}\right)\frac{\cos^2\theta_s}{\cos\theta_0}\big|\langle R(q|k)\rangle\big|^2 \tag{2.6.7}$$

enables us to determine the reflectivity of the surfaces we will design in this section, i.e., the model of a random surface we adopt is a stationary one.

In the Kirchhoff approximation, which we adopt here due to its simplicity, the mean scattering amplitude is given by (Eq. (2.2.13))

$$\big\langle R_\nu(q|k)\big\rangle = R_\nu\left(\frac{\theta_s+\theta_0}{2}\right)\frac{1}{\cos\theta_0}\frac{\cos\frac{1}{2}(\theta_s+\theta_0)}{\cos\frac{1}{2}(\theta_s-\theta_0)} \times \int_{-\infty}^{\infty} dx_1 \exp[-i(q-k)x_1]\big\langle \exp[-ia\zeta(x_1)]\big\rangle, \tag{2.6.8}$$

where $R_\nu(\frac{\theta_s+\theta_0}{2})$ has been defined in Eq. (2.2.12) and $a(\theta_s,\theta_0)$ has been defined in Eq. (2.1.61).

We choose for the surface profile function the form

$$\zeta(x_1) = d_n b_1, \quad nb < x_1 < (n+1)b,\ n = 0, \pm 1, \pm 2, \ldots, \tag{2.6.9}$$

where the $\{d_n\}$ are independent identically distributed random deviates, while b_1 and b are characteristic lengths. Because the $\{d_n\}$ are identically distributed random deviates, the probability density function (pdf) of d_n,

$$f(\gamma) = \langle \delta(\gamma - d_n) \rangle, \tag{2.6.10}$$

is independent of n.

The assumption of a surface profile function of the form given by Eq. (2.6.9) violates our initial assumptions that $\zeta(x_1)$ is a continuous, differentiable function of x_1. Nevertheless, as we will see, this assumption allows us to solve the problem of designing a surface that produces thermal emission with a specified spectrum at a specified angle of observation.

To simplify the notation, we use the definitions (2.5.16). The integral in Eq. (2.6.8) then becomes

$$\begin{aligned}
&\int_{-\infty}^{\infty} dx_1 \exp(-igx_1)\big\langle \exp[-ih\zeta(x_1)]\big\rangle \\
&= \sum_{n=-\infty}^{\infty} \int_{nb}^{(n+1)b} dx_1 \exp(-igx_1)\langle \exp(-ihb_1 d_n)\rangle \\
&= \sum_{n=-\infty}^{\infty} \int_{nb}^{(n+1)b} dx_1 \exp(-igx_1) \int_{-\infty}^{\infty} d\gamma\, f(\gamma)\exp(-ihb_1\gamma) \\
&= \int_{-\infty}^{\infty} d\gamma\, f(\gamma)\exp(-ihb_1\gamma) \int_{-\infty}^{\infty} dx_1 \exp(-igx_1) \\
&= 2\pi\delta(g)F(b_1 h),
\end{aligned} \tag{2.6.11}$$

where

$$F(\nu) = \int_{-\infty}^{\infty} d\gamma\, f(\gamma)\exp(-i\nu\gamma). \tag{2.6.12}$$

On combining Eqs. (2.6.8) and (2.6.11) we obtain from Eq. (2.6.7) the result that

$$\left\langle \frac{\partial R_\nu}{\partial \theta_s} \right\rangle_{\text{coh}} = \frac{1}{L_1}\left(\frac{\omega}{2\pi c}\right) \frac{\cos^2\theta_s \cos^2\frac{1}{2}(\theta_s+\theta_0)}{\cos^3\theta_0 \cos^2\frac{1}{2}(\theta_s-\theta_0)} \left| R_\nu\left(\frac{\theta_s+\theta_0}{2}\right)\right|^2$$
$$\times \left[2\pi\delta\big((\omega/c)(\sin\theta_s-\sin\theta_0)\big)\right]^2 \left|F_\nu\big((\omega b_1/c)(\cos\theta_s+\cos\theta_0)\big)\right|^2. \tag{2.6.13}$$

We now use the result that

$$\begin{aligned}\left[2\pi\delta\big((\omega/c)(\sin\theta_s-\sin\theta_0)\big)\right]^2 &= \left[2\pi\delta(q-k)\right]^2 \\ &= 2\pi\delta(0)2\pi\delta(q-k) \\ &= L_1 2\pi\delta(q-k) \\ &= L_1 2\pi\delta\big((\omega/c)(\sin\theta_s-\sin\theta_0)\big) \\ &= L_1\frac{2\pi c}{\omega}\frac{\delta(\theta_s-\theta_0))}{\cos\theta_0},\end{aligned} \tag{2.6.14}$$

where we have used the result that in one dimension $2\pi\delta(q=0)=L_1$. Equation (2.6.13) then becomes

$$\left\langle \frac{\partial R_\nu}{\partial \theta_s} \right\rangle_{\text{coh}} = \delta(\theta_s-\theta_0) r_\nu(\omega,\theta_0), \tag{2.6.15}$$

where the reflectivity $r_\nu(\omega,\theta_0)$ is given by

$$r_\nu(\omega,\theta_0) = \left|R_\nu(\theta_0)\right|^2 \left|F\big(2(\omega b_1/c)\cos\theta_0\big)\right|^2, \tag{2.6.16}$$

where

$$R_\nu(\theta_0) = \frac{\kappa_\nu\cos\theta_0 - [\epsilon-\sin^2\theta_0]^{\frac{1}{2}}}{\kappa_\nu\cos\theta_0 + [\epsilon-\sin^2\theta_0]^{\frac{1}{2}}}. \tag{2.6.17}$$

Thus, if we wish to obtain a randomly rough surface that produces a particular frequency dependence of the emissivity $e_\nu(\omega,\theta_0)$ at an angle of emission θ_0, we have to solve the equation

$$\left|F_\nu\big((2\omega b_1/c)\cos\theta_0\big)\right| = \frac{[1-e_\nu(\omega,\theta_0)]^{\frac{1}{2}}}{|R_\nu(\theta_0)|} \tag{2.6.18}$$

to obtain $f_\nu(\gamma)$. Note that $R_\nu(\theta_0)$ is a function of the frequency ω when the dielectric constant ϵ is in fact a function of frequency. The problem of determining $f_\nu(\gamma)$ from $|F_\nu(\nu)|$ is solved by the use of the modified Gerchberg–Saxton algorithm described in Section 2.5.1.

2.6.2. The Reflectivity

Once $f(\gamma)$ has been determined, a long sequence of $\{d_n\}$ is generated, for example by the rejection method [2.6], and a realization of the surface profile function $\zeta(x_1)$ is obtained by the use of Eq. (2.6.9). The scattered field produced when a ν-polarized plane wave is incident on this realization of the surface profile function is then calculated. The reflectivity, and hence the emissivity, of the surface are calculated from this result as functions of frequency, and compared with the input emissivity to assess the validity of the approach presented here.

To calculate the reflectivity from the results of a scattering calculation it is convenient to return to Eq. (2.6.7) and to use the result that $\langle R(q|k)\rangle = 2\pi\delta(q-k)R_\nu^{(0)}(k)$ for a stationary surface profile function, and that $2\pi\delta(q=0) = L_1$. In this way we obtain

$$\left\langle \frac{\partial R_\nu}{\partial \theta_s} \right\rangle_{\text{coh}} = \delta(\theta_s - \theta_0)\left|R_\nu^{(0)}\left((\omega/c)\sin\theta_0\right)\right|^2. \tag{2.6.19}$$

The reflectivity in general is therefore given by

$$r_\nu(\omega, \theta_0) = \left|R_\nu^{(0)}\left((\omega/c)\sin\theta_0\right)\right|^2. \tag{2.6.20}$$

To determine $R_\nu^{(0)}(k) = R_\nu^{(0)}((\omega/c)\sin\theta_0)$ through a scattering calculation we use the result that

$$R_\nu^{(0)}(k) = \frac{1}{L_1}\langle R_\nu(k|k)\rangle, \tag{2.6.21}$$

and proceed to determine the scattering amplitude $R_\nu(q|k)$ by a suitable computational approach.

2.6.3. Solution of the Scattering Problem

Because of the jump discontinuities $x_1 = nb$, $n = 0, \pm 1, \pm 2, \ldots$, possessed by the surface profile function (2.6.9), the scattering problem is solved here by the rigorous numerical approach described in Section 2.5.1.4.

2.6.4. Example

To illustrate the method developed in this section, we consider the design of a one-dimensional randomly rough surface that produces s-polarized thermal emission at a specified angle of emission θ_0 with a frequency dependence given by

$$e(\omega, \theta_0) = \begin{cases} (1/2) & 0 < \omega_1 < \omega < \omega_2, \\ 0 & \text{otherwise.} \end{cases} \tag{2.6.22}$$

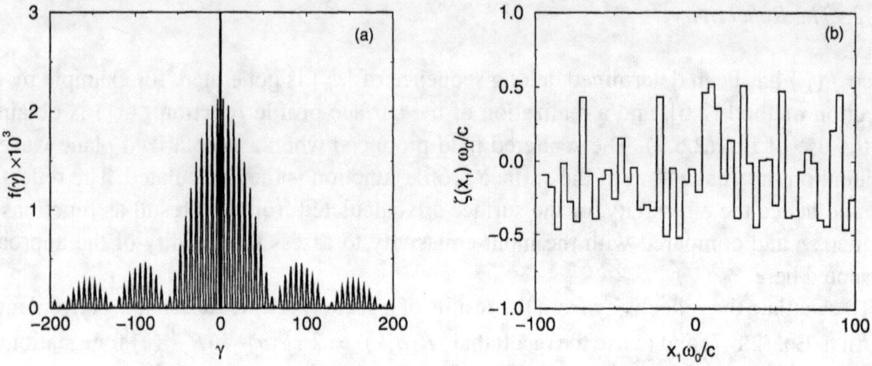

Figure 2.36. A plot of the pdf of d_n, $f(\gamma)$ (a) and a segment of one realization of the surface profile generated on the basis of this pdf (b).

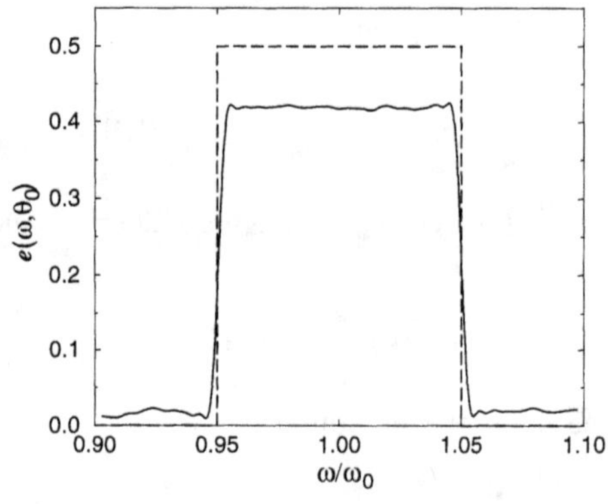

Figure 2.37. A plot of the calculated emissivity as a function of ω/ω_0 for an angle of emission $\theta_0 = 0°$, calculated for a single realization of the surface profile function. A plot of the input emissivity given by Eq. (2.6.22) is also presented for comparison.

In Fig. 2.36(a), we present a plot of the pdf of d_n, $f(\gamma)$, corresponding to this choice for $e(\omega, \theta_0)$. A segment of one realization of the surface profile generated on the basis of this result for $f(\gamma)$ is presented in Fig. 2.36(b). The values of the experimental and roughness parameters employed in these calculations were $\theta_0 = 0°$, $\omega_1/\omega_0 = 0.9$, $\omega_2/\omega_0 = 1.1$, $b\omega_0/c = 8$, and $b_1\omega_0/c = 0.1$. In Fig. 2.37 we present a plot of the calculated emissivity as a function of ω for an angle of emission $\theta_0 = 0°$, calculated for a single realization of the surface profile function. A plot of the input emissivity given by Eq. (2.6.22) is also presented for comparison. It is seen that the calculated emissivity is in good agreement with the input emissivity, giving us confidence in the reliability of the approach developed in this section.

Thus, we have shown that it is possible to design a one-dimensional randomly rough perfectly conducting surface that emits thermal radiation that at a specified emission angle has a prescribed frequency dependence. Such a surface can be fabricated by techniques such as those described in Refs. [2.65] and [2.9].

2.7. Control of the Coherence of the Light Scattered from a One-Dimensional Randomly Rough Surface that Acts as a Schell-Model Source

2.7.1. Coherence of Light

Coherence is the property of light that is most closely related to interference. It can be defined in the following way. Let $U(\mathbf{x}|\omega)$ be a component of a fluctuating electromagnetic field of frequency ω. Then the *spectral degree of coherence at frequency* ω of the field is defined by [2.90]

$$g(\mathbf{x}, \mathbf{x}'|\omega) = \frac{W(\mathbf{x}, \mathbf{x}'|\omega)}{[S(\mathbf{x}|\omega)]^{\frac{1}{2}}[S(\mathbf{x}'|\omega)]^{\frac{1}{2}}}. \tag{2.7.1}$$

In this expression $W(\mathbf{x}, \mathbf{x}'|\omega)$ is called the *cross-spectral density* of the field, and is defined by

$$W(\mathbf{x}, \mathbf{x}'|\omega) = \langle U(\mathbf{x}|\omega)U^*(\mathbf{x}'|\omega)\rangle, \tag{2.7.2}$$

where the angle brackets denote an average over the ensemble of realizations of the field $U(\mathbf{x}|\omega)$. The function $S(\mathbf{x}|\omega)$ represents the *spectral density* (intensity) of the field, and is given by

$$S(\mathbf{x}|\omega) = W(\mathbf{x}, \mathbf{x}|\omega) = \langle |U(\mathbf{x}|\omega)|^2\rangle. \tag{2.7.3}$$

The cross-spectral density function is clearly Hermitian,

$$W(\mathbf{x}', \mathbf{x}|\omega) = W^*(\mathbf{x}, \mathbf{x}'|\omega), \tag{2.7.4}$$

as is the spectral degree of coherence,

$$g(\mathbf{x}', \mathbf{x}|\omega) = g^*(\mathbf{x}, \mathbf{x}'|\omega). \tag{2.7.5}$$

In obtaining Eq. (2.7.5) we have used the result that from its definition (2.7.3) the spectral density $S(\mathbf{x}|\omega)$ is a real non-negative function of $\mathbf{x}$,

$$S(\mathbf{x}|\omega) \geqslant 0. \tag{2.7.6}$$

From Eqs. (2.7.1)–(2.7.3) we find that

$$g(\mathbf{x}, \mathbf{x}|\omega) = 1. \tag{2.7.7}$$

More generally, $W(\mathbf{x}, \mathbf{x}'|\omega)$ is a non-negative definite function, in the sense that for any n points $\mathbf{x}_1, \mathbf{x}_2, \ldots, \mathbf{x}_n$, where n is an arbitrary positive integer, for any n real or complex numbers $a_1, a_2, \ldots, a_n$, and for any frequency ω,

$$\sum_{j=1}^{n}\sum_{k=1}^{n} a_j a_k^* W(\mathbf{x}_j, \mathbf{x}_k|\omega) \geqslant 0. \tag{2.7.8}$$

This result follows directly from the obvious inequality

$$\left\langle \left| \sum_{j=1}^{n} a_j U(\mathbf{x}_j|\omega) \right|^2 \right\rangle \geqslant 0. \tag{2.7.9}$$

In the case $n = 2$, Eq. (2.7.8) can be rewritten in the form

$$(a_1, a_2) \begin{pmatrix} W(\mathbf{x}_1, \mathbf{x}_1|\omega) & W(\mathbf{x}_1, \mathbf{x}_2|\omega) \\ W(\mathbf{x}_2, \mathbf{x}_1|\omega) & W(\mathbf{x}_2, \mathbf{x}_2|\omega) \end{pmatrix} \begin{pmatrix} a_1^* \\ a_2^* \end{pmatrix} \geqslant 0. \tag{2.7.10}$$

In order for this inequality to be satisfied the principal minors of the matrix entering it should be non-negative. This condition together with Eqs. (2.7.3) and (2.7.4) yields the inequality

$$S(\mathbf{x}_1|\omega)S(\mathbf{x}_2|\omega) - \left|W(\mathbf{x}_1, \mathbf{x}_2|\omega)\right|^2 \geqslant 0. \tag{2.7.11}$$

It follows from this result and Eq. (2.7.1) that [2.90]

$$0 \leqslant \left|g(\mathbf{x}, \mathbf{x}'|\omega)\right| \leqslant 1. \tag{2.7.12}$$

Although a fully coherent beam, such as the output from a laser, has the desirable feature of a very narrow line width, its use in such fields as microdensitometry, [2.91,2.92], line width measurements [2.93,2.94], and lithography [2.95] sometimes has detrimental effects because its high spatial coherence gives rise to speckled images, which make it difficult to obtain good resolution. The reduction of the spatial coherence of a beam can increase the resolution of an image by decreasing the speckle it produces [2.90]. Moreover, partially coherent beams have been shown to propagate farther and to spread less in a random medium than does a fully coherent laser beam [2.96–2.98]. Thus, the ability to control the coherence of a light beam is important for a variety of optical applications.

2.7.2. *Schell-Model Sources*

Among the variety of fields to which the general formulas that we have just presented are applicable are those generated by secondary planar sources. Such a source is usually an aperture in an opaque planar screen illuminated, either directly or by means of an optical system, by a primary source. Chief among secondary planar sources are *Schell-model sources* [2.99]. These sources are characterized by the property that their spectral degree of coherence $g(\mathbf{x}_\parallel, 0; \mathbf{x}'_\parallel, 0)$ in the source plane $x_3 = 0$ depends on $\mathbf{x}_\parallel$ and $\mathbf{x}'_\parallel$ only through their difference $\mathbf{x}_\parallel - \mathbf{x}'_\parallel$. Here $\mathbf{x}_\parallel = (x_1, x_2, 0)$ is an arbitrary position vector in the plane $x_3 = 0$. The spectral degree of coherence of such a source therefore has the form

$$g(\mathbf{x}_\parallel, 0; \mathbf{x}'_\parallel, 0) = g^{(0)}(\mathbf{x}_\parallel - \mathbf{x}'_\parallel). \tag{2.7.13}$$

From Eq. (2.7.1) we see that the cross-spectral density in the source plane of a secondary, planar, Schell-model source has the form

$$W^{(0)}(\mathbf{x}_\parallel, 0|\mathbf{x}'_\parallel, 0) = \left[S^{(0)}(\mathbf{x}_\parallel)\right]^{\frac{1}{2}} g^{(0)}(\mathbf{x}_\parallel - \mathbf{x}'_\parallel)\left[S^{(0)}(\mathbf{x}'_\parallel)\right]^{\frac{1}{2}}. \tag{2.7.14}$$

Because everything is assumed to occur at the fixed frequency ω, we have suppressed mention of this frequency in writing Eqs. (2.7.13) and (2.7.14). The superscript (0) in these equations indicates that the functions involved are calculated in the source plane $x_3 = 0$. If $S^{(0)}(\mathbf{x}_\parallel)$ were a constant, an infinitely extended Schell-model source would be a homogeneous source due to its dependence on $\mathbf{x}_\parallel$ and $\mathbf{x}'_\parallel$ only through their difference. When $S^{(0)}(\mathbf{x}_\parallel)$, the spectral density of the light at a typical point in the source plane, is not a constant, the source defined by Eq. (2.7.14) is said to be a quasi-homogeneous, planar, secondary source.

Several methods of producing sources with different coherence properties have been described in the literature [2.100]. One such method [2.101] consists of placing in the path of a laser beam a rotating random phase screen created by spraying a finishing mist on a clear glass blank. With appropriate coating a quasi-homogeneous source is produced. Another method uses liquid crystals to scatter a laser beam [2.102,2.103]. The application of an electric field to the liquid crystal modifies the coherence properties of the scattered light. Other methods use transmission filters or holographic filters to change the coherence properties of light passing through them [2.104,2.105]. The interaction of light with sound waves has also been used to modify its coherence [2.106–2.109].

In this section we consider the possibility of designing a one-dimensional randomly rough surface that scatters a beam of light incident normally on it into a field that possesses a cross-spectral density in the mean scattering surface—the source plane—of the Schell-model form. A related problem for a two-dimensional randomly rough surface is discussed in Section 3.3.

2.7.3. An Incident Beam of General Form

Thus, in the one-dimensional case we consider here, we seek to design a surface defined by $x_3 = \zeta(x_1)$ which, when illuminated at normal incidence from vacuum ($x_3 > \zeta(x_1)$) by a suitable s-polarized beam of light of frequency ω, whose plane of incidence is the x_1x_3

plane, produces a scattered electric field whose only nonzero component $E_2(x_1, x_3|\omega)_{sc}$ possesses the cross-spectral density function

$$W^{(0)}(x_1, x_1'|\omega) = \langle E_2(x_1, 0|\omega)_{sc} E_2^*(x_1', 0|\omega)_{sc}\rangle$$
$$= [S^{(0)}(x_1)]^{\frac{1}{2}} g^{(0)}(|x_1 - x_1'|)[S^{(0)}(x_1')]^{\frac{1}{2}} \tag{2.7.15}$$

in the mean scattering plane $x_3 = 0$. In writing Eq. (2.7.15) we have assumed that the spectral degree of coherence $g^{(0)}(|x_1 - x_1'|)$ is a function of the magnitude of the difference $x_1 - x_1'$ instead of the difference itself. The consequence of this assumption is that $g^{(0)}(|x_1 - x_1'|)$ is real and symmetric in x_1 and x_1' rather than just Hermitian. This assumption is easily relaxed.

The physical system we consider here is the one studied in Section 2.1. The region $x_3 > \zeta(x_1)$ is vacuum, while the region $x_3 < \zeta(x_1)$ is a perfect conductor. The surface profile function has the properties described in Section 2.1.1.

The surface $x_3 = \zeta(x_1)$ is illuminated at normal incidence by an s-polarized beam of frequency ω. The single nonzero component of the incident electric field is written in the form

$$E_2(x_1, x_3|\omega)_{inc} = \int_{-\infty}^{\infty} \frac{dk}{2\pi} F(k) \exp[ikx_1 - i\alpha_0(k)x_3], \tag{2.7.16}$$

where the weight function $F(k)$ is not specified as yet. On invoking the Rayleigh hypothesis [2.28], we write the scattered field as a superposition of outgoing plane waves,

$$E_2(x_1, x_3|\omega)_{sc} = \int_{-\infty}^{\infty} \frac{dq}{2\pi} R(q) \exp[iqx_1 + i\alpha_0(q)x_3]. \tag{2.7.17}$$

The scattering amplitude $R(q)$ is seen to be given by

$$R(q) = \int_{-\infty}^{\infty} dx_1 \exp(-iqx_1) E_2(x_1, 0|\omega)_{sc}. \tag{2.7.18}$$

It follows from this result that

$$\langle R(q)R^*(q')\rangle = \int_{-\infty}^{\infty} dx_1 \int_{-\infty}^{\infty} dx_1' \exp(-iqx_1 + iq'x_1')\langle E_2(x_1, 0|\omega)_{sc} E_2^*(x_1', 0|\omega)_{sc}\rangle. \tag{2.7.19}$$

The substitution of Eq. (2.7.15) into this expression yields $\langle R(q)R^*(q')\rangle$ in the form

$$\langle R(q)R^*(q')\rangle = \int_{-\infty}^{\infty} dx_1 \int_{-\infty}^{\infty} dx_1' \exp(-iqx_1 + iq'x_1')\left[S^{(0)}(x_1)\right]^{\frac{1}{2}} \times g^{(0)}\left(|x_1 - x_1'|\right)\left[S^{(0)}(x_1')\right]^{\frac{1}{2}}. \tag{2.7.20}$$

The spectral density function $S^{(0)}(x_1)$ is a real non-negative function of x_1. Its square root $[S^{(0)}(x_1)]^{\frac{1}{2}}$ is therefore also a real, non-negative function of x_1. We introduce its Fourier integral representation by

$$\left[S^{(0)}(x_1)\right]^{\frac{1}{2}} = \int_{-\infty}^{\infty} \frac{dp}{2\pi} \hat{S}(p) \exp(ipx_1), \tag{2.7.21}$$

where the Fourier coefficient $\hat{S}(p)$ has the property

$$\hat{S}(-p) = S^*(p) \tag{2.7.22}$$

due to the reality of $[S^{(0)}(x_1)]^{\frac{1}{2}}$. The use of Eqs. (2.7.21) and (2.7.22) in Eq. (2.7.20) yields finally the result that

$$\langle R(q)R^*(q')\rangle = \int_{-\infty}^{\infty} du\, g^{(0)}\left(|u|\right) \exp\left[-i\left(\frac{q+q'}{2}\right)u\right] \times \int_{-\infty}^{\infty} \frac{dp}{2\pi} \exp(ipu)\hat{S}\left(p + \left(\frac{q-q'}{2}\right)\right)\hat{S}^*\left(p - \left(\frac{q-q'}{2}\right)\right). \tag{2.7.23}$$

2.7.3.1. Design of a Surface that Acts as a Schell-Model Source Since our goal is to design a surface that produces a scattered field whose cross-spectral density in the plane $x_3 = 0$ is given by Eq. (2.7.15), we now seek an alternative expression for $\langle R(q)R^*(q')\rangle$ that depends on the surface profile function, to equate to the expression given by Eq. (2.7.23). In this way we obtain an equation from which the surface profile function can be determined. To this end we return to the expression for the scattered field given by Eq. (2.7.17). Due to the linearity of the scattering problem, the scattering amplitude $R(q)$ can be written as

$$R(q) = \int_{-\infty}^{\infty} \frac{dk}{2\pi} R(q|k)F(k), \tag{2.7.24}$$

where $R(q|k)$ is the amplitude for the scattering of the plane wave $\exp[ikx_1 - i\alpha_0(k)x_3]$ into the plane wave $\exp[iqx_1 + i\alpha_0(q)x_3]$. It is the solution of Eq. (2.5.97), and to the lowest nonzero order in the surface profile function is given by

$$R(q|k) \cong -2\pi\delta(q-k) + 2i\alpha_0(k)\int_{-\infty}^{\infty} dx_1\,\zeta(x_1)\exp[-i(q-k)x_1]$$

$$= -\int_{-\infty}^{\infty} dx_1\,[1 - 2i\alpha_0(k)\zeta(x_1)]\exp[-i(q-k)x_1]. \tag{2.7.25}$$

By exponentiating the factor $1 - 2i\alpha_0(k)\zeta(x_1)$ in the integrand in this expression, and keeping only the contribution linear in $\zeta(x_1)$ in the exponent, we obtain $R(q|k)$ in the simplest version of phase perturbation theory,

$$R(q|k) \cong -\int_{-\infty}^{\infty} dx_1\,\exp[-i(q-k)x_1]\exp[-i2\alpha_0(k)\zeta(x_1)]. \tag{2.7.26}$$

The scattering amplitude $R(q)$, can now be written as

$$R(q) = -\int_{-\infty}^{\infty} dx_1\,\exp(-iqx_1)\int_{-\infty}^{\infty}\frac{dk}{2\pi}F(k)\exp[ikx_1 - i2\alpha_0(k)\zeta(x_1)]. \tag{2.7.27}$$

With this result the correlation function $\langle R(q)R^*(q')\rangle$ takes the form

$$\langle R(q)R^*(q')\rangle = \int_{-\infty}^{\infty} dx_1\,\exp(-iqx_1)\int_{-\infty}^{\infty}\frac{dk}{2\pi}F(k)\exp(ikx_1)$$

$$\times\int_{-\infty}^{\infty} dx_1'\,\exp(iq'x_1')\int_{-\infty}^{\infty}\frac{dk'}{2\pi}F^*(k')\exp(-ik'x_1')$$

$$\times\left\langle\exp[-i2\alpha_0(k)\zeta(x_1) + i2\alpha_0^*(k')\zeta(x_1')]\right\rangle. \tag{2.7.28}$$

At this point we make a major approximation. Namely, we replace $\alpha_0(k)$ and $\alpha_0^*(k')$ in this expression by (ω/c). The neglect of the k and k' dependence of these functions leads to only small errors in the results obtained. With this approximation Eq. (2.7.28) becomes

$$\langle R(q)R^*(q')\rangle = \int_{-\infty}^{\infty}\frac{dk}{2\pi}F(k)\int_{-\infty}^{\infty} dx_1\,\exp[-i(q-k)x_1]$$

$$\times \int_{-\infty}^{\infty} \frac{dk'}{2\pi} F^*(k') \int_{-\infty}^{\infty} dx_1' \exp[i(q'-k')x_1']$$
$$\times \langle \exp\{-i(2\omega/c)[\zeta(x_1)-\zeta(x_1')]\}\rangle. \qquad (2.7.29)$$

To evaluate the integral over x_1' in this expression we make an approximation that is equivalent to passing to the geometrical optics limit: we expand $\zeta(x_1)$ about $x_1 = x_1'$, $\zeta(x_1) = \zeta(x_1') + (x_1 - x_1')\zeta'(x_1') + \cdots$ and retain only the terms up to and including those linear in $(x_1 - x_1')$. We then use the representation of the surface profile function $\zeta(x_1)$ given by Eq. (2.1.83), and obtain

$$\int_{-\infty}^{\infty} dx_1' \exp[i(q'-k')x_1']\langle \exp[-i(2\omega/c)(x_1-x_1')\zeta'(x_1')]\rangle$$
$$= \sum_{n=-\infty}^{\infty} \int_{nb}^{(n+1)b} dx_1' \exp[i(q'-k')x_1']\langle \exp[-i(2\omega/c)(x_1-x_1')a_n]\rangle$$
$$= \sum_{n=-\infty}^{\infty} \int_{nb}^{(n+1)b} dx_1' \exp[i(q'-k')x_1'] \int_{-\infty}^{\infty} d\gamma\, f(\gamma) \exp[-i(2\omega/c)(x_1-x_1')\gamma]$$
$$= \int_{-\infty}^{\infty} d\gamma\, f(\gamma) \exp[-i(2\omega/c)x_1\gamma] \int_{-\infty}^{\infty} dx_1' \exp\{i[q'-k'+2(\omega\gamma/c)x_1']\}.$$
$$(2.7.30)$$

With this result Eq. (2.7.29) becomes

$$\langle R(q)R^*(q')\rangle = \int_{-\infty}^{\infty} d\gamma\, f(\gamma) \int_{-\infty}^{\infty} dk\, F(k)\delta(q-k+(2\omega/c)\gamma)$$
$$\times \int_{-\infty}^{\infty} dk'\, F^*(k')\delta(q'-k'+(2\omega/c)\gamma)$$
$$= \int_{-\infty}^{\infty} d\gamma\, f(\gamma) F(q+(2\omega/c)\gamma) F^*(q'+(2\omega/c)\gamma). \qquad (2.7.31)$$

We now introduce the new variables

$$q' - q = Q, \qquad \frac{1}{2}(q + q') = P, \tag{2.7.32}$$

and equate the right-hand sides of Eqs. (2.7.23) and (2.7.31):

$$\int_{-\infty}^{\infty} du\, g^{(0)}(|u|) \exp(-iPu) \int_{-\infty}^{\infty} \frac{dp}{2\pi} \exp(ipu) \hat{S}\left(p + \frac{Q}{2}\right) S^*\left(p - \frac{Q}{2}\right)$$

$$= \int_{-\infty}^{\infty} d\gamma\, f(\gamma) F\left(P + \frac{2\omega}{c}\gamma + \frac{Q}{2}\right) F^*\left(P + \frac{2\omega}{c}\gamma - \frac{Q}{2}\right). \tag{2.7.33}$$

This is the integral equation satisfied by the pdf $f(\gamma)$ of the slope a_n in the representation of the surface profile function $\zeta(x_1)$ given by Eq. (2.1.83).

To solve Eq. (2.7.33) we use the Fourier inversion theorem to rewrite it as

$$g^{(0)}(|u|) \int_{-\infty}^{\infty} \frac{dp}{2\pi} \exp(ipu) \hat{S}\left(p + \frac{Q}{2}\right) \hat{S}^*\left(p - \frac{Q}{2}\right)$$

$$= \int_{-\infty}^{\infty} d\gamma\, f(\gamma) \exp\left(-i\frac{2\omega}{c}u\gamma\right) \int_{-\infty}^{\infty} \frac{dp}{2\pi} \exp(ipu) F\left(p + \frac{Q}{2}\right) F^*\left(p - \frac{Q}{2}\right). \tag{2.7.34}$$

We see that this equation has a solution if we set

$$\hat{S}(k) = AF(k), \tag{2.7.35}$$

where A is a constant. With this identification Eq. (2.7.34) becomes

$$|A|^2 g^{(0)}(|u|) = \int_{-\infty}^{\infty} d\gamma\, f(\gamma) \exp[-i(2\omega u/c)\gamma], \tag{2.7.36}$$

which can be inverted to yield

$$f(\gamma) = \frac{|A|^2}{2\pi} \int_{-\infty}^{\infty} dx\, g^{(0)}\left(\frac{c}{2\omega}|x|\right) \exp(i\gamma x). \tag{2.7.37}$$

The pdf $f(\gamma)$ must be normalized to unity,

$$\int_{-\infty}^{\infty} d\gamma\, f(\gamma) = 1 = \frac{|A|^2}{2\pi} \int_{-\infty}^{\infty} d\gamma \int_{-\infty}^{\infty} dx\, g^{(0)}\left(\frac{c}{2\omega}|x|\right) \exp(i\gamma x)$$
$$= |A|^2 \int_{-\infty}^{\infty} dx\, \delta(x) g^{(0)}\left(\frac{c}{2\omega}|x|\right)$$
$$= |A|^2, \tag{2.7.38}$$

because $g^{(0)}(0) = 1$. With this result we obtain finally

$$f(\gamma) = \frac{1}{\pi} \int_{0}^{\infty} dx\, g^{(0)}\left(\frac{c}{2\omega}x\right) \cos\gamma x, \tag{2.7.39}$$

where it should be kept in mind that $g^{(0)}(|u|)$ is an even function of u.

From the result given by Eq. (2.7.39) we see that apart from the requirement that $g^{(0)}(0) = 1$ the function $g^{(0)}(|u|)$ is constrained only by the requirement that its Fourier cosine transform be a non-negative function of γ. This restriction follows from the fact that a probability density function is a non-negative function of its argument.

The preceding results also show that if we wish to have a one-dimensional random surface that produces a scattered field whose cross-spectral density function in the plane $x_3 = 0$ has the form given by Eq. (2.7.15), not only must the pdf of the slope a_n in the representation of the surface profile function given by Eq. (2.1.83) be obtained from Eq. (2.7.39), but the surface must also be illuminated by a beam given by

$$E_2(x_1, x_3|\omega)_{\text{inc}} = \int_{-\infty}^{\infty} \frac{dk}{2\pi} \hat{S}(k) \exp\left[ikx_1 - i\alpha_0(k)x_3\right], \tag{2.7.40}$$

where $\hat{S}(k)$ is defined in Eq. (2.7.21).

Once $f(\gamma)$ has been determined, a long sequence of $\{a_n\}$ is generated by the rejection method [2.6], and the corresponding sequence of $\{b_n\}$ is obtained from Eq. (2.1.85). A single realization of the surface profile function $\zeta(x_1)$ is then constructed on the basis of Eq. (2.1.83).

In this way we can generate a one-dimensional randomly rough surface that scatters a beam with a suitable intensity profile incident on it into a beam with a specified spectral density of the scattered light in the source plane (2.7.15), and a spectral degree of coherence of the scattered light in this plane of a rather general form.

2.7.3.2. Example To determine how well a surface generated in the manner described produces a scattered field that possesses the correlation property expressed by Eq. (2.7.15) we proceed as follows. The fields $E_2(x_1,0|\omega)_{sc}$ and $E_2^*(x_1',0|\omega)_{sc}$ scattered from this surface are calculated by the rigorous approach described in Section 2.1.3.1, and their product $E_2(x_1,0|\omega)_{sc}E_2^*(x_1',0|\omega)_{sc}$ is formed. This calculation is repeated for a total of N_p realizations of $\zeta(x_1)$ for the same values of x_1 and x_1', and an arithmetic average of the results obtained yields the correlation function $\langle E_2(x_1,0|\omega)_{sc}E_2^*(x_1',0|\omega)_{sc}\rangle$. A different pair of values of x_1 and x_1' then selected, and the correlation function $\langle E_2(x_1,0|\omega)_{sc}E_2^*(x_1',0|\omega)_{sc}\rangle$ is calculated for it in the manner described. These calculations are continued for enough pairs of values of x_1 and x_1' that a comparison of the correlation function obtained with the input function (2.7.15) can be made.

To illustrate the preceding results we generate a one-dimensional randomly rough surface that produces a scattered field whose spectral density in the plane $x_3=0$ is given by

$$S^{(0)}(x_1)=\frac{\sqrt{\pi}}{2}\theta\big(\sqrt{2}\sigma_s-|x_1|\big), \tag{2.7.41}$$

and a spectral degree of coherence in this plane given by

$$g^{(0)}\big(|u|\big)=J_0\big((\sqrt{2}/\sigma_g)|u|\big), \tag{2.7.42}$$

where $J_0(z)$ is a Bessel function of the first kind and zero order. These forms for $S^{(0)}(x_1)$ and $g^{(0)}(|u|)$ were chosen because they lead to a cross-spectral density function $W^{(0)}(x_1,x_1'|\omega)$ with considerable structure, which poses a challenge to the design procedure to reproduce. The Fourier coefficient $\hat{S}(p)$ obtained from Eqs. (2.7.21) and (2.7.41) is

$$\hat{S}(p)=2\pi^{\frac{1}{4}}\sigma_s\operatorname{sinc}(\sqrt{2}\sigma_s p). \tag{2.7.43}$$

The pdf $f(\gamma)$ obtained from Eqs. (2.7.39) and (2.7.42) is

$$f(\gamma)=\begin{cases}\frac{1}{\pi}\frac{1}{[(c/\sqrt{2}\sigma_g\omega)^2-\gamma^2]^{1/2}} & |\gamma|<(c/\sqrt{2}\sigma_g\omega),\\ 0 & |\gamma|>(c/\sqrt{2}\sigma_g\omega).\end{cases} \tag{2.7.44}$$

The values of the experimental and roughness parameters assumed in carrying out these calculations were $\omega=2\pi c/\lambda$, where $\lambda=612.7$ nm, $\sigma_s=12.5$ μm, $\sigma_g=3$ μm, $b=5$ μm, the length of the surface $L=100$ μm, and $N_p=3000$.

In Fig. 2.38(a) we present a comparison of the result for $W^{(0)}(x_1,x_1|\omega)=\langle E_2(x_1,0|\omega)_{sc}|^2\rangle=S^{(0)}(x_1)$ obtained for scattering from a perfectly conducting surface and the input expression given by Eq. (2.7.15). The analogous comparison between the computer simulation results for $W^{(0)}(x_1,-x|\omega)=\langle E_2(x_1,0|\omega)_{sc}E_2^*(-x_1,0|\omega)_{sc}\rangle=S^{(0)}(x_1)g^{(0)}(2|x_1|)$ is shown in Fig. 2.38(b). In each case the comparison between the computed and input results is very good. Thus, the results obtained show that the random

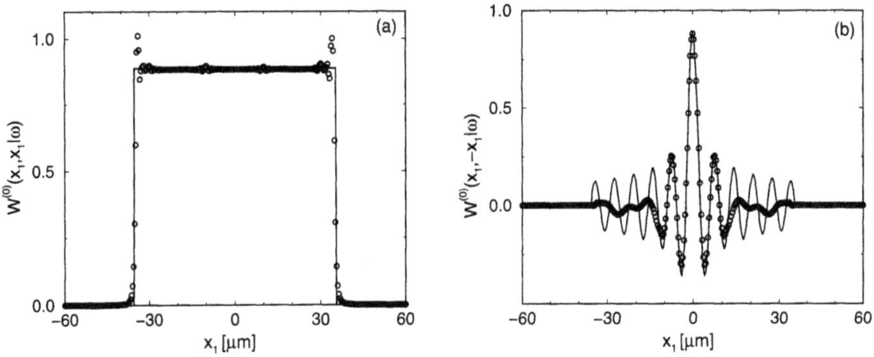

Figure 2.38. (a) The $x_1' = x_1$ cross section of the cross spectral density $W^{(0)}(x_1, x_1'|\omega)$ for scattering from a perfectly conducting surface designed to produce a scattered field with the spectral density and spectral degree of coherence given by Eqs. (2.7.41) and (2.7.42), respectively. The solid curve shows the shape of the input function, while the circles show the shape of $W^{(0)}(x_1, x_1|\omega)$ obtained by a rigorous computer simulation approach. (b) The same as (a) but for the $x_1' = -x_1$ cross section of $W^{(0)}(x_1, x_1'|\omega)$.

surfaces generated in the manner described in this section produce scattered fields that possess the correlation property expressed by Eq. (2.7.15).

2.7.3.3. Evolution of the Scattered Field To explore the manner in which the scattered field propagates in the vacuum region $x_3 > 0$ we use the result that the mean intensity of the scattered field in the far zone can be expressed in terms of $g^{(0)}(|u|)$ and $\hat{S}(p)$. To obtain this relation we substitute Eq. (2.7.18) into Eq. (2.7.17) and use the parabolic approximation

$$\alpha_0(q) \cong (\omega/c) - (c/2\omega)q^2 \tag{2.7.45}$$

in evaluating the integral over q. In this way we obtain

$$E_2(x_1, x_3|\omega)_{sc} = \exp\big[i(\omega/c)x_3 - i(\pi/4)\big]\left(\frac{\omega}{2\pi c x_3}\right)^{\frac{1}{2}} \times \int_{-\infty}^{\infty} dx_1' \exp\big[i(\omega/2cx_3)(x_1 - x_1')^2\big]E_2(x_1', 0|\omega)_{sc}. \tag{2.7.46}$$

It follows from this result that

$$\big\langle\big|E_2(x_1, x_3|\omega)_{sc}\big|^2\big\rangle = \frac{\omega}{2\pi c x_3}\int_{-\infty}^{\infty} dx_1' \int_{-\infty}^{\infty} dx_1'' \exp\big[i(\omega/2cx_3)(x_1 - x_1')^2\big] \times \exp\big[-i(\omega/2cx_3)(x_1 - x_1'')^2\big]\big\langle E_2(x_1', 0|\omega)_{sc} E_2^*(x_1'', 0|\omega)_{sc}\big\rangle. \tag{2.7.47}$$

When we substitute into this expression the representation of $\langle E_2(x_1', 0|\omega)_{sc} E_2^*(x_1'', 0|\omega)_{sc}\rangle$ given by Eq. (2.7.15) we obtain

$$\begin{aligned}\langle|E_2(x_1, x_3|\omega)_{sc}|^2\rangle \\ &= \frac{\omega}{2\pi c x_3}\int_{-\infty}^{\infty} dx_1' \int_{-\infty}^{\infty} dx_1'' \exp\left[-i(\omega/cx_3)(x_1' - x_1'')x_1 + i(\omega/2cx_3)(x_1'^2 - x_1''^2)\right] \\ &\quad \times \left[S^{(0)}(x_1')\right]^{\frac{1}{2}} g^{(0)}(|x_1' - x_1''|)\left[S^{(0)}(x_1'')\right]^{\frac{1}{2}}. \qquad (2.7.48)\end{aligned}$$

The use of the representation (2.7.21) and the change of variable $x_1' = u + x_1''$ leads to

$$\begin{aligned}\langle|E_2(x_1, x_3|\omega)_{sc}|^2\rangle &= \frac{\omega}{2\pi c x_3}\int_{-\infty}^{\infty} du\, g^{(0)}(|u|)\exp\left[-i(\omega x_1/cx_3)u\right] \\ &\quad \times \int_{-\infty}^{\infty} \frac{dp}{2\pi}\exp(ipu)\hat{S}\bigl(p - (\omega u/2cx_3)\bigr)\hat{S}^*\bigl(p + (\omega u/2cx_3)\bigr).\end{aligned} \qquad (2.7.49)$$

An alternative, equivalent, expression for $\langle|E_2(x_1, x_3|\omega)_{sc}|^2\rangle$ is obtained if we use the inverse of Eq. (2.7.21),

$$\hat{S}(p) = \int_{-\infty}^{\infty} dx_1 \left[S^{(0)}(x_1)\right]^{\frac{1}{2}} \exp(-ipx_1), \qquad (2.7.50)$$

in Eq. (2.7.49):

$$\begin{aligned}\langle|E_2(x_1, x_3|\omega)_{sc}|^2\rangle &= \frac{\omega}{2\pi c x_3}\int_{-\infty}^{\infty} du\, g^{(0)}(|u|)\exp\left[-i(\omega x_1/cx_3)u\right] \\ &\quad \times \int_{-\infty}^{\infty} dx \exp\left[i(\omega u/cx_3)x\right]\left[S^{(0)}\bigl(x + (u/2)\bigr)\right]^{\frac{1}{2}}\left[S^{(0)}\bigl(x - (u/2)\bigr)\right]^{\frac{1}{2}}.\end{aligned} \qquad (2.7.51)$$

In some cases the use of this expression results in integrals that are easier to evaluate.

To illustrate this result we consider the evolution of the beam produced by the surface for which $S^{(0)}(x_1)$ is given by Eq. (2.7.41) and $g^{(0)}(|u|)$ is given by Eq. (2.7.42). The

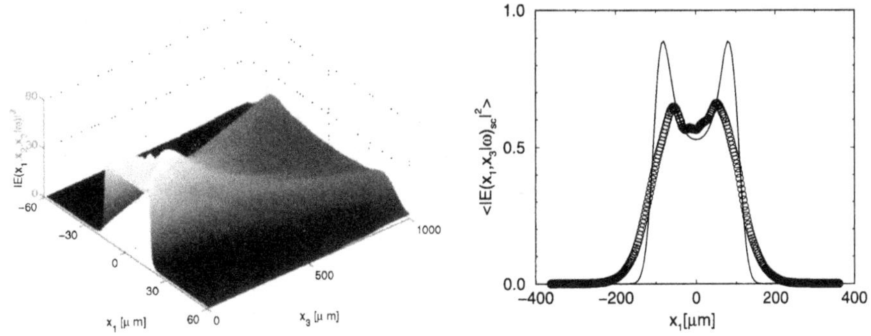

(a) (b)

Figure 2.39. (a) The profile of the beam scattered from a perfectly conducting surface designed to produce the spectral density and spectral degree of coherence given by Eqs. (2.7.41) and (2.7.42), respectively, as a function of the distance from the surface. (b) The profile of the beam scattered from the same perfectly conducting surface at a fixed distance $x_3 = 3500$ μm from the surface in the far field.

integral over x in Eq. (2.7.51) is readily calculated in this case, with the result that

$$\int_{-\infty}^{\infty} dx \exp\left[i(\omega u/cx_3)x\right]\left[S^{(0)}(x+(u/2))\right]^{\frac{1}{2}}\left[S^{(0)}(x-(u/2))\right]^{\frac{1}{2}}$$

$$= \sqrt{\pi}\frac{cx_3}{\omega}\frac{1}{u}\sin\left[\left(\frac{\omega u}{cx_3}\right)\left(\sqrt{2}\sigma_s - \frac{|u|}{2}\right)\right] \quad |u| < 2\sqrt{2}\sigma_s \tag{2.7.52a}$$

$$= 0 \quad |u| > 2\sqrt{2}\sigma_s. \tag{2.7.52b}$$

Equation (2.7.51) then takes the form

$$\left\langle\left|E_2(x_1, x_3|\omega)_{sc}\right|^2\right\rangle$$

$$= \frac{1}{2\sqrt{\pi}}\int_0^{2\sqrt{2}\sigma_s} du\, \frac{g^{(0)}(u)}{u}$$

$$\times\left[\sin\left(\frac{\omega u}{cx_3}\left(\sqrt{2}\sigma_s - \frac{u}{2} + x_1\right)\right) + \sin\left(\frac{\omega u}{cx_3}\left(\sqrt{2}\sigma_s - \frac{u}{2} - x_1\right)\right)\right], \tag{2.7.53}$$

where we have used the fact that the spectral degree of coherence is an even function of u.

In Fig. 2.39(a) we show the evolution of the beam scattered from a perfectly conducting surface designed to produce a scattered field whose cross-spectral density in the plane $x_3 = 0$ is given by Eqs. (2.7.15), (2.7.41), and (2.7.42) when it is illuminated at normal incidence by a beam defined by Eqs. (2.7.40) and (2.7.43). This result was calculated on the basis of Eq. (2.7.53). The values of the experimental and roughness parameters assumed in obtaining this result are those used in the calculations resulting in Fig. 2.38.

In Fig. 2.39(b) we plot the intensity profile of the scattered beam at a distance $x_3 =$ 3500 μm. The profile obtained from the numerical simulation is represented by open circles, while the one calculated on the basis of Eq. (2.7.53) is plotted as a solid curve. The two peaks that develop with increasing x_3 in the mean scattered intensity plotted in Figs. 2.39(a) and 2.39(b) arise from the square-root singularities in the expression for the pdf $f(\gamma)$ given by Eq. (2.7.44), on which the calculation of these figures is based (see Section 2.7.4.3 below). The reproduction of these peaks in the result of a rigorous scattering calculation of these mean intensity is a challenge to the corresponding surface design problem. Nevertheless, we see from Fig. 2.39(b) that the surface designed on the basis of the approach developed in this section is able to produce a mean intensity of the field scattered from it that displays peaks not far from the positions given by the calculation based on Eq. (2.7.53), although their heights are smaller than predicted by Eq. (2.7.53). We feel that the surface that produced the numerical simulation result plotted in Fig. 2.39(b) is a respectable solution to a difficult design problem.

2.7.4. *A Gaussian Incident Beam*

The preceding results are quite general, both as to the spectral density function of the scattered field in the plane $x_3 = 0$, which implies a particular intensity profile for the incident field, and as to the spectral degree of coherence in that plane. In this section we specialize these results to the practically important case of an incident field that has a Gaussian intensity profile, as in the case of a laser beam.

2.7.4.1. Evolution of the Scattered Field The case of a Gaussian incident beam is equivalent to choosing for the spectral density function of the scattered field in the plane $x_3 = 0$ the Gaussian form

$$S^{(0)}(x_1) = \exp(-x_1^2/2\sigma_s^2). \tag{2.7.54}$$

To see that this is the case we note first that the function $\hat{S}(k)$ defined by Eq. (2.7.50) now takes the form

$$\hat{S}(k) = 2\sqrt{\pi}\sigma_s \exp(-\sigma_s^2 k^2). \tag{2.7.55}$$

The incident field defined by Eq. (2.7.40) then becomes

$$E_2(x_1, x_3|\omega)_{\text{inc}} = \frac{\sigma_s}{\sqrt{\pi}} \int_{-\infty}^{\infty} dk \exp(-\sigma_s^2 k^2) \exp[ikx_1 - i\alpha_0(k)x_3]. \tag{2.7.56}$$

In the limit that $2\pi\sigma_s \gg \lambda$, where λ is the wavelength of the incident light, which is the case of interest here, we can use the parabolic approximation (2.7.45) to simplify the integrand in this expression, which then becomes

$$E_2(x_1, x_3|\omega)_{\text{inc}} \cong 2\frac{\sigma_s}{\sqrt{\pi}} \exp\bigl[-i(\omega/c)x_3\bigr] \int_0^\infty dk \, \cos kx_1 \exp\bigl\{-\bigl[\sigma_s^2 - i(cx_3/2\omega)\bigr]k^2\bigr\}$$

$$= \exp\bigl[-i(\omega/c)x_3\bigr] \frac{\exp\{-(x_1^2/w^2)[1 - i(2cx_3/\omega w^2)]\}}{[1 - i(2cx_3/\omega w^2)]^{\frac{1}{2}}}, \tag{2.7.57}$$

where we have defined

$$w = 2\sigma_s, \tag{2.7.58}$$

the $1/e$ half width of the incident beam. Under the condition assumed, $\lambda \ll 2\pi\sigma_s = \pi w$, and the fact that only the incident field or its normal derivative evaluated at $x_3 = \zeta(x_1)$ enters the rigorous scattering calculations, $2cx_3/\omega w^2$ can be neglected in comparison with unity, and the incident field in the vicinity of the surface is given by

$$E_2(x_1, x_3|\omega)_{\text{inc}} \cong \exp\bigl(-x_1^2/w^2\bigr) \exp\bigl[-i(\omega/c)x_3\bigr]. \tag{2.7.59}$$

The cross-spectral density function (2.7.15) of interest to us now has the Schell-model form

$$W^{(0)}(x_1, x_1'|\omega) = \exp\bigl(-x_1^2/4\sigma_s^2\bigr) g^{(0)}\bigl(|x_1 - x_1'|\bigr) \exp\bigl(-x_1'^2/4\sigma_s^2\bigr). \tag{2.7.60}$$

The pdf $f(\gamma)$ of the slope a_n in the representation (2.1.83) of the surface profile function $\zeta(x_1)$ is still related to the spectral degree of coherence $g^{(0)}(|u|)$ by Eq. (2.7.39). When we substitute the expression for the spectral density $S^{(0)}(x_1)$ given by Eq. (2.7.54) into Eq. (2.7.51), the mean intensity profile of the scattered field is obtained in the form

$$\bigl\langle\bigl|E_2(x_1, x_3|\omega)_{\text{sc}}\bigr|^2\bigr\rangle = \left(\frac{2}{\pi}\right)^{\frac{1}{2}} \sigma_s \frac{\omega}{cx_3} \int_0^\infty du \, g^{(0)}(u) \cos\left(\frac{\omega x_1 u}{cx_3}\right)$$

$$\times \exp\left[-\left(\frac{1}{8\sigma_s^2} + \frac{\sigma_s^2\omega^2}{2c^2x_3^2}\right)u^2\right]. \tag{2.7.61}$$

2.7.4.2. Examples To illustrate the result given by Eq. (2.7.61) we generate one-dimensional randomly rough surfaces that produce scattered fields whose spectral degrees of coherence in the plane $x_3 = 0$, and the corresponding pdfs $f(\gamma)$, are given by

$$g_a^{(0)}\bigl(|u|\bigr) = \exp\bigl(-u^2/2\sigma_g^2\bigr), \qquad f_a(\gamma) = \left(\frac{2}{\pi}\right)\sigma_g\left(\frac{\omega}{c}\right)\exp\left[-2\sigma_g^2\left(\frac{\omega}{c}\right)\gamma^2\right], \tag{2.7.62a}$$

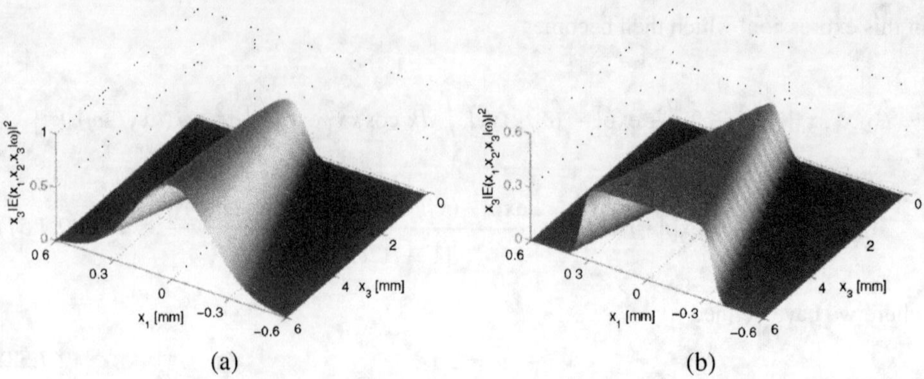

Figure 2.40. The profiles of the beams scattered from perfectly conducting surfaces designed to produce scattered fields with the cross-spectral density $W^{(0)}(x_1, x_1'|\omega)$ defined by Eq. (2.7.60) and Eqs. (2.7.62a)–(2.7.62b), respectively, as functions of the distance from the surface.

$$g_b^{(0)}(|u|) = \frac{\sin(\sqrt{3}u/\sigma_g)}{(\sqrt{3}u/\sigma_g)}, \qquad f_b(\gamma) = \frac{\sigma_g}{\sqrt{3}}\frac{\omega}{c}\theta\left(\frac{\sqrt{3}}{2}\frac{c}{\sigma_g\omega} - |\gamma|\right). \tag{2.7.62b}$$

In Figs. 2.40(a) and 2.40(b) we show the evolution of the beam scattered from a perfectly conducting surface designed to produce a scattered field with the cross-spectral density $W^{(0)}(x_1, x_1'|\omega)$ defined by Eq. (2.7.60) and Eqs. (2.7.62a) and (2.7.62b), respectively, when it is illuminated at normal incidence by an s-polarized Gaussian beam defined by Eqs. (2.7.40) and (2.7.55). These results were calculated on the basis of Eq. (2.7.61) for the values of the parameters used in obtaining Fig. 2.38. In the case (a), where the spectral degree of coherence in the plane $x_3 = 0$ is a Gaussian, Eq. (2.7.62a), the scattered beam is also a Gaussian beam. It is characterized by a mean half width given by $\mu_2(x_3, \omega) = \sigma_s^2 + (cx_3/\omega)^2[(1\sigma_g^2) + (1/4\sigma_s^2)]$ and an angular divergence $\Delta\theta = (c/\omega)[(1/\sigma_g^2) + (1/4\sigma_s^2)]^{\frac{1}{2}}$. In contrast, in the case (b), where the spectral degree of coherence in the plane $x_3 = 0$ is a sinc function, Eq. (2.7.62b), the scattered beam has a nearly constant mean intensity profile as it propagates. These features of the scattered beam are clearly seen in the results presented in Figs. 2.41(a) and 2.41(b), where we have plotted the intensity profiles of the beams scattered from perfectly conducting surfaces designed to produce the spectral degree of coherence of the scattered field in the plane $x_3 = 0$ given by Eqs. (2.7.62), at a distance $x_3 = 2000$ μm from that plane. The profiles obtained from the numerical simulations are represented by open circles, while those calculated on the basis of Eq. (2.7.61) are plotted as solid curves. The agreement between the results of the two types of calculations is seen to be excellent.

2.7.4.3. Transformations of the Incident Beam There has been considerable interest in recent years in the design of optical elements that transform a beam with one intensity distribution into a beam with a different intensity distribution [2.110–2.115]. Much of this work has been devoted to the transformation of a laser beam, with a Gaussian intensity

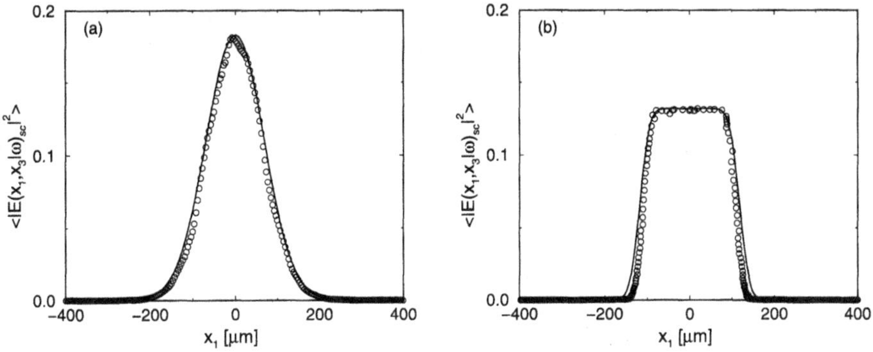

Figure 2.41. The profiles of the beams scattered from perfectly conducting surfaces designed to produce scattered fields with the cross-spectral density $W^{(0)}(x_1, x_1|\omega)$ defined by Eq. (2.7.60) and Eqs. (2.7.62a)–(2.7.62b), respectively, at a fixed distance from the surface $x_3 = 3500$ μm in the far field.

profile, into a beam with a constant intensity—a flat top beam. The results presented in Figs. 2.40(b) and 2.41(b) show that a one-dimensional randomly rough surface fabricated on the basis of Eqs. (2.7.15) and (2.7.62b) effects such a transformation.

The results we have obtained here, however, are more general than this. We see from the plots in Figs. 2.41 that the intensity profile of the scattered beam at a fixed distance x_3 in the far field resembles the pdf $f(\gamma)$ used in generating the surface from which the beam is scattered. This is not accidental. In the limit $x_3 \to \infty$, Eq. (2.7.61) together with the inverse of Eq. (2.7.37), with $|A|^2 = 1$,

$$g^{(0)}(|u|) = \int_{-\infty}^{\infty} d\gamma\, f(\gamma) \exp[-i(2\omega/c)u\gamma], \tag{2.7.63}$$

yields the result that

$$\langle |E_2(x_1, x_3|\omega)_{\mathrm{sc}}|^2 \rangle = 2\sigma_s^2 \frac{\omega}{cx_3} \int_{-\infty}^{\infty} d\gamma\, f(\gamma) \exp\{-8\sigma_s^2(\omega/c)^2[\gamma - (x_1/2x_3)]^2\}. \tag{2.7.64}$$

In the limit $\sigma_s/\lambda \gg 1$, we have the equivalence

$$\exp\{-8\sigma_s^2(\omega/c)^2[\gamma - (x_1/2x_3)]^2\} = \frac{1}{2}\left(\frac{\pi}{2}\right)^{\frac{1}{2}} \frac{c}{\omega} \frac{1}{\sigma_s} \delta(\gamma - (x_1/2x_3)). \tag{2.7.65}$$

With this result we find from Eq. (2.7.64) that

$$\langle |E_2(x_1, x_3|\omega)_{\mathrm{sc}}|^2 \rangle = \left(\frac{\pi}{2}\right)^{\frac{1}{2}} \frac{\sigma_s}{x_3} f\left(\frac{x_1}{2x_3}\right). \tag{2.7.66}$$

This relationship shows how to produce a one-dimensional randomly rough surface that transforms an incident laser beam with a Gaussian intensity profile (2.7.59), obtained from Eqs. (2.7.40) and (2.7.55) into a beam with an arbitrary intensity profile $\langle |E_2(x_1, x_3|\omega)_{\mathrm{sc}}|^2\rangle$ at a distance x_3 from it, when the laser beam is reflected from it: one generates the surface by means of Eq. (2.1.83) and the rejection method [2.6] on the basis of the pdf $f(\gamma)$ defined by Eq. (2.7.66),

$$f(\gamma) = \left(\frac{2}{\pi}\right)^{\frac{1}{2}} \frac{x_3}{\sigma_s} \langle |E_2(2\gamma x_3, x_3|\omega)_{\mathrm{sc}}|^2\rangle. \tag{2.7.67}$$

Although the preceding discussion has been based on the assumption that the incident beam has a Gaussian intensity profile, the approach used is not limited to incident fields with this intensity distribution. For if we return to the expression for the mean scattered intensity given by Eq. (2.7.49), we see that in the far field it becomes

$$\begin{aligned}&\langle |E_2(x_1, x_3|\omega)_{\mathrm{sc}}|^2\rangle \\ &\quad = \frac{\omega}{2\pi c x_3} \int_{-\infty}^{\infty} du\, g^{(0)}(|u|) \exp[-i(\omega x_1/c x_3)u] \int_{-\infty}^{\infty} \frac{dp}{2\pi} \exp(ipu) |\hat{S}(p)|^2.\end{aligned} \tag{2.7.68}$$

If we replace $g^{(0)}(|u|)$ by its Fourier representation (2.7.36), with $|A|^2 = 1$, we obtain

$$\begin{aligned}\langle |E_2(x_1, x_3|\omega)_{\mathrm{sc}}|^2\rangle &= \frac{\omega}{2\pi c x_3} \int_{-\infty}^{\infty} du \int_{-\infty}^{\infty} d\gamma \int_{-\infty}^{\infty} \frac{dp}{2\pi} f(\gamma) \\ &\quad \times \exp[-i(2\omega\gamma/c)u - i(\omega x_1/c x_3)u + ipu]|\hat{S}(p)|^2 \\ &= \frac{\omega}{2\pi c x_3} \int_{-\infty}^{\infty} d\gamma\, f(\gamma) \int_{-\infty}^{\infty} \frac{dp}{2\pi} |\hat{S}(p)|^2 2\pi\delta\left(p - 2\frac{\omega}{c}\left(\gamma + \frac{x_1}{2x_3}\right)\right) \\ &= \frac{\omega}{2\pi c x_3} \int_{-\infty}^{\infty} d\gamma\, f(\gamma) \left|\hat{S}\left(\frac{2\omega}{c}\left(\gamma + \frac{x_1}{2x_3}\right)\right)\right|^2.\end{aligned} \tag{2.7.69}$$

Thus, if the function $|\hat{S}(\frac{2\omega}{c}(\gamma + \frac{x_1}{2x_3}))|^2$ is sufficiently narrowly peaked as a function of γ that it can be replaced by a delta function,

$$\left|\hat{S}\left(\frac{2\omega}{c}\left(\gamma + \frac{x_1}{2x_3}\right)\right)\right|^2 \cong B\delta\left(\gamma + \frac{x_1}{2x_3}\right), \tag{2.7.70}$$

we find that

$$\langle |E_2(x_1, x_3|\omega)_{sc}|^2\rangle = \frac{\omega}{2\pi c x_3} Bf\left(-\frac{x_1}{2x_3}\right). \tag{2.7.71}$$

Because we have assumed that the spectral degree of coherence is an even function of u, it follows from Eq. (2.7.39) that $f(\gamma)$ is an even function of γ. Thus we have that

$$\langle |E_2(x_1, x_3|\omega)_{sc}|^2\rangle = \frac{\omega}{2\pi c x_3} Bf\left(\frac{x_1}{2x_3}\right). \tag{2.7.72}$$

The constant B is obtained from the normalization of the delta function as

$$B = \int_{-\infty}^{\infty} d\gamma \left|\hat{S}\left(\frac{2\omega}{c}\left(\gamma + \frac{x_1}{2x_3}\right)\right)\right|^2$$

$$= \frac{c}{2\omega} \int_{-\infty}^{\infty} dp |\hat{S}(p)|^2 \tag{2.7.73a}$$

$$= \pi \frac{c}{\omega} \int_{-\infty}^{\infty} dx_1\, S^{(0)}(x_1). \tag{2.7.73b}$$

In general, the more slowly varying $S^{(0)}(x_1)$ is as a function of x_1, the more sharply peaked the function $|\hat{S}(\frac{2\omega}{c}(\gamma + \frac{x_1}{2x_3}))|^2$ is, and the better the approximation (2.7.70) is.

As an example, we consider the case where

$$S^{(0)}(x_1) = \frac{2\sigma_s^2}{(x_1^2 + 2\sigma_s^2)}. \tag{2.7.74}$$

In this case we find that

$$\hat{S}(p) = 2\sqrt{2}\sigma_s K_0(\sqrt{2}\sigma_s p), \tag{2.7.75}$$

where $K_0(x)$ is a modified Bessel function of the second kind of zero order. The squared modulus of this function is indeed more sharply and narrowly peaked the larger σ_s is, i.e., the more slowly $S^{(0)}(x)$ varies with x_1.

In Figs. 2.42(a) and 2.42(b) we show the evolution of the beam scattered from a perfectly conducting surface designed to produce a scattered field with the spectral density $S^{(0)}(x_1)$ defined by Eqs. (2.7.54) and (2.7.74) and the spectral degree of coherence given by Eq. (2.7.62b) when it is illuminated at normal incidence by s-polarized beams defined by Eqs. (2.7.54) and (2.7.74), respectively. These results were calculated on the basis of Eq. (2.7.51) for the values of the parameters used in obtaining Fig. 2.38. Both incident beams, Gaussian (a) and Lorentzian (b) evolve into a flat top beam at large distances from

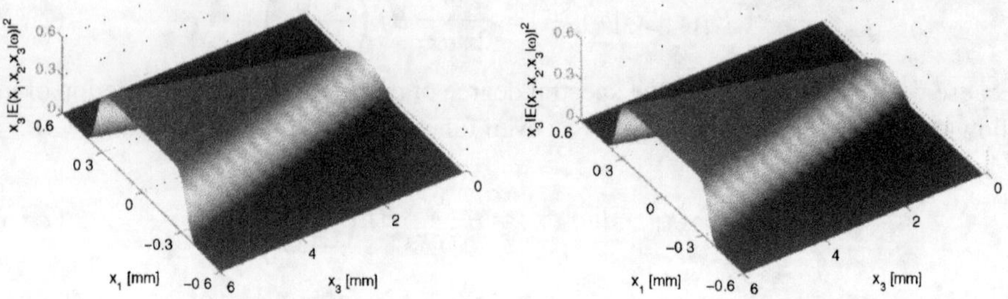

(a) (b)

Figure 2.42. The profiles of the beam scattered from perfectly conducting surfaces designed to produce scattered fields with the spectral densities defined by Eqs. (2.7.54) (a) and (2.7.74) (b) and the spectral degree of coherence defined by Eq. (2.7.62b) as functions of the distance from the surface.

the surface. However, the flat top beam forms at considerably shorter distances in the case of the incident Lorentzian beam.

Thus, in this section we have first presented an approach to the design of a one-dimensional randomly rough perfectly conducting surface that, when illuminated at normal incidence by an s-polarized beam with a suitable intensity profile obtained from Eq. (2.7.40) produces a scattered field whose cross-spectral density $W^{(0)}(x_1, x_1'|\omega)$ in the mean scattering plane $x_3 = 0$ has the Schell-model form given by Eq. (2.7.15). We have then shown that when the incident field is one that produces a scattered field with an intensity distribution in the plane $x_3 = 0$, $S^{(0)}(x_1)$, that is a slowly varying function of x_1, the surfaces designed by the approach described in this section can transform that beam into a beam with a different, arbitrary, intensity profile, including a flat top beam. These results could be useful for producing beams with specified coherence properties.

2.8. Surfaces that Produce a Prescribed Angular Dependence of the Mean Intensity of the Field Transmitted through Them

Thus far in this chapter we have considered only the design of one-dimensional surfaces that possess specified scattering properties. However, one can envision situations in which it would be useful to have an optical diffuser that produces *transmitted* light with a specified angular dependence of the mean differential transmission coefficient, or with a specified dependence of its mean intensity on the wavelength of the incident light, for example. It is desirable, therefore, to extend the theory developed in the preceding sections of this chapter for the scattering problem to the transmission problem. This we do in the present section for the case of light transmitted through a free-standing or supported dielectric film with a planar illuminated surface and a randomly rough back surface.

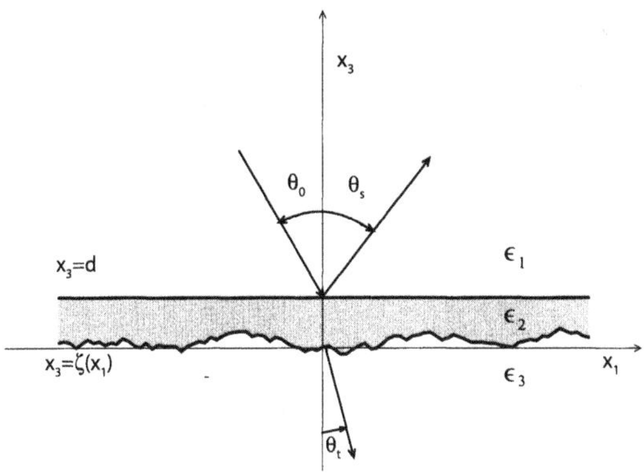

Figure 2.43. The structure studied in the present section.

The first step in designing a random surface on a film that transmits light in a prescribed fashion is to obtain the transmission amplitude, in terms of which the transmitted field is expressed. Because a film has two surfaces or interfaces this is a somewhat more complicated task than it is in the single interface systems we have considered so far. Nevertheless, it can be carried out, and yields a result in a form that allows the design problem to be solved.

2.8.1. Scattering Theory

The structure we consider in this work consists of a dielectric medium characterized by a real dielectric constant ϵ_1 in the region $x_3 > d$; a dielectric medium characterized by a real dielectric constant ϵ_2 in the region $\zeta(x_1) < x_3 < d$; and a dielectric medium characterized by a real dielectric constant ϵ_3 in the region $x_3 < \zeta(x_1)$ (Fig. 2.43). The surface profile function $\zeta(x_1)$ is assumed to be a single-valued function of x_1 that is differentiable and constitutes a random process, but not necessarily a stationary one.

This system is illuminated from the region $x_3 > d$ by a p- or s-polarized plane wave of frequency ω, whose plane of incidence is the x_1x_3 plane. With these assumptions there is no cross-polarized scattering.

To deal simultaneously with fields of p and s polarization we again introduce a function $F_\nu(x_1, x_3|\omega)$ that is the single nonzero component of the magnetic field in the system, $H_2(x_1, x_3|\omega)$, when $\nu = p$, and is the single nonzero component of the electric field, $E_2(x_1, x_3|\omega)$ when $\nu = s$. This function is the solution of the Helmholtz equation

$$\left(\frac{\partial^2}{\partial x_1^2} + \frac{\partial^2}{\partial x_3^2} + \epsilon_1 \frac{\omega^2}{c^2}\right) F_\nu^{(1)}(x_1, x_3|\omega) = 0 \tag{2.8.1}$$

in the region $x_3 > d$, the equation

$$\left(\frac{\partial^2}{\partial x_1^2}+\frac{\partial^2}{\partial x_3^2}+\epsilon_2\frac{\omega^2}{c^2}\right)F_\nu^{(2)}(x_1,x_3|\omega)=0 \tag{2.8.2}$$

in the region $\zeta(x_1) < x_3 < d$, and the equation

$$\left(\frac{\partial^2}{\partial x_1^2}+\frac{\partial^2}{\partial x_3^2}+\epsilon_3\frac{\omega^2}{c^2}\right)F_\nu^{(3)}(x_1,x_3|\omega)=0 \tag{2.8.3}$$

in the region $x_3 < \zeta(x_1)$. It satisfies the boundary conditions

$$F_\nu^{(1)}(x_1,x_3|\omega)\big|_{x_3=d}=F_\nu^{(2)}(x_1,x_3|\omega)\big|_{x_3=d}, \tag{2.8.4a}$$

$$\frac{1}{\kappa_\nu}\frac{\partial}{\partial x_3}F_\nu^{(1)}(x_1,x_3|\omega)\bigg|_{x_3=d}=\frac{1}{\lambda_\nu}\frac{\partial}{\partial x_3}F_\nu^{(2)}(x_1,x_3|\omega)\bigg|_{x_3=d} \tag{2.8.4b}$$

at the interface $x_3 = d$, and the conditions

$$F_\nu^{(2)}(x_1,x_3|\omega)\big|_{x_3=\zeta(x_1)}=F_\nu^{(3)}(x_1,x_3|\omega)\big|_{x_3=\zeta(x_1)}, \tag{2.8.5a}$$

$$\frac{1}{\lambda_\nu}\frac{\partial}{\partial n}F_\nu^{(2)}(x_1,x_3|\omega)\bigg|_{x_3=\zeta(x_1)}=\frac{1}{\mu_\nu}\frac{\partial}{\partial n}F_\nu^{(3)}(x_1,x_3|\omega)\bigg|_{x_3=\zeta(x_1)} \tag{2.8.5b}$$

at the interface $x_3 = \zeta(x_1)$. In these equations $\kappa_p = \epsilon_1$ and $\kappa_s = 1$, $\lambda_p = \epsilon_2$ and $\lambda_s = 1$, and $\mu_p = \epsilon_3$ and $\mu_s = 1$. The normal derivative $\partial/\partial n$ is defined by Eq. (2.1.4). In addition, $F_\nu(x_1, x_3|\omega)$ consists of an incoming incident plane wave and outgoing scattered waves as $x_3 \to \infty$, and of outgoing transmitted waves as $x_3 \to -\infty$.

We next introduce three Green's functions $G_j(x_1, x_3|x_1', x_3')$ $(j = 1, 2, 3)$ that satisfy the equations

$$\left(\frac{\partial^2}{\partial x_1^2}+\frac{\partial^2}{\partial x_3^2}+\epsilon_j\frac{\omega^2}{c^2}\right)G_j(x_1,x_3|x_1',x_3')=-4\pi\delta(x_1-x_1')\delta(x_3-x_3') \tag{2.8.6}$$

in all space, subject to outgoing wave or vanishing boundary conditions at infinity. These functions have the representations

$$G_j(x_1,x_3|x_1',x_3')=i\pi H_0^{(1)}\left(n_j(\omega/c)\left[(x_1-x_1')^2+(x_3-x_3')^2\right]^{\frac{1}{2}}\right) \tag{2.8.7a}$$

$$=\int_{-\infty}^{\infty}\frac{dq}{2\pi}\frac{2\pi i}{\alpha_j(q)}\exp\left[iq(x_1-x_1')+i\alpha_j(q)|x_3-x_3'|\right], \tag{2.8.7b}$$

where $H_0^{(1)}(z)$ is a Hankel function of the first kind and zero order, n_j is the index of refraction of medium j,

$$n_j = (\epsilon_j)^{\frac{1}{2}}, \quad \mathrm{Re}\, n_j > 0,\ \mathrm{Im}\, n_j > 0, \tag{2.8.8}$$

and

$$\alpha_j(q) = \left[\epsilon_j(\omega/c)^2 - q^2\right]^{\frac{1}{2}}, \quad \mathrm{Re}\,\alpha_j(q) > 0,\ \mathrm{Im}\,\alpha_j(q) > 0. \tag{2.8.9}$$

2.8.1.1. The Scattered and Transmitted Fields We now use the preceding results in Green's second integral identity in the plane, Eq. (2.1.8), which we apply to the regions $x_3 > d$, $\zeta(x_1) < x_3 < d$, and $x_3 < \zeta(x_1)$ in turn. In the first region we set $u(x_1, x_3) = F_\nu^{(1)}(x_1, x_3|\omega)$ and $v(x_1, x_3) = G_1(x_1, x_3|x_1', x_3')$. In the second region we set $u(x_1, x_3) = F_\nu^{(2)}(x_1, x_3|\omega)$ and $v(x_1, x_3) = G_2(x_1, x_3|x_1', x_3')$. Finally, in the third region we set $u(x_1, x_3) = F_\nu^{(3)}(x_1, x_3|\omega)$ and $v(x_1, x_3) = G_3(x_1, x_3|x_1', x_3')$. With the use of Eq. (2.1.8) together with Eqs. (2.8.1)–(2.8.3) and (2.8.6), and the boundary conditions at infinity satisfied by the fields and the Green's functions, we obtain the following three equations, respectively,

$$\begin{aligned}
&\theta(x_3 - d)F_\nu^{(1)}(x_1, x_3|\omega) \\
&\quad = F_\nu(x_1, x_3|\omega)_{\mathrm{inc}} + \frac{1}{4\pi}\int_{-\infty}^{\infty} dx_1' \left\{\left[\frac{\partial}{\partial x_3'} G_1(x_1, x_3|x_1', x_3')\right] F_\nu^{(1)}(x_1', x_3'|\omega)\right. \\
&\quad \left. - G_1(x_1, x_3|x_1', x_3')\frac{\partial}{\partial x_3'} F_\nu^{(1)}(x_1', x_3'|\omega)\right\}_{x_3'=d},
\end{aligned} \tag{2.8.10}$$

$$\begin{aligned}
&\theta\big(x_3 - \zeta(x_1)\big)\theta(d - x_3)F_\nu^{(2)}(x_1, x_3|\omega) \\
&\quad = -\frac{1}{4\pi}\int_{-\infty}^{\infty} dx_1' \left\{\left[\frac{\partial}{\partial x_3'} G_2(x_1, x_3|x_1', x_3')\right] F_\nu^{(2)}(x_1', x_3'|\omega)\right. \\
&\quad \left. - G_2(x_1, x_3|x_1', x_3')\frac{\partial}{\partial x_3'} F_\nu^{(2)}(x_1', x_3'|\omega)\right\}_{x_3'=d} \\
&\quad + \frac{1}{4\pi}\int_S ds' \left\{\left[\frac{\partial}{\partial n'} G_2(x_1, x_3|x_1', x_3')\right] F_\nu^{(2)}(x_1', x_3'|\omega)\right. \\
&\quad \left. - G_2(x_1, x_3|x_1', x_3')\frac{\partial}{\partial n'} F_\nu^{(2)}(x_1', x_3'|\omega)\right\}_{x_3'=\zeta(x_1')},
\end{aligned} \tag{2.8.11}$$

$$\begin{aligned}
&\theta\big(\zeta(x_1) - x_3\big)F_\nu^{(3)}(x_1, x_3|\omega) \\
&\quad = -\frac{1}{4\pi}\int_S ds' \left\{\left[\frac{\partial}{\partial n'} G_3(x_1, x_3|x_1', x_3')\right] F_\nu^{(3)}(x_1', x_3'|\omega)\right. \\
&\quad \left. - G_3(x_1, x_3|x_1', x_3')\frac{\partial}{\partial n'} F_\nu^{(3)}(x_1', x_3'|\omega)\right\}_{x_3'=\zeta(x_1')}.
\end{aligned} \tag{2.8.12}$$

In these equations $\theta(z)$ is the Heaviside unit step function, and ds is the element of arc length along the curve $x_3 = \zeta(x_1)$, which we denote by s. The incident field $F_\nu(x_1, x_3|\omega)_{\rm inc}$ is a solution of Eq. (2.8.1), and we assume that it has the plane wave form

$$F_\nu(x_1, x_3|\omega)_{\rm inc} = \exp\bigl[ikx_1 - i\alpha_1(k)(x_3 - d)\bigr]. \tag{2.8.13}$$

Since we have assumed that the surface profile function $\zeta(x_1)$ is a single-valued function of x_1, we can express the element of arc length ds in Eqs. (2.8.11) and (2.8.12) as in Eq. (2.1.15). If we also use the boundary conditions (2.8.4b) and (2.8.5b) at the interfaces $x_3 = d$ and $x_3 = \zeta(x_1)$, then Eqs. (2.8.10), (2.8.11), and (2.8.12) can be rewritten as

$$\begin{aligned}
&\theta(x_3 - d)F_\nu^{(1)}(x_1, x_3|\omega) \\
&\quad = F_\nu(x_1, x_3|\omega)_{\rm inc} + \frac{1}{4\pi}\int_{-\infty}^{\infty} dx_1'\biggl[\left(\frac{\partial}{\partial x_3'} G_1(x_1, x_3|x_1', x_3')\right)_{x_3'=d} H_\nu^{(1)}(x_1'|\omega) \\
&\qquad - \bigl(G_1(x_1, x_3|x_1', x_3')\bigr)_{x_3'=d} L_\nu^{(1)}(x_1'|\omega)\biggr],
\end{aligned} \tag{2.8.14}$$

$$\begin{aligned}
&\theta\bigl(x_3 - \zeta(x_1)\bigr)\theta(d - x_3)F_\nu^{(2)}(x_1, x_3|\omega) \\
&\quad = -\frac{1}{4\pi}\int_{-\infty}^{\infty} dx_1'\biggl[\left(\frac{\partial}{\partial x_3'} G_2(x_1, x_3|x_1', x_3')\right)_{x_3'=d} H_\nu^{(1)}(x_1'|\omega) \\
&\qquad - \frac{\lambda_\nu}{\kappa_\nu}\bigl(G_2(x_1, x_3|x_1', x_3')\bigr)_{x_3'=d} L_\nu^{(1)}(x_1'|\omega)\biggr] \\
&\qquad + \frac{1}{4\pi}\int_{-\infty}^{\infty} dx_1'\biggl[\left(\frac{\partial}{\partial N'} G_2(x_1, x_3|x_1', x_3')\right)_{x_3'=\zeta(x_1')} H_\nu^{(2)}(x_1'|\omega) \\
&\qquad - \bigl(G_2(x_1, x_3|x_1', x_3')\bigr)_{x_3'=\zeta(x_1')} L_\nu^{(2)}(x_1'|\omega)\biggr],
\end{aligned} \tag{2.8.15}$$

$$\begin{aligned}
&\theta\bigl(\zeta(x_1) - x_3\bigr)F_\nu^{(3)}(x_1, x_3|\omega) \\
&\quad = -\frac{1}{4\pi}\int_{-\infty}^{\infty} dx_1'\biggl[\left(\frac{\partial}{\partial N'} G_3(x_1, x_3|x_1', x_3')\right)_{x_3'=\zeta(x_1')} H_\nu^{(2)}(x_1'|\omega) \\
&\qquad - \frac{\mu_\nu}{\lambda_\nu}\bigl(G_3(x_1, x_3|x_1', x_3')\bigr)_{x_3'=\zeta(x_1')} L_\nu^{(2)}(x_1'|\omega)\biggr],
\end{aligned} \tag{2.8.16}$$

where we have introduced the source functions

$$H_\nu^{(1)}(x_1|\omega) = F_\nu^{(1)}(x_1, x_3|\omega)\big|_{x_3=d}, \tag{2.8.17a}$$

$$L_\nu^{(1)}(x_1|\omega) = \frac{\partial}{\partial x_3} F_\nu^{(1)}(x_1, x_3|\omega)\bigg|_{x_3=d}, \tag{2.8.17b}$$

$$H_\nu^{(2)}(x_1|\omega) = F_\nu^{(2)}(x_1, x_3|\omega)\big|_{x_3=\zeta(x_1)}, \tag{2.8.17c}$$

$$L_\nu^{(2)}(x_1|\omega) = \frac{\partial}{\partial N} F_\nu^{(2)}(x_1, x_3|\omega)\bigg|_{x_3=\zeta(x_1)}. \tag{2.8.17d}$$

The scattered field in the region $x_3 > d$ is given by the second term on the right-hand side of Eq. (2.8.14),

$$F_\nu(x_1, x_3|\omega)_{sc} = \int_{-\infty}^{\infty} \frac{dq}{2\pi} R_\nu(q|k) \exp\big[iqx_1 + i\alpha_1(q)(x_3 - d)\big], \tag{2.8.18a}$$

where

$$R_\nu(q|k) = \frac{1}{2\alpha_1(q)} \int_{-\infty}^{\infty} dx_1 \left[\alpha_1(q) H_\nu^{(1)}(x_1|\omega) - iL_\nu^{(1)}(x_1|\omega)\right] \exp(-iqx_1). \tag{2.8.18b}$$

The dependence of the scattering amplitude $R_\nu(q|k)$ on the wave number k arises from the dependence of the source functions $H_\nu^{(1)}(x_1|\omega)$ and $L_\nu^{(1)}(x_1|\omega)$ on the incident field, Eq. (2.8.13). In obtaining Eq. (2.8.18b) we have used the representation of $G_1(x_1, x_3|x_1', x_3')$ given by Eq. (2.8.7b) in the case that $x_3 > d$.

The transmitted field in the region $x_3 < \zeta(x_1)$ is given by the right-hand side of Eq. (2.8.16), and can be represented in the form

$$F_\nu(x_1, x_3|\omega)_{tr} = \int_{-\infty}^{\infty} \frac{dq}{2\pi} T_\nu(q|k) \exp\big[iqx_1 - i\alpha_3(q)x_3\big], \tag{2.8.19a}$$

where

$$T_\nu(q|k) = \frac{1}{2\alpha_3(q)} \int_{-\infty}^{\infty} dx_1 \exp\big[-iqx_1 + i\alpha_3(q)\zeta(x_1)\big] \times \left\{\big[q\zeta'(x_1) + \alpha_3(q)\big] H_\nu^{(2)}(x_1|\omega) + i\frac{\mu_\nu}{\lambda_\nu} L_\nu^{(2)}(x_1|\omega)\right\}. \tag{2.8.19b}$$

We have used the representation of $G_3(x_1, x_3|x_1', x_3')$ given by Eq. (2.8.7b) in the case that $x_3 \to -\infty$ in writing this result.

2.8.1.2. The Mean Differential Reflection and Transmission Coefficients The 3-component of the complex Poynting vector in the region $x_3 > d$, when the field in it is p-polarized, is

$$\begin{aligned}\left(S_3^c\right)_p &= \frac{c}{8\pi} E_1 H_2^* \\ &= -i\frac{c^2}{8\pi\omega\epsilon_1}\frac{\partial H_2}{\partial x_3} H_2^*. \end{aligned} \tag{2.8.20}$$

The 3-component of the complex Poynting vector in the region $x_3 > d$, when the field in it is s-polarized, is

$$\begin{aligned}\left(S_3^c\right)_s &= -\frac{c}{8\pi} E_2 H_1^* \\ &= i\frac{c^2}{8\pi\omega}\frac{\partial E_2^*}{\partial x_3} E_2. \end{aligned} \tag{2.8.21}$$

Since the time-averaged Poynting vector is given by the real part of the complex Poynting vector, we can combine the results given by Eqs. (2.8.20) and (2.8.21) into

$$\operatorname{Re}\left(S_3^c\right)_\nu = \operatorname{Re}\left(-i\frac{c^2}{8\pi\omega\kappa_\nu}\frac{\partial F_\nu}{\partial x_3} F_\nu^*\right). \tag{2.8.22}$$

We assume for the incident field the plane wave (2.8.13). The magnitude of the total time-averaged flux incident on the region $|x_1| < L_1/2$, $|x_2| < L_2/2$ of the plane $x_3 = d$ is then given by

$$(P_{\text{inc}})_\nu = -\int_{-\frac{L_1}{2}}^{\frac{L_1}{2}} dx_1 \int_{-\frac{L_2}{2}}^{\frac{L_2}{2}} dx_2 \operatorname{Re}\left\{-i\frac{c^2}{8\pi\omega\kappa_\nu}\big(-i\alpha_1(k)\big)\right\} \tag{2.8.23a}$$

$$= L_1 L_2 \frac{c^2\alpha_1(k)}{8\pi\omega\kappa_\nu}. \tag{2.8.23b}$$

The minus sign in front of the integral on the right-hand side of Eq. (2.8.23a) compensates for the fact that the 3-component of the Poynting vector of the incident field is negative. We introduce the angle of incidence θ_0, measured counterclockwise from the x_3 axis, by

$$k = \sqrt{\epsilon_1}(\omega/c)\sin\theta_0. \tag{2.8.24}$$

The total time-averaged incident flux then becomes

$$(P_{\text{inc}})_\nu = L_1 L_2 \frac{c}{8\pi}\frac{\sqrt{\epsilon_1}}{\kappa_\nu}\cos\theta_0. \tag{2.8.25}$$

The scattered field in the region $x_3 > d$ has the form (Eq. (2.8.18a))

$$F_\nu(x_1, x_3|\omega)_{\text{sc}} = \int_{-\infty}^{\infty} \frac{dq}{2\pi} R_\nu(q|k)\exp\bigl[iqx_1 + i\alpha_1(q)(x_3 - d)\bigr]. \tag{2.8.26}$$

The total time-averaged flux scattered from the region $|x_1| < L_1/2$, $|x_2| < L_2/2$ of the plane $x_3 = d$ is then

$$\begin{aligned}(P_{\text{sc}})_\nu &= \int_{-\frac{L_1}{2}}^{\frac{L_1}{2}} dx_1 \int_{-\frac{L_2}{2}}^{\frac{L_2}{2}} dx_2 \,\text{Re}\Biggl\{-i\frac{c^2}{8\pi\omega\kappa_\nu}\int_{-\infty}^{\infty}\frac{dq}{2\pi}\int_{-\infty}^{\infty}\frac{dq'}{2\pi} \\ &\quad \times i\alpha_1(q)R_\nu(q|k)R_\nu^*(q'|k)e^{i(q-q')x_1}e^{i[\alpha_1(q)-\alpha_1^*(q')](x_3-d)}\Biggr\} \\ &= L_2\frac{c^2}{8\pi\omega\kappa_\nu}\,\text{Re}\int_{-\infty}^{\infty}\frac{dq}{2\pi}\alpha_1(q)\bigl|R_\nu(q|k)\bigr|^2 e^{i[\alpha_1(q)-\alpha_1^*(q)](x_3-d)} \\ &= L_2\frac{c^2}{8\pi\omega\kappa_\nu}\int_{-\sqrt{\epsilon_1}\frac{\omega}{c}}^{\sqrt{\epsilon_1}\frac{\omega}{c}}\frac{dq}{2\pi}\alpha_1(q)\bigl|R_\nu(q|k)\bigr|^2,\end{aligned} \tag{2.8.27}$$

where we have let L_1 tend to infinity.

We introduce the scattering angle θ_s, measured clockwise from the x_3 axis, by

$$q = \sqrt{\epsilon_1}(\omega/c)\sin\theta_s. \tag{2.8.28}$$

The total time-averaged scattered flux then becomes

$$(P_{\text{sc}})_\nu = L_2\frac{\omega\epsilon_1}{16\pi^2\kappa_\nu}\int_{-\frac{\pi}{2}}^{\frac{\pi}{2}} d\theta_s\,\cos^2\theta_s\bigl|R_\nu(q|k)\bigr|^2 \tag{2.8.29a}$$

$$\equiv \int_{-\frac{\pi}{2}}^{\frac{\pi}{2}} d\theta_s \left(P_{\rm sc}(\theta_s)\right)_\nu. \tag{2.8.29b}$$

The differential reflection coefficient by definition is given by

$$\frac{\partial R_\nu}{\partial \theta_s} = \frac{(P_{\rm sc}(\theta_s))_\nu}{(P_{\rm inc}(\theta_0))_\nu} = \frac{\sqrt{\epsilon_1}}{L_1}\frac{\omega}{2\pi c}\frac{\cos^2\theta_s}{\cos\theta_0}\left|R_\nu(q|k)\right|^2, \tag{2.8.30}$$

where q and k are given by Eqs. (2.8.28) and (2.8.24), respectively. The mean differential reflection coefficient is therefore given by

$$\left\langle\frac{\partial R_\nu}{\partial \theta_s}\right\rangle = \frac{\sqrt{\epsilon_1}}{L_1}\frac{\omega}{2\pi c}\frac{\cos^2\theta_s}{\cos\theta_0}\left\langle\left|R_\nu(q|k)\right|^2\right\rangle. \tag{2.8.31}$$

We now turn to a determination of the mean differential transmission coefficient. The 3-component of the complex Poynting vector in the region $x_3 < \zeta(x_1)$, when the field in it is p-polarized is

$$\begin{aligned}\left(S_3^c\right)_p &= \frac{c}{8\pi}E_1 H_2^* \\ &= -i\frac{c^2}{8\pi\omega\epsilon_3}\frac{\partial H_2}{\partial x_3}H_2^*.\end{aligned} \tag{2.8.32}$$

The 3-component of the complex Poynting vector in the region $x_3 < \zeta(x_1)$ when the field in it is s-polarized is

$$\begin{aligned}\left(S_3^c\right)_s &= -\frac{c}{8\pi}E_2 H_1^* \\ &= i\frac{c^2}{8\pi\omega}\frac{\partial E_2^*}{\partial x_3}E_2.\end{aligned} \tag{2.8.33}$$

We can therefore write $\mathrm{Re}(S_3^c)_\nu$ in the form

$$\mathrm{Re}\left(S_3^c\right)_\nu = \mathrm{Re}\left(-i\frac{c^2}{8\pi\omega\mu_\nu}\frac{\partial F_\nu}{\partial x_3}F_\nu^*\right). \tag{2.8.34}$$

The transmitted field in the region $x_3 < \zeta(x_1)$ has the form (Eq. (2.8.19a))

$$F_\nu^{(3)}(x_1, x_2|\omega)_{\rm tr} = \int_{-\infty}^{\infty}\frac{dq}{2\pi}T_\nu(q|k)\exp\left[iqx_1 - i\alpha_3(q)x_3\right]. \tag{2.8.35}$$

The total time-averaged flux transmitted through the region $|x_1| < L_1/2$, $|x_2| < L_2/2$ is then given by

$$(P_{\rm tr})_\nu = -\int_{-\frac{L_1}{2}}^{\frac{L_1}{2}} dx_1 \int_{-\frac{L_2}{2}}^{\frac{L_2}{2}} dx_2 \,\mathrm{Re}\left(-i\frac{c^2}{8\pi\omega\mu_\nu}\right)$$
$$\times \int_{-\infty}^{\infty}\frac{dq}{2\pi}\int_{-\infty}^{\infty}\frac{dq'}{2\pi}\big(-i\alpha_3(q)\big)T_\nu(q|k)T_\nu^*(q'|k)e^{-i(q-q')x_1}e^{-i[\alpha_3(q)-\alpha_3^*(q')]x_3}$$
$$\tag{2.8.36a}$$

$$= L_2\frac{c^2}{8\pi\omega\mu_\nu}\,\mathrm{Re}\int_{-\infty}^{\infty}\frac{dq}{2\pi}\alpha_3(q)\big|T_\nu(q|k)\big|^2 e^{-i[\alpha_3(q)-\alpha_3^*(q)]x_3}$$

$$= L_2\frac{c^2}{16\pi^2\omega\mu_\nu}\int_{-\sqrt{\epsilon_3}\frac{\omega}{c}}^{\sqrt{\epsilon_3}\frac{\omega}{c}} dq\,\alpha_3(q)\big|T_\nu(q|k)\big|^2. \tag{2.8.36b}$$

We now introduce the angle of transmission θ_t, measured counterclockwise from the $-x_3$ axis,

$$q = \sqrt{\epsilon_3}(\omega/c)\sin\theta_t. \tag{2.8.37}$$

The total time-averaged transmitted flux then takes the form

$$(P_{\rm tr})_\nu = L_2\frac{\omega\epsilon_3}{16\pi^2\mu_\nu}\int_{-\frac{\pi}{2}}^{\frac{\pi}{2}} d\theta_t\,\cos^2\theta_t\big|T_\nu(q|k)\big|^2 \tag{2.8.38a}$$

$$\equiv \int_{-\frac{\pi}{2}}^{\frac{\pi}{2}} d\theta_t\,\big(P_{\rm tr}(\theta_t)\big)_\nu. \tag{2.8.38b}$$

The differential transmission coefficient $(\partial T_\nu/\partial\theta_t)$ is defined such that $(\partial T_\nu/\partial\theta_t)\,d\theta_t$ is the fraction of the total time-averaged flux transmitted into the angular interval $(\theta_t, \theta_t + d\theta_t)$. It is therefore given by

$$\frac{\partial T_\nu}{\partial\theta_t} = \frac{(P_{\rm tr}(\theta_t))_\nu}{(P_{\rm inc}(\theta_0))_\nu} = \frac{1}{L_1}\frac{\epsilon_3}{\sqrt{\epsilon_1}}\frac{\kappa_\nu}{\mu_\nu}\frac{\omega}{2\pi c}\frac{\cos^2\theta_t}{\cos\theta_0}\big|T_\nu(q|k)\big|^2, \tag{2.8.39}$$

where q and k are now given by Eqs. (2.8.37) and (2.8.24), respectively. The mean differential transmission coefficient is

$$\left\langle \frac{\partial T_\nu}{\partial \theta_t} \right\rangle = \frac{1}{L_1} \frac{\epsilon_3}{\sqrt{\epsilon_1}} \frac{\kappa_\nu}{\mu_\nu} \frac{\omega}{2\pi c} \frac{\cos^2\theta_t}{\cos\theta_0} \langle |T_\nu(q|k)|^2 \rangle. \tag{2.8.40}$$

2.8.1.3. The Equations Satisfied by the Source Functions To obtain the equations satisfied by the source functions $H_\nu^{(1)}(x_1|\omega)$, $L_\nu^{(1)}(x_1|\omega)$, $H_\nu^{(2)}(x_1|\omega)$, and $L_\nu^{(2)}(x_1|\omega)$, we set $x_3 = d + \eta$ in Eqs. (2.8.14) and (2.8.15), and set $x_3 = \zeta(x_1) + \eta$ in Eqs. (2.8.15) and (2.8.16), where η is a positive infinitesimal. The result is the following set of four coupled inhomogeneous integral equations.

$$H_\nu^{(1)}(x_1|\omega) = H_\nu(x_1|\omega)_{\mathrm{inc}} + \int_{-\infty}^{\infty} dx_1' \big[H^{(1)}(x_1|x_1') H_\nu^{(1)}(x_1'|\omega) - L^{(1)}(x_1|x_1') L_\nu^{(1)}(x_1'|\omega) \big], \tag{2.8.41}$$

$$0 = \int_{-\infty}^{\infty} dx_1' \Big[H^{(11)}(x_1|x_1') H_\nu^{(1)}(x_1'|\omega) - \frac{\lambda_\nu}{\kappa_\nu} L^{(11)}(x_1|x_1') L_\nu^{(1)}(x_1'|\omega) - H^{(12)}(x_1|x_1') H_\nu^{(2)}(x_1'|\omega) + L^{(12)}(x_1|x_1') L_\nu^{(2)}(x_1'|\omega) \Big], \tag{2.8.42}$$

$$H_\nu^{(2)}(x_1|\omega) = \int_{-\infty}^{\infty} dx_1' \Big[-H^{(21)}(x_1|x_1') H_\nu^{(1)}(x_1'|\omega) + \frac{\lambda_\nu}{\kappa_\nu} L^{(21)}(x_1|x_1') L_\nu^{(1)}(x_1'|\omega) + H^{(22)}(x_1|x_1') H_\nu^{(2)}(x_1'|\omega) - L^{(22)}(x_1|x_1') L_\nu^{(2)}(x_1'|\omega) \Big],$$

$$0 = \int_{-\infty}^{\infty} dx_1' \Big[H^{(3)}(x_1|x_1') H_\nu^{(2)}(x_1'|\omega) - \frac{\mu_\nu}{\lambda_\nu} L^{(2)}(x_1|x_1') L_\nu^{(2)}(x_1'|\omega) \Big], \tag{2.8.43}$$

where

$$H_\nu(x_1|\omega)_{\mathrm{inc}} = F_\nu(x_1, x_3|\omega)_{\mathrm{inc}}\big|_{x_3=d}. \tag{2.8.44}$$

The kernels entering these equations are tabulated in Appendix A.

We will return to Eqs. (2.8.41)–(2.8.45) in Section 2.8.2.1.

2.8.1.4. The Kirchhoff Approximation and its Geometrical Optics Limit The rigorous approach to the scattering of light from, and its transmission through, a film with a rough boundary presented in the preceding three sections cannot be used as the basis for the determination of a surface profile function $\zeta(x_1)$ that produces a scattered or transmitted field with a prescribed dependence of the mean differential reflection or transmission coefficient on the angles of scattering or transmission. An exact solution of Eqs. (2.8.41)–(2.8.44) for the source functions $H_\nu^{(1)}(x_1|\omega)$, $L_\nu^{(1)}(x_1|\omega)$, $H_\nu^{(2)}(x_1|\omega)$, and $L_\nu^{(2)}(x_1|\omega)$ in an analytic form that is simple enough for the scattering and transmission amplitudes to be calculated from Eqs. (2.8.18b) and (2.8.19b) does not exist. Only numerical solutions can be obtained. We are therefore forced to use approximate single-scattering versions of the rigorous theory developed in the preceding three sections as the basis for the solution of the inverse problem. The approximate theory that we use is the Kirchhoff approximation, otherwise known as the tangent plane approximation. The rigorous theory derived in the preceding three sections will be used later in this chapter to validate the results obtained by the use of the Kirchhoff approximation. We now turn to a derivation of this approximation for transmission through the film.

Our starting point is the expression for the transmission amplitude $T_\nu(q|k)$ given by Eqs. (2.8.17d) and (2.8.19b):

$$T_\nu(q|k) = \frac{1}{2\alpha_3(q)} \int_{-\infty}^{\infty} dx_1 \exp\bigl[-iqx_1 + i\alpha_3(q)\zeta(x_1)\bigr] \times \Biggl\{\bigl[q\zeta'(x_1) + \alpha_3(q)\bigr]F_\nu^{(2)}(x_1, x_3|\omega)\big|_{x_3=\zeta(x_1)} + i\frac{\mu_\nu}{\lambda_\nu}\frac{\partial}{\partial N}F_\nu^{(2)}(x_1, x_3|\omega)\bigg|_{x_3=\zeta(x_1)}\Biggr\}. \tag{2.8.45}$$

To obtain the source functions $F_\nu^{(2)}(x_1, x_3|\omega)|_{x_3=\zeta(x_1)}$ and $(\partial F_\nu^{(2)}(x_1, x_3|\omega)/\partial N)_{x_3=\zeta(x_1)}$ we proceed as follows. When the plane wave (2.8.13) is incident on the planar surface $x_3 = d$, the field $F^{(1)}(x_1, x_3|\omega)$ in the region $x_3 > d$ can be written as

$$F_\nu^{(1)}(x_1, x_3|\omega) = \exp\bigl[ikx_1 - i\alpha_1(k)(x_3 - d)\bigr] + R_\nu \exp\bigl[ikx_1 + i\alpha_1(k)(x_3 - d)\bigr]. \tag{2.8.46}$$

The field transmitted into the region $x_3 < d$, assumed for the moment to extend to $x_3 = -\infty$, can be written as

$$F_\nu^{(2)}(x_1, x_3|\omega) = A_\nu \exp\bigl[ikx_1 - i\alpha_2(k)x_3\bigr]. \tag{2.8.47}$$

The satisfaction of the boundary conditions at the interface $x_3 = d$, Eqs. (2.8.4b), yields the results that

$$R_\nu = \frac{\lambda_\nu \alpha_1(k) - \kappa_\nu \alpha_2(k)}{\lambda_\nu \alpha_1(k) + \kappa_\nu \alpha_2(k)}, \tag{2.8.48a}$$

$$A_\nu = \frac{2\lambda_\nu \alpha_1(k)}{\lambda_\nu \alpha_1(k) + \kappa_\nu \alpha_2(k)} \exp[i\alpha_2(k)d]. \tag{2.8.48b}$$

We next assume that the plane wave $F_\nu^{(2)}(x_1, x_3|\omega) = A_\nu \exp[ikx_1 - i\alpha_2(k)x_3]$ is incident on the surface $x_3 = \zeta(x_1)$, and that it is reflected from the plane tangent to this surface at each point. It follows from the discussion in Section 2.1.1.4 that with this assumption we have

$$F_\nu^{(2)}(x_1, x_3|\omega)\big|_{x_3=\zeta(x_1)} = \big[1 + R_\nu\big(\alpha_\ell(k|x_1)\big)\big] A_\nu \exp\big[ikx_1 - i\alpha_2(k)\zeta(x_1)\big], \tag{2.8.49a}$$

$$\frac{\partial}{\partial N} F_\nu^{(2)}(x_1, x_3|\omega)\bigg|_{x_3=\zeta(x_1)}$$
$$= -i\big[1 + \big(\zeta'(x_1)\big)^2\big]^{\frac{1}{2}} \alpha_\ell(k|x_1)\big[1 - R_\nu\big(\alpha_\ell(k|x_1)\big)\big] A_\nu \exp\big[ikx_1 - i\alpha_2(k)\zeta(x_1)\big]. \tag{2.8.49b}$$

In these expressions $\alpha_\ell(k|x_1)$ is the modulus of the projection of the wave vector of the incident wave $(k, 0, -\alpha_2(k))$ on a local normal to the surface, $\hat{\mathbf{n}} = (-\zeta'(x_1), 0, 1)/[1 + (\zeta'(x_1))^2]^{\frac{1}{2}}$, so that

$$\alpha_\ell(k|x_1) = \frac{\alpha_2(k) + k\zeta'(x_1)}{[1 + (\zeta'(x_1))^2]^{\frac{1}{2}}}. \tag{2.8.50}$$

The function $R_\nu(\alpha_\ell(k|x_1))$ is the Fresnel reflection coefficient $R_\nu(\alpha_2(k))$ for light of polarization ν and wave vector $(k, 0, -\alpha_2(k))$ incident from the medium with a dielectric constant ϵ_2 onto its planar interface $x_3 = 0$ with the medium whose dielectric constant is ϵ_3,

$$R_\nu\big(\alpha_2(k)\big) = \frac{\mu_\nu \alpha_2(k) - \lambda_\nu \alpha_3(k)}{\mu_\nu \alpha_2(k) + \lambda_\nu \alpha_3(k)}, \tag{2.8.51}$$

but with $\alpha_2(k)$ replaced everywhere by $\alpha_\ell(k|x_1)$. Thus we have that

$$R_\nu\big(\alpha_\ell(k|x_1)\big) = \frac{\mu_\nu \alpha_\ell(k|x_1) - \lambda_\nu[(\epsilon_3 - \epsilon_2)(\omega/c)^2 + \alpha_\ell^2(k|x_1)]^{\frac{1}{2}}}{\mu_\nu \alpha_\ell(k|x_1) + \lambda_\nu[(\epsilon_3 - \epsilon_2)(\omega/c)^2 + \alpha_\ell^2(k|x_1)]^{\frac{1}{2}}}. \tag{2.8.52}$$

When the results given by Eqs. (2.8.49a) and (2.8.49b) are substituted into Eq. (2.8.45), the transmission amplitude $T_\nu(q|k)$ takes the form

$$T_\nu(q|k) = \frac{A_\nu}{2\alpha_3(q)} \int_{-\infty}^{\infty} dx_1 \exp[-i(q-k)x_1] \exp[i(\alpha_3(q) - \alpha_2(k))\zeta(x_1)]$$
$$\times \left\{ \left(q + \frac{\mu_\nu}{\lambda_\nu}k\right)\zeta'(x_1) + \left(\alpha_3(q) + \frac{\mu_\nu}{\lambda_\nu}\alpha_2(k)\right) + \left[\left(q - \frac{\mu_\nu}{\lambda_\nu}k\right)\zeta'(x_1) + \left(\alpha_3(q) - \frac{\mu_\nu}{\lambda_\nu}\alpha_2(k)\right)\right] R_\nu\left(\alpha_\ell(k|x_1)\right) \right\}. \tag{2.8.53}$$

We again evaluate the integral in this equation by the method of stationary phase. The ith stationary point is obtained from

$$\zeta'(x_{1i}) = \frac{q-k}{\alpha_3(q) - \alpha_2(k)}. \tag{2.8.54}$$

At each stationary point $\zeta'(x_1)$ has the value given by the right-hand side of this equation, and at each stationary point

$$\alpha_\ell(k|x_{1i}) = \frac{|\alpha_3(q) - \alpha_2(k)|}{\alpha_3(q) - \alpha_2(k)} \frac{\alpha_3(q)\alpha_2(k) + qk - \epsilon_2(\omega/c)^2}{\sqrt{2}[\frac{1}{2}(\epsilon_2 + \epsilon_3)(\omega/c)^2 - \alpha_3(q)\alpha_2(k) - qk]^{\frac{1}{2}}}. \tag{2.8.55}$$

We approximate the factor multiplying the exponential in the integrand on the right-hand side of Eq. (2.8.53) by its value calculated at the stationary points with the result that

$$T_\nu(q|k) = T_\nu^{(0)}(q|k) \int_{-\infty}^{\infty} dx_1 \exp[-i(q-k)x_1] \exp\left\{ i[\alpha_3(q) - \alpha_2(k)]\zeta(x_1) \right\}, \tag{2.8.56}$$

where

$$T_\nu^{(0)}(q|k) = \frac{A_\nu(k)}{2\alpha_3(q)[\alpha_3(q) - \alpha_2(k)]}$$
$$\times \left\{ \left[\left(\epsilon_3 - \frac{\mu_\nu}{\lambda_\nu}\epsilon_2\right)\frac{\omega^2}{c^2} - \left(1 - \frac{\mu_\nu}{\lambda_\nu}\right)\left(qk + \alpha_3(q)\alpha_2(k)\right)\right] + \left[\left(\epsilon_3 + \frac{\mu_\nu}{\lambda_\nu}\epsilon_2\right)\frac{\omega^2}{c^2} - \left(1 + \frac{\mu_\nu}{\lambda_\nu}\right)\left(qk + \alpha_3(q)\alpha_2(k)\right)\right] R_\nu\left(\alpha_\ell(k|x_{1i})\right) \right\}. \tag{2.8.57}$$

It should be kept in mind that $k = \sqrt{\epsilon_1}(\omega/c)\sin\theta_0$ and $q = \sqrt{\epsilon_3}(\omega/c)\sin\theta_t$.

From Eqs. (2.8.40) and (2.8.56) we find that the mean differential transmission coefficient is given by

$$\left\langle \frac{\partial T_\nu}{\partial \theta_t} \right\rangle = \frac{1}{L_1}\left(\frac{\omega}{2\pi c}\right)\frac{\epsilon_3}{\sqrt{\epsilon_1}}\frac{\kappa_\nu}{\mu_\nu}\frac{\cos^2\theta_t}{\cos\theta_0}\left|T_\nu^{(0)}(q|k)\right|^2 \times \int_{-\infty}^{\infty} dx_1 \int_{-\infty}^{\infty} dx_1'\, e^{-i(q-k)(x_1-x_1')}\left\langle e^{i[\alpha_3(q)-\alpha_2(k)][\zeta(x_1)-\zeta(x_1')]}\right\rangle. \quad (2.8.58)$$

In writing this result we have used the assumption that $\alpha_2(k) = (\omega/c)[\epsilon_2 - \epsilon_1 \sin^2\theta_0]^{\frac{1}{2}}$ is real. This requires that $|\theta_0| < \sin^{-1}(\epsilon_2/\epsilon_1)^{\frac{1}{2}}$. The latter condition of course, is automatically satisfied when $\epsilon_2 > \epsilon_1$, which we will assume to be the case. However, in the case where $\epsilon_2 < \epsilon_1$, the condition $|\theta_0| < \sin^{-1}(\epsilon_2/\epsilon_1)^{\frac{1}{2}}$ implies that the angle of incidence should be smaller than the critical angle for total internal reflection $\theta_c = \sin^{-1}(\epsilon_2/\epsilon_1)^{\frac{1}{2}}$. Since we are interested here in the radiation transmitted through the film, this condition does not introduce any restriction on our analysis. We recall that $\alpha_3(q) = \sqrt{\epsilon_3}(\omega/c)\sin\theta_t$ is real because we have assumed that ϵ_3 is real.

Our goal is to obtain a surface profile function $\zeta(x_1)$ that produces a specified form for $\langle \partial T_\nu/\partial\theta_t\rangle$. As it stands, Eq. (2.8.58) is too difficult to invert, and we pass to its geometrical optics limit by making the change of variable $x_1' = x_1 + u$, and the approximation that $\zeta(x_1) - \zeta(x_1+u) \cong -u\zeta'(x_1)$. In this way we obtain the result that

$$\left\langle \frac{\partial T_\nu}{\partial \theta_t} \right\rangle = \frac{1}{L_1}\left(\frac{\omega}{2\pi c}\right)\frac{\epsilon_3}{\sqrt{\epsilon_1}}\frac{\kappa_\nu}{\mu_\nu}\frac{\cos^2\theta_t}{\cos\theta_0}\left|T_\nu^{(0)}(q|k)\right|^2 \times \int_{-\infty}^{\infty} dx_1 \int_{-\infty}^{\infty} du\, e^{i(q-k)u}\left\langle e^{-i[\alpha_3(q)-\alpha_2(k)]u\zeta'(x_1)}\right\rangle. \quad (2.8.59)$$

It is this equation that we will invert to obtain the desired surface profile function.

2.8.1.5. The Inverse Problem To evaluate the double integral in Eq. (2.8.59) we again use the representation of the surface profile function $\zeta(x_1)$ given by Eq. (2.1.83), and obtain

$$\int_{-\infty}^{\infty} dx_1 \int_{-\infty}^{\infty} du\, e^{i(q-k)u}\left\langle e^{-i[\alpha_3(q)-\alpha_2(k)]u\zeta'(x_1)}\right\rangle$$
$$= \int_{-\infty}^{\infty} du\, e^{i(q-k)u} \sum_{n=-N_s}^{N_s-1} \int_{nb}^{(n+1)b} dx_1 \left\langle e^{-i[\alpha_3(q)-\alpha_2(k)]ua_n}\right\rangle$$

$$= \int_{-\infty}^{\infty} du\, e^{i(q-k)u} \sum_{n=-N_s}^{N_s-1} \int_{nb}^{(n+1)b} dx_1 \int_{-\infty}^{\infty} d\gamma\, f(\gamma) e^{-i[\alpha_3(q)-\alpha_2(k)]u\gamma}$$

$$= 2N_s b \int_{-\infty}^{\infty} d\gamma\, f(\gamma) 2\pi \delta\big(q - k - [\alpha_3(q) - \alpha_2(k)]\gamma\big)$$

$$= \frac{4\pi N_s b}{|\alpha_3(q) - \alpha_2(k)|} f\left(\frac{q-k}{\alpha_3(q) - \alpha_2(k)}\right), \qquad (2.8.60)$$

where $f(\gamma)$ is the pdf of a_n. On combining this result with Eq. (2.8.59) we obtain for the mean differential transmission coefficient

$$\left\langle \frac{\partial T_\nu}{\partial \theta_t} \right\rangle (q, k) = \frac{\epsilon_3}{\sqrt{\epsilon_1}} \frac{\kappa_\nu}{\mu_\nu} \frac{\cos^2 \theta_t}{\cos \theta_0} \frac{|T_\nu^{(0)}(q|k)|^2}{|\sqrt{\epsilon_3} \cos \theta_t - \sqrt{\epsilon_2 - \epsilon_1 \sin^2 \theta_0}|} \times f\left(\frac{\sqrt{\epsilon_3} \sin \theta_t - \sqrt{\epsilon_1} \sin \theta_0}{\sqrt{\epsilon_3} \cos \theta_t - \sqrt{\epsilon_2 - \epsilon_1 \sin^2 \theta_0}}\right). \qquad (2.8.61)$$

If we maintain complete generality beyond this point, and consider non-normal incidence, the inversion of Eq. (2.8.61) to obtain $f(\gamma)$ in terms of $\langle \partial T_\nu / \partial \theta_t \rangle$, while it can be carried out analytically, leads to such a cumbersome expression that it is simpler to effect the inversion numerically. In order to avoid this, in what follows we restrict our attention to the common case of normal incidence, $\theta_0 = 0$, $k = 0$, for which the inversion can be carried out analytically, and produces a rather simple result.

Thus, in this case Eq. (2.8.61) becomes

$$\left\langle \frac{\partial T_\nu}{\partial \theta_t} \right\rangle (q, 0) = \frac{\epsilon_3}{\sqrt{\epsilon_1}} \frac{\kappa_\nu}{\mu_\nu} \cos^2 \theta_t \frac{|T_\nu^{(0)}(q|0)|^2}{|\sqrt{\epsilon_3} \cos \theta_t - \sqrt{\epsilon_2}|} f\left(\frac{\sqrt{\epsilon_3} \sin \theta_t}{\sqrt{\epsilon_3} \cos \theta_t - \sqrt{\epsilon_2}}\right) \qquad (2.8.62)$$

To be specific in what follows, we will assume that $\epsilon_2 > \epsilon_3$. A simple expression for $T_\nu^{(0)}(q|0)$ can be obtained from Eq. (2.8.57) with the aid of the results

$$A_\nu(0) = \frac{2\lambda_\nu \sqrt{\epsilon_1}}{\lambda_\nu \sqrt{\epsilon_1} + \kappa_\nu \sqrt{\epsilon_2}} \exp\big[i\sqrt{\epsilon_2}(\omega/c)d\big] \qquad (2.8.63)$$

$$\alpha_\ell(0|x_{1i}) = \frac{\omega}{c} \frac{\epsilon_2 - \sqrt{\epsilon_2 \epsilon_3} \cos \theta_t}{[\epsilon_2 + \epsilon_3 - 2\sqrt{\epsilon_2 \epsilon_3} \cos \theta_t]^{\frac{1}{2}}} \qquad (2.8.64)$$

and

$$\big[(\epsilon_3 - \epsilon_2)(\omega/c)^2 + \alpha_\ell^2(0|x_{1i})\big]^{\frac{1}{2}} = \frac{\omega}{c} \frac{|\epsilon_3 - \sqrt{\epsilon_2 \epsilon_3} \cos \theta_t|}{[\epsilon_2 + \epsilon_3 - 2\sqrt{\epsilon_2 \epsilon_3} \cos \theta_t]^{\frac{1}{2}}}. \qquad (2.8.65)$$

The modulus present in the expression on the right-hand side of Eq. (2.8.65) is significant. We also note that $\epsilon_2 + \epsilon_3 - 2\sqrt{\epsilon_2\epsilon_3}\cos\theta_t$ is positive independently of whether $\epsilon_2 > \epsilon_3$ or $\epsilon_2 < \epsilon_3$. With the use of Eqs. (2.8.64) and (2.8.65) we obtain

$$R_\nu\big(\alpha_\ell(0|x_{1i})\big) = \frac{\mu_\nu(\epsilon_2 - \sqrt{\epsilon_2\epsilon_3}\cos\theta_t) - \lambda_\nu|\epsilon_3 - \sqrt{\epsilon_2\epsilon_3}\cos\theta_t|}{\mu_\nu(\epsilon_2 - \sqrt{\epsilon_2\epsilon_3}\cos\theta_t) + \lambda_\nu|\epsilon_3 - \sqrt{\epsilon_2\epsilon_3}\cos\theta_t|}. \tag{2.8.66}$$

We obtain from Eqs. (2.8.57) and (2.8.66) the result that

$$T_\nu^{(0)}(q|0) = 2\sqrt{\frac{\epsilon_2}{\epsilon_3}}\mu_\nu \frac{A_\nu(0)}{\cos\theta_t}$$
$$\times \frac{\cos\theta_t - \sqrt{\epsilon_3/\epsilon_2}}{\mu_\nu(\sqrt{\epsilon_2/\epsilon_3} - \cos\theta_t) + \lambda_\nu(\cos\theta_t - \sqrt{\epsilon_3/\epsilon_2})} \qquad \cos\theta_t > \sqrt{\epsilon_3/\epsilon_2} \tag{2.8.67a}$$

$$= 0 \qquad \cos\theta_t < \sqrt{\epsilon_3/\epsilon_2}. \tag{2.8.67b}$$

The restriction on $\cos\theta_t$ arises from the existence of a critical angle for total internal reflection due to our assumption that $\epsilon_2 > \epsilon_3$.

We now make the change of variable

$$\frac{\sqrt{\epsilon_3}\sin\theta_t}{\sqrt{\epsilon_2} - \sqrt{\epsilon_3}\cos\theta_t} = \gamma. \tag{2.8.68}$$

Keeping in mind that the left-hand side of this equation is an odd function of θ_t, we find that it increases from zero as θ_t increases from zero until it reaches a maximum at $\theta_t = \cos^{-1}\sqrt{\epsilon_3/\epsilon_2}$, where it equals $\sqrt{\epsilon_3}/\sqrt{\epsilon_2 - \epsilon_3}$, and then decreases with further increase of θ_t to a value of $\sqrt{\epsilon_3/\epsilon_2}$ at $\theta_t = \pi/2$. However, from Eq. (2.8.67b) we see that $T_\nu^{(0)}(q|0)$ is nonzero only for $0 \leqslant \theta_t \leqslant \cos^{-1}\sqrt{\epsilon_3/\epsilon_2}$. Consequently the range of variation of γ is restricted to

$$|\gamma| \leqslant \left(\frac{\epsilon_3}{\epsilon_2 - \epsilon_3}\right)^{\frac{1}{2}}. \tag{2.8.69}$$

From Eq. (2.8.68) we obtain the useful results

$$\cos\theta_t = \frac{1}{1+\gamma^2}\left[\sqrt{\frac{\epsilon_2}{\epsilon_3}}\gamma^2 + \sqrt{1 - \left(\frac{\epsilon_2 - \epsilon_3}{\epsilon_3}\right)\gamma^2}\right], \tag{2.8.70}$$

$$\sin\theta_t = \frac{\gamma}{1+\gamma^2}\left[\sqrt{\frac{\epsilon_2}{\epsilon_3}} - \sqrt{1 - \left(\frac{\epsilon_2 - \epsilon_3}{\epsilon_3}\right)\gamma^2}\right]. \tag{2.8.71}$$

Because of the condition (2.8.69), both $\cos\theta_t$ and $\sin\theta_t$ are real.

With the preceding results we obtain

$$\sqrt{\epsilon_2/\epsilon_3} - \cos\theta_t = \sin\theta_t/\gamma, \tag{2.8.72}$$

$$\cos\theta_t - \sqrt{\epsilon_3/\epsilon_2} = \sqrt{\epsilon_3/\epsilon_2}\sqrt{1 - \left(\frac{\epsilon_2 - \epsilon_3}{\epsilon_3}\right)\gamma^2}\frac{\sin\theta_t}{\gamma}, \tag{2.8.73}$$

which allows us to simplify Eq. (2.8.67b) further:

$$T_\nu^{(0)}(q|0) = 2\mu_\nu \frac{A_\nu(0)}{\cos\theta_t} \frac{\sqrt{1 - (\frac{\epsilon_2-\epsilon_3}{\epsilon_3})\gamma^2}}{\mu_\nu + \sqrt{\epsilon_3/\epsilon_2}\lambda_\nu\sqrt{1 - (\frac{\epsilon_2-\epsilon_3}{\epsilon_3})\gamma^2}} \quad |\gamma| < \left(\frac{\epsilon_3}{\epsilon_2 - \epsilon_3}\right)^{\frac{1}{2}} \tag{2.8.74a}$$

$$= 0 \quad |\gamma| > \left(\frac{\epsilon_3}{\epsilon_2 - \epsilon_3}\right)^{\frac{1}{2}}. \tag{2.8.74b}$$

The mean differential transmission coefficient (2.8.62) can then be put into the form

$$\left\langle \frac{\partial T_\nu}{\partial \theta_t} \right\rangle(\gamma) = 4\kappa_\nu\mu_\nu\sqrt{\frac{\epsilon_3}{\epsilon_1}}\left(\frac{2\lambda_\nu\sqrt{\epsilon_1}}{\lambda_\nu\sqrt{\epsilon_1} + \kappa_\nu\sqrt{\epsilon_2}}\right)^2 \times \frac{(1+\gamma^2)[1 - (\frac{\epsilon_2-\epsilon_3}{\epsilon_3})\gamma^2]f(-\gamma)}{[\sqrt{\epsilon_2/\epsilon_3} - \sqrt{1 - (\frac{\epsilon_2-\epsilon_3}{\epsilon_3})\gamma^2}\,][\mu_\nu + \lambda_\nu\sqrt{\epsilon_3/\epsilon_2}\sqrt{1 - (\frac{\epsilon_2-\epsilon_3}{\epsilon_3})\gamma^2}\,]^2}, \tag{2.8.75}$$

where $\langle \partial T_\nu/\partial\theta_t\rangle(\gamma)$ is the mean differential transmission coefficient in which its dependence on the scattering angle θ_t has been replaced by a dependence on γ through the use of Eqs. (2.8.70) and (2.8.71).

We invert Eq. (2.8.75) and obtain for the pdf of a_n

$$f(\gamma) = \frac{1}{4\kappa_\nu\mu_\nu}\sqrt{\frac{\epsilon_1}{\epsilon_3}}\left(\frac{\lambda_\nu\sqrt{\epsilon_1} + \kappa_\nu\sqrt{\epsilon_2}}{2\lambda_\nu\sqrt{\epsilon_1}}\right)^2 \times \frac{[\sqrt{\epsilon_2/\epsilon_3} - \sqrt{1 - (\frac{\epsilon_2-\epsilon_3}{\epsilon_3})\gamma^2}\,][\mu_\nu + \lambda_\nu\sqrt{\epsilon_3/\epsilon_2}\sqrt{1 - (\frac{\epsilon_2-\epsilon_3}{\epsilon_3})\gamma^2}\,]^2}{(1+\gamma^2)[1 - (\frac{\epsilon_2-\epsilon_3}{\epsilon_3})\gamma^2]}\left\langle \frac{\partial T_\nu}{\partial \theta_t} \right\rangle(-\gamma). \tag{2.8.76}$$

2.8.2. *Solution of the Transmission Problem*

Two approaches to the solution of the transmission problem will be used to validate the method developed in Sections 2.8.1.4 and 2.8.1.5 for designing one-dimensional randomly rough surfaces that transmit light with a specified angular dependence of its mean intensity. The first is a rigorous computer simulation approach based on the results obtained in Section 2.8.1.3; the second is the Kirchhoff approximation developed in Section 2.8.1.4, but without passing to its geometrical optics limit.

2.8.2.1. Computer Simulations The starting point for the rigorous solution of the transmission problem is the set of four coupled integral equations for the source functions $H_\nu^{(1)}(x_1|\omega)$, $L_\nu^{(1)}(x_1|\omega)$, $H_\nu^{(2)}(x_1|\omega)$, and $L_\nu^{(2)}(x_1|\omega)$ given by Eqs. (2.8.41)–(2.8.44). They are solved by being converted into $N \times N$ matrix equations by the approach described in Sections 2.1.3.2 and 2.2.2.1. The matrix equations that are obtained in this way are $(m = 1, 2, \ldots, N)$

$$H_\nu^{(1)}(x_m|\omega) = H_\nu(x_n|\omega)_{\text{inc}} + \sum_{n=1}^{N}\left[H_{mn}^{(1)} H_\nu^{(1)}(x_n|\omega) - L_{mn}^{(1)} L_\nu^{(1)}(x_n|\omega)\right], \quad (2.8.77)$$

$$0 = \sum_{n=1}^{N}\left[H_{mn}^{(11)} H_\nu^{(1)}(x_n|\omega) - \frac{\lambda_\nu}{\kappa_\nu} L_{mn}^{(11)} L_\nu^{(1)}(x_n|\omega) - H_{mn}^{(12)} H_\nu^{(2)}(x_n|\omega) + L_{mn}^{(12)} L_\nu^{(2)}(x_n|\omega)\right], \quad (2.8.78)$$

$$H_\nu^{(2)}(x_m|\omega) = \sum_{n=1}^{N}\left[-H_{mn}^{(21)} H_\nu^{(1)}(x_n|\omega) + \frac{\lambda_\nu}{\kappa_\nu} L_{mn}^{(21)} L_\nu^{(1)}(x_n|\omega) + H_{mn}^{(22)} H_\nu^{(2)}(x_n|\omega) - L_{mn}^{(22)} L_\nu^{(2)}(x_n|\omega)\right], \quad (2.8.79)$$

$$0 = \sum_{n=1}^{N}\left[H_{mn}^{(3)} H_\nu^{(2)}(x_n|\omega) - \frac{\mu_\nu}{\lambda_\nu} L_{mn}^{(3)} L_\nu^{(2)}(x_n|\omega)\right]. \quad (2.8.80)$$

Explicit expressions for the matrix elements entering these equations are presented in Appendix B.

The mean differential transmission coefficient given by Eqs. (2.8.19b) and (2.8.40) can be written as

$$\left\langle \frac{\partial T_\nu}{\partial \theta_t} \right\rangle = \frac{1}{L_1} \frac{c}{8\pi\omega} \frac{\kappa_\nu}{\mu_\nu} \frac{1}{\sqrt{\epsilon_1}} \frac{\langle |t_\nu(\theta_t|\theta_0)|^2 \rangle}{\cos\theta_0}, \quad (2.8.81)$$

where

$$t_\nu(\theta_t|\theta_0) = \int_{-\infty}^{\infty} dx_1 \exp\{-i\sqrt{\epsilon_3}(\omega/c)[x_1 \sin\theta_t - \zeta(x_1)\cos\theta_t]\} \times \left\{i\sqrt{\epsilon_3}(\omega/c)[\zeta'(x_1)\sin\theta_t + \cos\theta_t]H_\nu^{(2)}(x_1|\omega) - \frac{\mu_\nu}{\lambda_\nu}L_\nu^{(2)}(x_1|\omega)\right\}. \tag{2.8.82}$$

In terms of the solutions of Eqs. (2.8.77)–(2.8.80), we can write $t_\nu(\theta_t|\theta_0)$ alternatively as

$$t_\nu(\theta_t|\theta_s) \cong \Delta x \sum_{n=1}^{N} \exp\{-i\sqrt{\epsilon_3}(\omega c)[x_n \sin\theta_t - \zeta(x_n)\cos\theta_t]\} \times \left\{i\sqrt{\epsilon_3}(\omega/c)[\zeta'(x_n)\sin\theta_t + \cos\theta_t]H_\nu^{(2)}(x_n|\omega) - \frac{\mu_\nu}{\lambda_\nu}L_\nu^{(2)}(x_n|\omega)\right\}. \tag{2.8.83}$$

In a similar fashion, the mean differential reflection coefficient given by Eqs. (2.8.18b) and (2.8.31) can be rewritten as

$$\left\langle \frac{\partial R_\nu}{\partial \theta_s} \right\rangle = \frac{1}{L_1} \frac{c}{8\pi\omega} \frac{1}{\sqrt{\epsilon_1}} \frac{\langle |r_\nu(\theta_s|\theta_0)|^2\rangle}{\cos\theta_0}, \tag{2.8.84}$$

where

$$r_\nu(\theta_s|\theta_0) = \int_{-\infty}^{\infty} dx_1 \exp[-i\sqrt{\epsilon_1}(\omega/c)x_1 \sin\theta_s] \times \{[i\sqrt{\epsilon_1}(\omega/c)\cos\theta_s]H_\nu^{(1)}(x_1|\omega) + L_\nu^{(1)}(x_1|w)\}. \tag{2.8.85}$$

In terms of the solutions of Eqs. (2.8.77)–(2.8.80), the scattering amplitude $r_\nu(\theta_s|\theta_0)$ can be written in the form

$$r_\nu(\theta_s|\theta_0) \cong \Delta x \sum_{n=1}^{N} \exp[-i\sqrt{\epsilon_1}(\omega/c)x_n \sin\theta_s] \times \{[i\sqrt{\epsilon_1}(\omega/c)\cos\theta_s]H_\nu^{(1)}(x_n|\omega) + L_\nu^{(1)}(x_n|\omega)\}. \tag{2.8.86}$$

2.8.2.2. The Kirchhoff Approximation The expression for the transmission amplitude $T_\nu(q|k)$ obtained in the Kirchhoff approximation is (Eq. (2.8.56))

$$T_\nu(q|k) = T_\nu^{(0)}(q|k) \int_{-\infty}^{\infty} dx_1 \exp[-i(q-k)x_1] \exp\{i[\alpha_3(q) - \alpha_2(k)]\zeta(x_1)\}, \tag{2.8.87}$$

where $T_\nu^{(0)}(q|k)$ is given by Eq. (2.8.57). When the representation of the surface profile function $\zeta(x_1)$ given by Eq. (2.1.83) is introduced into Eq. (2.8.87), the latter becomes

$$\begin{aligned} T_\nu(q|k) &= T_\nu^{(0)}(q|k) \sum_{n=-N_s}^{N_s-1} \exp\{i[\alpha_3(q) - \alpha_2(k)]b_n\} \\ &\quad \times \int_{nb}^{(n+1)b} dx_1 \exp\{-i[q - k - (\alpha_3(q) - \alpha_2(k))a_n]x_1\} \\ &= bT_\nu^{(0)}(q|k) t_\nu(\theta_t|\theta_0), \end{aligned} \tag{2.8.88}$$

where

$$\begin{aligned} t_\nu(\theta_t|\theta_0) = &\sum_{n=-N_s}^{N_s-1} \exp\{i(\omega/c)[\sqrt{\epsilon_3}\cos\theta_t - \sqrt{\epsilon_2 - \epsilon_1 \sin^2\theta_0}\,]b_n\} \\ &\times \exp\Big\{-i(\omega/c)[\sqrt{\epsilon_3}\sin\theta_t - \sqrt{\epsilon_1}\sin\theta_0 \\ &- (\sqrt{\epsilon_3}\cos\theta_t - \sqrt{\epsilon_2 - \epsilon_1 \sin^2\theta_0}\,)a_n]\Big(n + \frac{1}{2}\Big)b\Big\} \\ &\times \operatorname{sinc}(\omega/c)\{\sqrt{\epsilon_3}\sin\theta_t - \sqrt{\epsilon_1}\sin\theta_0 - [\sqrt{\epsilon_3}\cos\theta_t - \sqrt{\epsilon_2 - \epsilon_1 \sin^2\theta_0}\,]a_n\}\frac{b}{2}. \end{aligned} \tag{2.8.89}$$

The mean differential transmission coefficient is now obtained from Eqs. (2.8.40) and (2.8.88) in the form

$$\left\langle \frac{\partial T_\nu}{\partial \theta_t} \right\rangle = \frac{b}{2N_s} \frac{\epsilon_3}{\sqrt{\epsilon_1}} \frac{\kappa_\nu}{\mu_\nu} \frac{\omega}{2\pi c} \frac{\cos^2\theta_t}{\cos\theta_0} \left|T_\nu^{(0)}(q|k)\right|^2 \left\langle \left|t_\nu(\theta_t|\theta_0)\right|^2 \right\rangle. \tag{2.8.90}$$

The amplitude $T_\nu^{(0)}(q|k)$, defined by Eq. (2.8.57), is given explicitly in terms of the angles of incidence and transmission by

$$\begin{aligned} T_\nu^{(0)}(q|k) &= \frac{\mu_\nu}{\sqrt{\epsilon_3}\cos\theta_t} \frac{[\sqrt{\epsilon_3}\cos\theta_t\sqrt{\epsilon_2-\epsilon_1\sin^2\theta_0}+\sqrt{\epsilon_1\epsilon_3}\sin\theta_t\sin\theta_0-\epsilon_2]}{\sqrt{\epsilon_3}\cos\theta_t-\sqrt{\epsilon_2-\epsilon_1\sin^2\theta_0}} \\ &\times \frac{2\lambda_\nu\sqrt{\epsilon_1}\cos\theta_0\exp[i(\omega d/c)\sqrt{\epsilon_2-\epsilon_1\sin^2\theta_0}\,]}{\lambda_\nu\sqrt{\epsilon_1}\cos\theta_0+\kappa_\nu\sqrt{\epsilon_2-\epsilon_1\sin^2\theta_0}} \frac{N(\theta_t,\theta_0)}{D(\theta_t,\theta_0)}, \end{aligned} \tag{2.8.91}$$

where

$$\begin{aligned} N(\theta_t,\theta_0) &= \operatorname{sgn}\Big(\sqrt{\epsilon_3}\cos\theta_t-\sqrt{\epsilon_2-\epsilon_1\sin^2\theta_0}\Big) \\ &\quad\times\Big[\epsilon_3-\sqrt{\epsilon_3}\cos\theta_t\sqrt{\epsilon_2-\epsilon_1\sin^2\theta_0}-\sqrt{\epsilon_1\epsilon_3}\sin\theta_t\sin\theta_0\Big] \\ &\quad+\Big|\epsilon_3-\sqrt{\epsilon_3}\cos\theta_t\sqrt{\epsilon_2-\epsilon_1\sin^2\theta_0}-\sqrt{\epsilon_1\epsilon_3}\sin\theta_t\sin\theta_0\Big|, \end{aligned} \tag{2.8.92}$$

$$\begin{aligned} D(\theta_t,\theta_0) &= \mu_\nu\operatorname{sgn}\Big(\sqrt{\epsilon_3}\cos\theta_t-\sqrt{\epsilon_2-\epsilon_1\sin^2\theta_0}\Big) \\ &\quad\times\Big[\sqrt{\epsilon_3}\cos\theta_t\sqrt{\epsilon_2-\epsilon_1\sin^2\theta_0}+\sqrt{\epsilon_1\epsilon_3}\sin\theta_t\sin\theta_0-\epsilon_2\Big] \\ &\quad+\lambda_\nu\Big|\sqrt{\epsilon_3}\cos\theta_t\sqrt{\epsilon_2-\epsilon_1\sin^2\theta_0}+\sqrt{\epsilon_1\epsilon_3}\sin\theta_t\sin\theta_0-\epsilon_3\Big|. \end{aligned} \tag{2.8.93}$$

2.8.3. *Example: A Band-Limited Uniform Diffuser*

If we wish to design a surface that produces transmitted light with a constant mean differential transmission coefficient within the interval $|\theta_t| < \theta_m$ of angles of transmission, and produces no transmission for θ_t outside this interval, the form of the mean dtc we wish to reproduce is

$$\left\langle\frac{\partial T_\nu}{\partial\theta_t}\right\rangle = A\theta\big(\theta_m-|\theta_t|\big). \tag{2.8.94}$$

Since $\sin\theta_t$ is a monotonically increasing function of θ_t in the interval $(-\pi/2,\pi/2)$, we can rewrite Eq. (2.8.94) as

$$\begin{aligned} \left\langle\frac{\partial T_\nu}{\partial\theta_t}\right\rangle &= A\theta\big(\sin\theta_m-|\sin\theta_t|\big) \\ &= A\theta\left(\sin\theta_m-\frac{|\gamma|}{1+\gamma^2}\left[\sqrt{\frac{\epsilon_2}{\epsilon_3}}-\sqrt{1-\left(\frac{\epsilon_2-\epsilon_3}{\epsilon_3}\right)\gamma^2}\right]\right). \end{aligned} \tag{2.8.95}$$

This is an awkward expression to deal with. However, for small values of γ we can approximate it by

$$\left\langle \frac{\partial T_\nu}{\partial \theta_t} \right\rangle = A\theta\left(\sin\theta_m - \frac{2b|\gamma|}{1+\gamma^2}\right), \tag{2.8.96}$$

where

$$2b = \sqrt{\epsilon_2/\epsilon_3} - 1. \tag{2.8.97}$$

Equation (2.8.96) can be written in a more convenient form as

$$\left\langle \frac{\partial T_\nu}{\partial \theta_t} \right\rangle = A\theta\big(\gamma_m - |\gamma|\big), \tag{2.8.98}$$

where

$$\gamma_m = \frac{b - \sqrt{b^2 - \sin^2\theta_m}}{\sin\theta_m}. \tag{2.8.99}$$

It follows from Eq. (2.8.76) that the pdf of a_n is given by

$$f(\gamma) = A\frac{1}{4\kappa_\nu\mu_\nu}\sqrt{\frac{\epsilon_1}{\epsilon_3}}\left(\frac{\lambda_\nu\sqrt{\epsilon_1} + \kappa_\nu\sqrt{\epsilon_2}}{2\lambda_\nu\sqrt{\epsilon_1}}\right)^2 \times \frac{[\sqrt{\epsilon_2/\epsilon_3} - \sqrt{1-(\frac{\epsilon_2-\epsilon_3}{\epsilon_3})\gamma^2}\,][\mu_\nu + \lambda_\nu\sqrt{\epsilon_3/\epsilon_2}\sqrt{1-(\frac{\epsilon_2-\epsilon_3}{\epsilon_3})\gamma^2}\,]^2}{(1+\gamma^2)[1-(\frac{\epsilon_2-\epsilon_3}{\epsilon_3})\gamma^2]}\theta\big(\gamma_m - |\gamma|\big). \tag{2.8.100}$$

Of course, γ_m must be smaller than $(\epsilon_3/(\epsilon_2-\epsilon_3))^{\frac{1}{2}}$, in view of Eq. (2.8.69). The constant A is determined from the normalization of $f(\gamma)$.

An ensemble of N_p one-dimensional randomly rough surfaces was generated on the basis of the preceding results. For each member of the ensemble a sequence of coefficients $\{a_n\}$ was obtained by the rejection method [2.6], and the surface profile function generated by the use of Eqs. (2.1.83) and (2.1.85). The film structure depicted in Fig. 2.43 was characterized by the values $d = 20$ μm, $\epsilon_1 = 2.25$ (glass), $\epsilon_2 = 2.69$ (photoresist), and $\epsilon_3 = 1$ (vacuum). The value of θ_m was chosen to be $\theta_m = 7°$.

This system was illuminated at normal incidence by both an s-polarized and a p-polarized plane wave of wavelength $\lambda = 612.7$ nm (in vacuum). The mean differential transmission coefficients was then calculated for incident light of each polarization by the rigorous numerical approach described in Section 2.8.2.1. The characteristic length b assumed in these calculations was $b = 20$ μm $= 31.6\lambda$. The length of the x_1 axis covered by the random surface was $L_1 = 85.8$ μm $= 140\lambda$, and the sampling length was $\Delta x = \lambda/10$. The results from $N_p = 30\,000$ realizations of the random surface were averaged in obtaining the mean differential transmission coefficient. We note that the surfaces studied

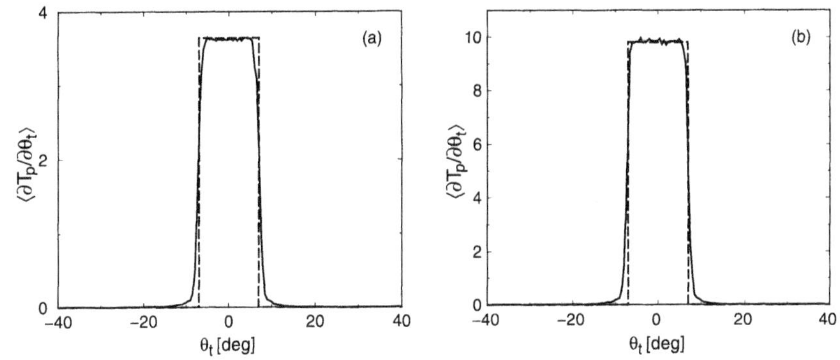

Figure 2.44. The mean differential transmission coefficient $\langle \partial T_s/\partial\theta_t\rangle$ for s-polarized light of wavelength $\lambda = 612.7$ nm incident normally on the structure depicted in Fig. 2.43. The dielectric constants of the layers constituting this structure are $\epsilon_1 = 2.25$, $\epsilon_2 = 2.69$, $\epsilon_3 = 1$, and the thickness parameter $d = 20$ μm. The solid curve is the result of rigorous computer simulation calculations, while the dashed curve is the result given by Eq. (2.8.94). The length of the surface used in the simulations was $L_1 = 85.8$ μm $= 140\lambda$, and the number of surface realizations over which $\partial T_s/\partial\theta_s$ was averaged was $N_p = 30\,000$. The number of sampling points used was $N = 1400$ ($\Delta x = \lambda/10$). The random surface was characterized by the parameters $b = 20$ μm and $\theta_m = 7°$. (b) The same as (a) but for p-polarized incident light.

were sufficiently rough that the contribution to the mean differential transmission coefficient from the light transmitted coherently was negligible compared to the contribution from the light transmitted incoherently. Consequently, it sufficed to calculate only the mean differential transmission coefficient. We also note that the characteristic length b of the surface structure is quite large and, although the maximal slope of the surface is small ($\gamma_m = 0.19784$), in some realizations of the surface profile function the film could be pierced by the structure. Such realizations were discarded in the calculations.

The result for the mean differential transmission coefficient $\langle \partial T_s/\partial\theta_t\rangle$ for s-polarized incident light is plotted in Fig. 2.44(a) as a function of the angle of transmission θ_t. The corresponding result, $\langle \partial T_p/\partial\theta_t\rangle$ for p-polarized incident light, is plotted in Fig. 2.44(b). For incident light of both polarizations the mean differential transmission coefficient is seen to be very well band limited. It is nonzero only for $-7° < \theta_t < 7°$. It is also quite uniform within this interval of angles of transmission.

In Figs. 2.45(a) and 2.45(b) we present plots of the mean differential transmission coefficient calculated for the same system for which Figs. 2.44(a) and 2.44(b), respectively, in the Kirchhoff approximation derived in Section 2.8.2.1. In these calculations a much larger value of the characteristic length b was used, namely $b = 50$ μm $\approx 81.6\lambda$. We have noted in the discussion of the results presented in Section 2.1.4.1, that the use of a large value of b in calculations of the mean differential reflection coefficient on the basis of the Kirchhoff approximation for a surface that acts as a band-limited uniform diffuser in reflection, sharpens the corners of the mean differential reflection coefficient and makes its sides more vertical. The use of a large value of b in obtaining the results plotted in Figs. 2.45(a) and 2.45(b) is seen to have the same effects in calculations of the mean differ-

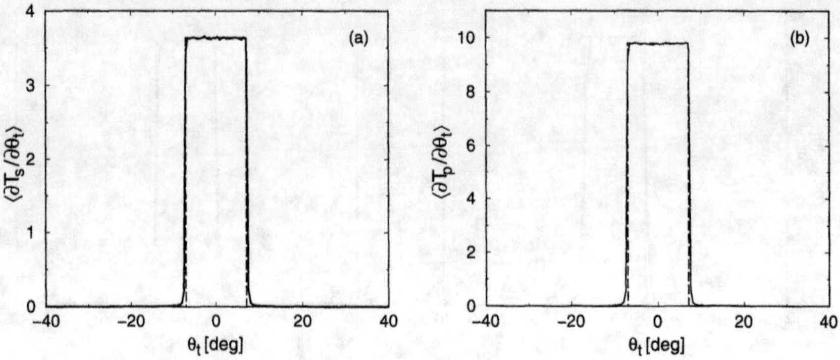

Figure 2.45. The mean differential transmission coefficient $\langle \partial T_s/\partial\theta_t \rangle$ for s-polarized light of wavelength $\lambda = 612.7$ nm incident normally on the structure depicted in Fig. 2.43. The dielectric constants of the layers constituting this structure are $\epsilon_1 = 2.25$, $\epsilon_2 = 2.69$, $\epsilon_3 = 1$, and $d = 20$ μm. The solid curve is the result of calculations based on the Kirchhoff approximation, while the dashed curve is the result given by Eq. (2.8.94). The length of the surface used in this calculation was $L_1 = 1$ mm $\approx 1632\lambda$, and the number of surface realizations over which $\partial T_s/\partial\theta_t$ was averaged was $N_p = 30\,000$. The random surface was characterized by the parameters $b = 50$ μm and $\theta_m = 7°$. (b) The same as (a) but for p-polarized incident light.

ential transmission coefficient for a band-limited uniform diffuser. The use of larger values of the characteristic length b and considerably longer surfaces in calculations based on the Kirchhoff approximation makes it even more important to check every realization of the surface profile function and to discard those that pierce the film. Such a selection of the surface profile functions, however, does not change the statistics of the surface roughness.

The results obtained in Section 2.8.2.1 enable us to calculate rigorously the mean differential reflection coefficient $\langle \partial R_\nu/\partial\theta_s \rangle$ for the structure depicted in Fig. 2.43. In Fig. 2.46(a) we present a plot of $\langle \partial R_p/\partial\theta_s \rangle$ as a function of the scattering angle θ_s. In Fig. 2.46(b) we present a plot of $\langle \partial R_s/\partial\theta_s \rangle$ as a function of θ_s. The experimental, material, and computational parameters assumed in obtaining these results are exactly the same as the ones assumed in obtaining the results plotted in Figs. 2.44(a) and 2.44(b), respectively. The results for $\langle \partial R_\nu/\partial\theta_s \rangle$ plotted in Figs. 2.46(a) and 2.46(b) are seen to be band-limited. The result for $\langle \partial R_p/\partial\theta_s \rangle$ is also nonzero only for $|\theta_s|$ smaller than approximately 24°; the result for $\langle \partial R_s/\partial\theta_s \rangle$ is nonzero only for $|\theta_s|$ smaller than approximately 24°. Both mean differential reflection coefficients are reasonably constant within these intervals. The narrow peak present in the results for $\langle R_{p,s}(\theta_s) \rangle$ at $\theta_s = 0°$ is due to the specular reflection of the incident light from the planar interface at $x_3 = d$. An argument based on the application of Snell's law at each interface yields the following relations between the maximum angle defining the angular region within which the mean differential reflection coefficient is nonzero and θ_m:

$$(\sin\theta_s)_{\max} = 2\left(\frac{\epsilon_2}{\epsilon_1}\right)^{\frac{1}{2}} \frac{(\sqrt{\epsilon_2/\epsilon_3} - \cos\theta_m)\sin\theta_m}{[(\epsilon_2/\epsilon_3) - 2\sqrt{\epsilon_2/\epsilon_3}\cos\theta_m + 1)]} \tag{2.8.101}$$

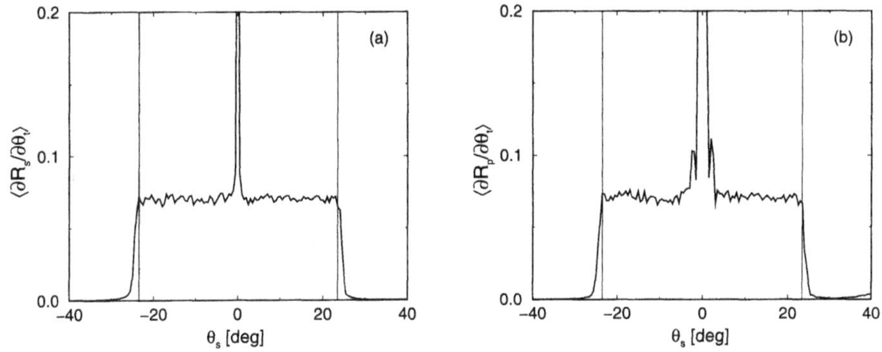

Figure 2.46. (a) The mean differential reflection coefficient $\langle \partial R_s / \partial \theta_s \rangle$ for s-polarized light incident normally on the structure depicted in Fig. 2.43, calculated by means of a rigorous computer simulation approach. The values of the experimental, material, and computational parameters used in carrying out this calculation were the same as those used in obtaining the results presented in Fig. 2.44(a). (b) The same as (a) but for p-polarized incident light. The vertical dashed lines in these figures are the predictions of Eq. (2.8.101).

for incident light of both polarizations. For $\theta_m = 7°$, Eq. (2.8.101) yields $\theta_s \approx 23.5°$, which agrees with the results depicted in Figs. 2.46(a) and 2.46(b). In view of the fact that the surface profile function $\zeta(x_1)$ was not designed to produce a band limited uniform diffuser in reflection, but only in transmission, the results for $\langle \partial R_\nu / \partial \theta_s \rangle$ presented in Fig. 2.46 are quite satisfactory. Thus, the results of this section show that the approach used in the earlier sections of this chapter to generate one-dimensional randomly rough surfaces that act as one-dimensional band-limited uniform diffusers in reflection can be applied to solve the same problem in transmission as well.

2.9. Replacement of Ensemble Averaging by the Use of a Broadband Incident Field in Calculations of Scattering from Randomly Rough Surfaces

In the preceding sections of this chapter we have assumed that the rough surfaces being designed were illuminated by a monochromatic plane wave (2.1.11). In addition, in those cases where the rough surface was assumed to be a randomly rough surface, to determine the pdf $f(\gamma)$ of the slope a_n in the representation of the surface profile function given by Eq. (2.1.83) we carried out an average over the ensemble of realizations of the surface profile function of the property of the scattered field that the surface was designed to produce, its intensity, for example. This averaging is inevitable in a stochastic approach to the design of surfaces with specified scattering properties, such as the one adopted in this book. However, in validating the predictions of the resulting algorithms, the property of the scattered field of interest is calculated for each member of an ensemble of N_p realizations of the surface profile function by a rigorous computational approach or by the Kirchhoff approximation, and an arithmetic average of the N_p results obtained in this way yields the ensemble average of this property. Without this latter ensemble average the intensity, for example, of the field scattered from a single realization of the randomly rough surface would consist of an array of bright spots on a dark background—a speckle pattern—as a function of the transverse coordinates on a screen at any distance from the mean scattering surface, rather than the smooth function of these coordinates produced by the ensemble averaging. In contrast with the ensemble averaging used in obtaining the pdf $f(\gamma)$, the latter ensemble averaging is not inevitable.

An experimentalist after all cannot fabricate an ensemble of several thousand realizations of a random surface drawn from a given pdf $f(\gamma)$, measure the intensity of the field scattered from each realization, and average the results of all of these measurements. He/she measures the intensity of the field scattered from a single realization of a randomly rough surface. Although the illuminated area of the rough surface can be quite large it is not of the infinite size needed for a single measurement to reproduce the results obtained by ensemble averaging. To obtain the kind of smooth curves produced by ensemble averaging the experimentalist has to average over the resulting speckles in some way. One way of doing this is to move the random surface, e.g., to rotate or dither the sample. However, in some applications moving the surface may not be an option.

A speckle pattern depends on the wavelength of a monochromatic incident beam. This suggests that it may be possible to average over speckles by using a polychromatic, broadband, beam rather than a monochromatic beam to illuminate the surface. In this section we consider this option for the example of scattering from a one-dimensional, randomly rough surface.

2.9.1. The Incident Field

The scattering system we consider is the one described in Section 2.1.1. It consists of vacuum in the region $x_3 > \zeta(x_1)$ and a perfect conductor in the region $x_3 < \zeta(x_1)$. The surface profile function is assumed to have the properties described in Section 2.1.1.

We assume that the surface $x_3 = \zeta(x_1)$ is illuminated at normal incidence by an s-polarized, two-dimensional, broadband Gaussian beam rather than by the plane wave (2.1.11). The single nonzero component of the electric vector of the incident field is assumed to have the form

$$E_2(x_1, x_3; t)_{\text{inc}} = \int_{-\infty}^{\infty} \frac{d\omega}{2\pi} F(\omega) \int_{-\frac{\omega}{c}}^{\frac{\omega}{c}} \frac{dk}{2\pi} G(k) \exp\{i[kx_1 - \alpha_0(k,\omega)x_3 - \omega t]\}, \tag{2.9.1}$$

where $\alpha_0(k,\omega) = [(\omega/c)^2 - k^2]^{\frac{1}{2}}$, with $\operatorname{Re}\alpha_0(k,\omega), \operatorname{Im}\alpha_0(k,\omega) > 0$. The function $G(k)$ is

$$G(k) = \frac{\omega}{c} \frac{\sqrt{\pi}\, w}{\alpha_0(k,\omega)} \exp\bigl[-(w^2\omega^2/4c^2)\bigl(\sin^{-1}(ck/\omega)\bigr)^2\bigr]. \tag{2.9.2}$$

The function $F(\omega)$ is a random function that possesses the properties

$$\big\langle F(\omega)F^*(\omega')\big\rangle_F = 2\pi\,\delta(\omega-\omega')S_0(\omega), \tag{2.9.3a}$$

$$\big\langle F(\omega)F(\omega')\big\rangle_F = 0, \tag{2.9.3b}$$

where the angle brackets $\langle\cdots\rangle_F$ denote an average over the ensemble of realizations of the field. The fluctuation of $|F(\omega)|^2$ as a function of ω, which depends on the integration time of the detector, is below the resolution of a normal spectrograph, so for practical purposes $F(\omega)$ can be considered to be delta correlated [2.116]. An incident field of this nature is produced, for example, by a superluminescent diode [2.117]. For definiteness, we assume that the spectral density of the incident light $S_0(\omega)$ is described by a Gaussian form with central frequency ω_0 and $1/e$ half width $\Delta\omega$,

$$S_0(\omega) = \frac{1}{\sqrt{\pi}\,\Delta\omega}\exp\big[-(\omega-\omega_0)^2/(\Delta\omega)^2\big]. \tag{2.9.4}$$

The half width $\Delta\omega$ is assumed to be small enough that the spectral density of the incident light can be regarded as zero when $\omega < 0$. It is also convenient to regard the function $F(\omega)$ as zero when $\omega < 0$.

To see that Eq. (2.9.1) describes a broadband Gaussian beam that is incident normally on the surface $x_3 = \zeta(x_1)$, we make the change of variable $k = (\omega/c)\sin\theta$, which yields for the incident field

$$E_2(x_1, x_3; t)_{\mathrm{inc}} = \frac{w}{2\sqrt{\pi}c}\int_{-\infty}^{\infty}\frac{d\omega}{2\pi}\omega F(\omega)\exp(-i\omega t)$$
$$\times\int_{-\frac{\pi}{2}}^{\frac{\pi}{2}} d\theta\,\exp\big[-(w^2\omega^2/4c^2)\theta^2\big]\exp\big[i(\omega/c)(x_1\sin\theta - x_3\cos\theta)\big]. \tag{2.9.5}$$

In the limit $(w\omega/2c) \gg 1$ only small values of θ contribute to the integral over θ. Therefore, by approximating $(x_1\sin\theta - x_3\cos\theta)$ by $(x_1\theta - x_3)$, and replacing the finite limits of integration by infinite limits, we obtain the incident beam in the form

$$E_2(x_1, x_3; t)_{\mathrm{inc}} \cong \int_{-\infty}^{\infty}\frac{d\omega}{2\pi}F(\omega)\exp\big[-(x_1^2/w^2) - i(\omega/c)x_3 - i\omega t\big]. \tag{2.9.6}$$

For future reference we also provide the nonzero components of the magnetic vector of the incident field:

$$H_1(x_1, x_3; t)_{\text{inc}} = \int_{-\infty}^{\infty} \frac{d\omega}{2\pi} \frac{c}{\omega} F(\omega) \int_{-\frac{\omega}{c}}^{\frac{\omega}{c}} \frac{dk}{2\pi} \alpha_0(k,\omega) G(k) \exp\{i[kx_1 - \alpha_0(k,\omega)x_3 - \omega t]\}$$

(2.9.7a)

$$\cong \int_{-\infty}^{\infty} \frac{d\omega}{2\pi} F(\omega) \exp[-(x_1^2/w^2) - i(\omega/c)x_3 - i\omega t] \quad (2.9.7b)$$

and

$$H_3(x_1, x_3; t)_{\text{inc}} = \int_{-\infty}^{\infty} \frac{d\omega}{2\pi} \frac{c}{\omega} F(\omega) \int_{-\frac{\omega}{c}}^{\frac{\omega}{c}} \frac{dk}{2\pi} k\, G(k) \exp\{i[kx_1 - \alpha_0(k,\omega)x_3 - \omega t]\}$$

(2.9.8a)

$$\cong i2\frac{cx_1}{w^2} \int_{-\infty}^{\infty} \frac{d\omega}{2\pi} \frac{F(\omega)}{\omega} \exp[-(x_1^2/w^2) - i(\omega/c)x_3 - i\omega t]. \quad (2.9.8b)$$

The approximate expressions obtain in the limit $(w\omega/2c) \gg 1$.

2.9.2. The Scattered Field

Due to the linearity of the scattering problem the scattered field produced by the incident field (2.9.1) can be rewritten as

$$E_2(x_1, x_3; t)_{\text{sc}} = \int_{-\infty}^{\infty} \frac{d\omega}{2\pi} F(\omega) \int_{-\infty}^{\infty} \frac{dq}{2\pi} \int_{-\frac{\omega}{c}}^{\frac{\omega}{c}} \frac{dk}{2\pi} R(q|k) G(k)$$

$$\times \exp\{i[qx_1 + \alpha_0(q,\omega)x_3] - i\omega t\}. \quad (2.9.9)$$

In this expression $R(q|k)$ is the amplitude for the scattering of an incident plane wave of frequency ω whose wavevector has a component k parallel to the mean scattering surface into a plane wave whose wavevector has a component q parallel to the mean scattering surface. It is the solution of Eq. (2.5.97). It has been shown in Section 2.7.3 that in the simplest version of phase perturbation theory $R(q|k)$ is given by

$$R(q|k) = -\int_{-\infty}^{\infty} dx_1 \exp[-i(q-k)x_1] \exp[-i2\alpha_0(k,\omega)\zeta(x_1)]. \quad (2.9.10)$$

We will use it here due to its simplicity.

When the results given by Eqs. (2.9.2) and (2.9.10) are substituted into Eq. (2.9.9), and the change of variable $k = (\omega/c)\sin\theta$ is made, the latter equation becomes

$$E_2(x_1, x_3; t)_{sc} = -\frac{\omega}{2\sqrt{\pi}c} \int_{-\infty}^{\infty} \frac{d\omega}{2\pi} \omega F(\omega) \exp(-i\omega t)$$
$$\times \int_{-\infty}^{\infty} dx_1' \int_{-\frac{\pi}{2}}^{\frac{\pi}{2}} d\theta \exp\bigl[-\bigl(w^2\omega^2/4c^2\bigr)\theta^2\bigr]$$
$$\times \exp\bigl\{i(\omega/c)\bigl[x_1' \sin\theta - 2\zeta(x_1')\cos\theta\bigr]\bigr\}$$
$$\times \int_{-\infty}^{\infty} \frac{dq}{2\pi} \exp\bigl[iq(x_1 - x_1') + i\alpha_0(q, \omega)x_3\bigr]. \tag{2.9.11}$$

We now invoke the parabolic approximation,

$$\alpha_0(q, \omega) \cong (\omega/c) - (c/2\omega)q^2, \tag{2.9.12}$$

which enables us to evaluate the integral over q in Eq. (2.9.19) analytically,

$$\int_{-\infty}^{\infty} \frac{dq}{2\pi} \exp\bigl[iq(x_1 - x_1') + i\alpha_0(q, \omega)x_3\bigr]$$
$$\cong \left(\frac{|\omega|}{2\pi c x_3}\right)^{\frac{1}{2}} \exp\bigl[i(\omega/c)x_3 - i(\pi/4)\,\mathrm{sgn}\,\omega + i(\omega/2cx_3)(x_1 - x_1')^2\bigr]. \tag{2.9.13}$$

In the limit $(w\omega/2c) \gg 1$ the integral over θ can be approximated by

$$\int_{-\frac{\pi}{2}}^{\frac{\pi}{2}} d\theta \exp\bigl[-\bigl(w^2\omega^2/4c^2\bigr)\theta^2\bigr] \exp\bigl\{i(\omega/c)\bigl[x_1' \sin\theta - 2\zeta(x_1')\cos\theta\bigr]\bigr\}$$
$$\cong \int_{-\infty}^{\infty} d\theta \exp\bigl[-\bigl(w^2\omega^2/4c^2\bigr)\theta^2\bigr] \exp\bigl[i(\omega/c)\bigl(x_1'\theta - 2\zeta(x_1')\bigr)\bigr]$$
$$= \frac{2\sqrt{\pi}c}{w|\omega|} \exp\bigl(-x_1'^2/w^2\bigr) \exp\bigl[-i(2\omega/c)\zeta(x_1')\bigr]. \tag{2.9.14}$$

On combining the results given by Eqs. (2.9.10), (2.9.13) and (2.9.14), we obtain the scattered field in the form

$$E_2(x_1, x_3; t)_{sc} = -\frac{1}{(2\pi c x_3)^{\frac{1}{2}}} \int_{-\infty}^{\infty} \frac{d\omega}{2\pi} \operatorname{sgn}\omega |\omega|^{\frac{1}{2}} F(\omega)$$

$$\times \exp\bigl[-i\omega t + i(\omega/c)x_3 - i(\pi/4)\operatorname{sgn}\omega\bigr] G(x_1, x_3; \omega), \quad (2.9.15)$$

where

$$G(x_1, x_3; \omega) = \int_{-\infty}^{\infty} dx_1' \exp\bigl[-(x_1'/w)^2 + i(\omega/2cx_3)(x_1 - x_1')^2 - i(2\omega/c)\zeta(x_1')\bigr]. \quad (2.9.16)$$

For completeness we also provide the nonzero components of the magnetic vector of the scattered field, and their expressions in the limit $(w\omega/2c) \gg 1$:

$$H_1(x_1, x_3; t)_{sc} = -\int_{-\infty}^{\infty} \frac{d\omega}{2\pi} \frac{c}{\omega} F(\omega) \int_{-\infty}^{\infty} \frac{dq}{2\pi} \int_{-\frac{\omega}{c}}^{\frac{\omega}{c}} \frac{dk}{2\pi} \alpha_0(q, \omega) R(q|k)$$

$$\times G(k) \exp\bigl\{i\bigl[qx_1 + \alpha_0(q, \omega)x_3 - \omega t\bigr]\bigr\} \quad (2.9.17a)$$

$$\cong \frac{1}{(2\pi c x_3)^{\frac{1}{2}}} \int_{-\infty}^{\infty} \frac{d\omega}{2\pi} \operatorname{sgn}\omega |\omega|^{\frac{1}{2}} F(\omega)$$

$$\times \exp\bigl[i(\omega/c)x_3 - i(\pi/4)\operatorname{sgn}\omega - i\omega t\bigr] G(x_1, x_3; \omega) \quad (2.9.17b)$$

and

$$H_3(x_1, x_3; t)_{sc} = \int_{-\infty}^{\infty} \frac{d\omega}{2\pi} \frac{c}{\omega} F(\omega) \int_{-\infty}^{\infty} \frac{dq}{2\pi} \int_{-\frac{\omega}{c}}^{\frac{\omega}{c}} \frac{dk}{2\pi} q R(q|k) G(k) \exp\bigl\{i\bigl[qx_1 + \alpha_0(q, \omega)x_3 - \omega t\bigr]\bigr\} \quad (2.9.18a)$$

$$\cong -\frac{1}{(2\pi c x_3)^{\frac{1}{2}}} \int_{-\infty}^{\infty} \frac{d\omega}{2\pi} \operatorname{sgn}\omega |\omega|^{\frac{1}{2}} F(\omega) \exp\bigl[i(\omega/c)x_3 - i(\pi/4)\operatorname{sgn}\omega - i\omega t\bigr]$$

$$\times \int_{-\infty}^{\infty} dx_1' \left(\frac{x_1 - x_1'}{x_3}\right) \exp\bigl[-(x_1'^2/w^2) - i(2\omega/c)\zeta(x_1') + i(\omega/2cx_3)(x_1 - x_1')^2\bigr]. \quad (2.9.18b)$$

2.9.3. Examples

To illustrate the quality of the results obtained when a single realization of a designed surface is illuminated by a broadband beam, we consider two properties of the scattered field, namely its intensity and the differential reflection coefficient. For each property three cases will be calculated: (i) the speckle pattern obtained when a single realization of the surface is illuminated by a monochromatic Gaussian beam of frequency ω_0 ($F(\omega) = 2\pi\,\delta(\omega - \omega_0)$ in Eq. (2.9.1)), calculated on the basis of the Kirchhoff approximation; (ii) the result obtained when the same realization of the surface is illuminated by a broadband beam centered at the frequency ω_0; (iii) the result obtained by averaging the property over the ensemble of realizations of the surface profile function when the surfaces are illuminated by a monochromatic Gaussian beam of frequency ω_0. These three results illustrate the extent to which speckle is suppressed by the use of a broadband Gaussian beam instead of a monochromatic Gaussian beam in scattering from a single realization of a designed surface. They also show how well the use of a broadband Gaussian beam illuminating a single realization of such a surface reproduces the result obtained by averaging the results obtained for an ensemble of realizations of the surface when a monochromatic Gaussian beam is used for illuminating the ensemble. The Kirchhoff approximation is used here because it allows to use considerably larger surfaces than are possible to use in the rigorous computer simulation calculations.

2.9.3.1. The Intensity of the Scattered Field The intensity profile of the scattered field at a distance x_3 from the plane $x_3 = 0$, averaged over the ensemble of realizations of the incident field, is obtained from Eqs. (2.9.3) and (2.9.15) in the form

$$\left\langle \left| E_2(x_1, x_3; t)_{\mathrm{sc}} \right|^2 \right\rangle_F = \frac{1}{2\pi c x_3} \int_{-\infty}^{\infty} \frac{d\omega}{2\pi} |\omega| S_0(\omega) \left| G(x_1, x_3; \omega) \right|^2. \tag{2.9.19}$$

We emphasize that there has been no averaging over the ensemble of realizations of the surface profile function involved in obtaining this result. The only averaging is over the ensemble of realizations of the incident field, and that is provided by the use of a broadband beam for illuminating the surface.

The procedure now is to generate a single realization of the surface profile function $\zeta(x_1)$ of the randomly rough surface on the basis of the pdf $f(\gamma)$ of the slope a_n in the representation of $\zeta(x_1)$ given by Eq. (2.1.83) and to use it in calculating the speckle pattern produced at a distance x_3 from it when it is illuminated at normal incidence by a monochromatic Gaussian beam of frequency ω_0. In Fig. 2.47(a) we present a plot of this speckle pattern. In Fig. 2.47(b) we present the intensity profile given by Eq. (2.9.19) when the same realization of the surface profile is illuminated at normal incidence by a broadband Gaussian beam centered at the frequency ω_0. Finally, in Fig. 2.47(c) we plot the intensity profile obtained by averaging the intensities of the fields produced by scattering

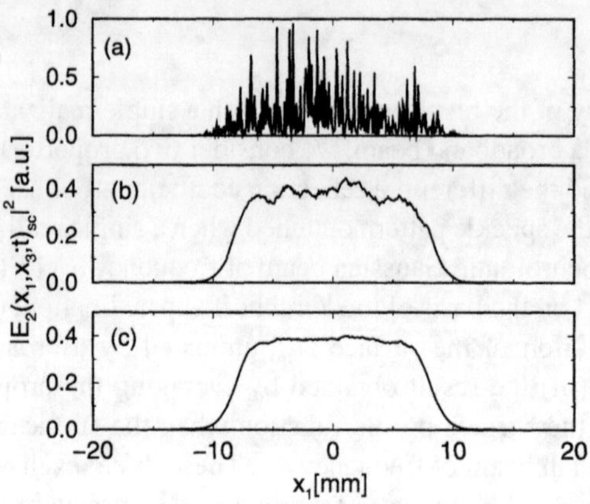

Figure 2.47. (a) The speckle pattern produced by the scattering of a monochromatic Gaussian beam of frequency $\omega_0 = 2\pi c/\lambda_0$ with $\lambda_0 = 1.55$ μm, from a single realization of a perfectly conducting randomly rough surface generated by the use of the pdf (2.7.62b); (b) the intensity profile of the scattered field given by Eq. (2.9.19) when this realization of the surface profile is illuminated by a broadband Gaussian beam with central frequency $\omega_0 = 2\pi c/\lambda_0$, where $\lambda_0 = 1.55$ μm; (c) the mean intensity of the scattered field averaged over the ensemble of realizations of the surface profile function generated by the use of the pdf (2.7.62b), when the perfectly conducting surfaces are illuminated by a monochromatic Gaussian beam of frequency ω_0.

a monochromatic Gaussian beam of frequency ω_0 from an ensemble of N_p realizations of the surface profile function, all drawn from the same pdf $f(\gamma)$.

The pdf $f(\gamma)$ assumed in carrying out all of these calculations was the one given by Eq. (2.7.62b). The values of the theoretical and experimental parameters used were $\omega_0 = 2\pi c/\lambda_0$, where $\lambda_0 = 1.55$ μm, $\Delta\omega = 0.4\omega_0$, $\omega = 1000$ μm, $\sigma_s = 500$ μm, $\sigma_g = 20$ μm, $b = 20$ μm, the length of the surface $L = 8$ mm, $N_p = 10\,000$, and $x_3 = 8$ cm.

From the results presented in Fig. 2.47 we see that the use of a broadband beam in illuminating a single realization of the randomly rough surface averages over the speckles produced by a monochromatic incident beam. It thereby produces an intensity profile of the scattered field that closely matches the one produced by a monochromatic beam when the intensity of the scattered field is averaged over the ensemble of realizations of the surface profile function. An example of the averaging effect of a polychromatic source can be clearly seen in Fig. 2.52(c) of Section 2.11.

2.9.3.2. The Differential Reflection Coefficient To calculate the differential reflection coefficient we require the 3-component of the total time-averaged Poynting vector of the incident field and of the scattered field. For an s-polarized electromagnetic field the 3-component of the complex Poynting vector is given by

$$S_3^c = -\frac{c}{8\pi} E_2(x_1, x_3; t) H_1^*(x_1, x_3; t). \tag{2.9.20}$$

The use of the expressions given by Eqs. (2.9.1) and (2.9.7a) and the result given by Eq. (2.9.3a) yields the result that the average of $(S_3^c)_{\text{inc}}$ over the ensemble of realizations of the incident field is

$$\langle (S_3^c)_{\text{inc}} \rangle_F = -\frac{c^2}{8\pi} \int_{-\infty}^{\infty} \frac{d\omega}{2\pi} \frac{S_0(\omega)}{\omega} \int_{-\frac{\omega}{c}}^{\frac{\omega}{c}} \frac{dk}{2\pi} G(k) \exp\bigl[ikx_1 - i\alpha_0(k,\omega)x_3\bigr]$$

$$\times \int_{-\frac{\omega}{c}}^{\frac{\omega}{c}} \frac{dk'}{2\pi} \alpha_0(k',\omega) G^*(k') \exp\bigl[-ik'x_1 + i\alpha_0(k',\omega)x_3\bigr]. \tag{2.9.21}$$

The magnitude of the total time-averaged flux illuminating the region $-L_1/2 < x_1 < L_1/2$, $-L_2/2 < x_2 < L_2/2$ of the plane $x_3 = 0$ is then

$$P_{\text{inc}} = -\text{Re} \int_{-\frac{L_1}{2}}^{\frac{L_1}{2}} dx_1 \int_{-\frac{L_2}{2}}^{\frac{L_2}{2}} dx_2 \langle (S_3^c)_{\text{inc}} \rangle_F$$

$$= L_2 \frac{c^2}{8\pi} \int_{-\infty}^{\infty} \frac{d\omega}{2\pi} \frac{S_0(\omega)}{\omega} \int_{-\frac{\omega}{c}}^{\frac{\omega}{c}} \frac{dk}{2\pi} \alpha_0(k,\omega) \bigl|G(k)\bigr|^2, \tag{2.9.22}$$

in the limit $L_1 \to \infty$. The reversal of sign in going from Eq. (2.9.21) to Eq. (2.9.22) compensates for the fact that S_3^c is negative for the incident field. The change of variable $k = (\omega/c)\sin\theta$ reduces Eq. (2.9.22) to

$$P_{\text{inc}} = L_2 \frac{w^2}{16\pi^2} \int_{-\infty}^{\infty} d\omega\, \omega S_0(\omega) \int_0^{\frac{\pi}{2}} d\theta \exp\bigl[-(w^2\omega^2/2c^2)\theta^2\bigr]$$

$$= L_2 \frac{cw}{8(2\pi)^{3/2}} \int_{-\infty}^{\infty} d\omega\, S_0(\omega) \,\text{Erf}\left(\frac{\pi}{\sqrt{2}} \frac{w|\omega|}{2c}\right), \tag{2.9.23}$$

where $\text{Erf}(z)$ is the error function,

$$\text{Erf}(z) = \frac{2}{\sqrt{\pi}} \int_0^z dt \exp(-t^2). \tag{2.9.24}$$

In the limit $w\omega/2c \gg 1$, $\mathrm{Erf}(\pi w|\omega|/2\sqrt{2}c)$ can be replaced by unity, and we find that

$$P_{\mathrm{inc}} = L_2 \frac{cw}{8(2\pi)^{3/2}}, \tag{2.9.25}$$

due to the normalization of $S_0(\omega)$.

The use of the expressions (2.9.8a) and (2.9.17a) together with the result given by Eq. (2.9.3a) gives the result that the average of $(S_3^c)_{\mathrm{sc}}$ over the ensemble of realizations of the incident field is

$$\begin{aligned}\langle (S_3^c)_{\mathrm{sc}}\rangle_F = \frac{c^2}{8\pi}\int_{-\infty}^{\infty}\frac{d\omega}{2\pi}\frac{S_0(\omega)}{\omega}\int_{-\infty}^{\infty}\frac{dq}{2\pi}R(q,\omega)\exp\bigl[iqx_1 + i\alpha_0(q,\omega)x_3\bigr]\\ \times\int_{-\infty}^{\infty}\frac{dq'}{2\pi}\alpha_0^*(q',\omega)R^*(q',\omega)\exp\bigl[-iq'x_1 - i\alpha_0^*(q',\omega)x_3\bigr],\end{aligned} \tag{2.9.26}$$

where the function $R(q,\omega)$ is defined by

$$R(q,\omega) = \int_{-\omega/c}^{\omega/c}\frac{dk}{2\pi}R(q|k)F(k). \tag{2.9.27}$$

The magnitude of the total time-averaged scattered flux leaving the region $-L_1/2 < x_1 < L_1/2$, $-L_2/2 < x_2 < L_2/2$ of the plane $x_3 = 0$ is then

$$\begin{aligned}P_{\mathrm{sc}} &= \mathrm{Re}\int_{-\frac{L_1}{2}}^{\frac{L_1}{2}}dx_1\int_{-\frac{L_2}{2}}^{\frac{L_2}{2}}dx_2\langle (S_3^c)_{\mathrm{sc}}\rangle_F\\ &= L_2\frac{c^2}{8\pi}\,\mathrm{Re}\int_{-\infty}^{\infty}\frac{d\omega}{2\pi}\frac{S_0(\omega)}{\omega}\int_{-\infty}^{\infty}\frac{dq}{2\pi}\alpha_0^*(q,\omega)\bigl|R(q,\omega)\bigr|^2\exp\bigl\{i\bigl[\alpha_0(q,\omega)-\alpha_0^*(q,\omega)\bigr]x_3\bigr\}\\ &= L_2\frac{c^2}{16\pi^2}\int_{-\infty}^{\infty}d\omega\frac{S_0(\omega)}{\omega}\int_{-\frac{\omega}{c}}^{\frac{\omega}{c}}\frac{dq}{2\pi}\alpha_0(q,\omega)\bigl|R(q,\omega)\bigr|^2,\end{aligned} \tag{2.9.28}$$

where we have let L_1 tend to infinity. The change of variable $q = (\omega/c)\sin\theta_s$ transforms Eq. (2.9.28) into

$$P_{\rm sc} = L_2 \frac{1}{16\pi^2} \int_{-\frac{\pi}{2}}^{\frac{\pi}{2}} d\theta_s \cos^2\theta_s \int_{-\infty}^{\infty} \frac{d\omega}{2\pi} \omega S_0(\omega) \left| R\big((\omega/c)\sin\theta_s, \omega\big)\right|^2$$

$$\equiv \int_{-\frac{\pi}{2}}^{\frac{\pi}{2}} d\theta_s\, P_{\rm sc}(\theta_s). \qquad (2.9.29)$$

By definition, the differential reflection coefficient, averaged over the ensemble of realizations of the incident field, is given by

$$\left\langle \frac{\partial R_s}{\partial \theta_s} \right\rangle_F = \frac{P_{\rm sc}(\theta_s)}{P_{\rm inc}}$$

$$= \left(\frac{2}{\pi}\right)^{\frac{1}{2}} \frac{\cos^2\theta_s}{cw} \int_{-\infty}^{\infty} \frac{d\omega}{2\pi} \omega S_0(\omega) \left| R\big((\omega/c)\sin\theta_s, \omega\big)\right|^2. \qquad (2.9.30)$$

Again, no averaging over the ensemble of realizations of the surface profile function was used in obtaining this result.

A simple result for the scattering amplitude $R(q,\omega)$ is obtained when we substitute into Eq. (2.9.27) the expression for $R(q|k)$ given by Eq. (2.9.10) and the expression for $G(k)$ given by Eq. (2.9.2), and then pass to the limit $(w\omega/2c) \gg 1$. The result is

$$R(q,\omega) = -\int_{-\infty}^{\infty} dx_1 \exp\big[-(x_1^2/w^2) - iqx_1 - i(2\omega/c)\zeta(x_1)\big]. \qquad (2.9.31)$$

The mean differential reflection coefficient, obtained by averaging the differential reflection coefficient over the ensemble of realizations of the surface profile function can be obtained from this result according to

$$\left\langle \frac{\partial R}{\partial \theta_s} \right\rangle = \frac{1}{\sqrt{2\pi}\,\pi} \frac{\omega}{cw} \cos^2\theta_s \left\langle \left| R\big((\omega/c)\sin\theta_s, \omega\big)\right|^2 \right\rangle. \qquad (2.9.32)$$

We apply the preceding results to the calculations of the differential reflection coefficients for a band-limited uniform diffuser, for which the pdf of the slope a_n in the representation (2.1.83) of the surface profile function $\zeta(x_1)$ is given by Eq. (2.1.139). In Fig. 2.48(a) we present a plot of the speckle pattern produced when a single realization of the surface

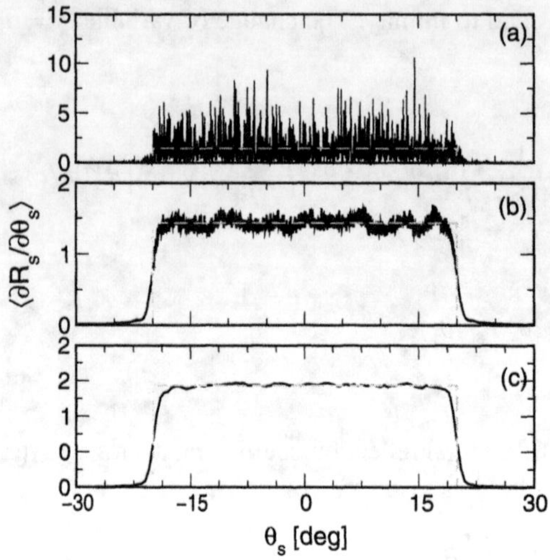

Figure 2.48. (a) The differential reflection coefficient produced by the scattering of a monochromatic Gaussian beam of frequency $\omega_0 = 2\pi c/\lambda_0$ with $\lambda_0 = 632.8$ nm, from a single realization of a perfectly conducting randomly rough surface generated by the use of the pdf (2.1.139); (b) the differential reflection coefficient when the same realization of the surface profile is illuminated by a broadband Gaussian beam centered at the frequency ω_0; (c) the mean differential reflection coefficient averaged over the ensemble of realizations of the surface profile function generated by the use of the pdf (2.1.139), when the perfectly conducting surfaces are illuminated by a monochromatic Gaussian beam of frequency ω_0.

profile function is illuminated at normal incidence by a monochromatic Gaussian beam of frequency ω_0. This result is produced by the use of the expression

$$\frac{\partial R_s}{\partial \theta_s} = \frac{1}{\sqrt{2\pi}\pi} \frac{\omega_0}{cw} \cos^2\theta_0 \left| R\left((\omega_0/c)\sin\theta_s, \omega_0\right)\right|^2, \tag{2.9.33}$$

which follows from Eq. (2.9.32) in the absence of ensemble averaging.

In Fig. 2.48(b) we present the differential reflection coefficient when the same realization of the surface profile is illuminated at normal incidence by a broadband Gaussian beam centered at the frequency ω_0. Finally, in Fig. 2.48(c) we plot the mean differential reflection coefficient obtained on the basis of Eq. (2.9.32) that is produced by averaging the differential reflection coefficients produced by scattering a monochromatic Gaussian beam of frequency ω_0 from an ensemble of N_p realizations of the surface profile function, all drawn from the same pdf $f(\gamma)$.

The values of the theoretical and experimental parameters employed in these calculations were $\omega_0 = 2\pi c/\lambda_0$, with $\lambda_0 = 632.8$ nm, $\Delta\omega = 0.1\omega_0$, $w = 1$ mm, $\theta_m = 20°$, $b =$ 20 μm, the length of the surface $L = 8$ mm and $N_p = 10\,000$.

The results presented in Fig. 2.48 demonstrate that the use of a broadband source in illuminating a single realization of the randomly rough surface averages over the speckles produced by a monochromatic incident beam. It consequently produces a differential reflection coefficient that matches well the one produced by a monochromatic beam when the resulting differential reflection coefficient is averaged over the ensemble of realizations of the surface profile function.

2.10. Fabrication of One-Dimensional Surfaces

Surfaces of the type considered in this chapter can be fabricated using optical techniques to selectively expose and etch photoresist-coated plates. Photoresists are light sensitive materials used in the microelectronics industry to create patterned coatings. They absorb light from approximately 300–450 nm, having a highly nonlinear response in the ultraviolet. A more linear relationship between exposure and resulting surface height can be obtained by using visible wavelengths and adjusting the coating and development processes.

In optics, photoresists are used to write the masters of two-layer or binary diffractive optical elements. They have also been used for the fabrication of continuous level structures like sinusoidal diffraction gratings [2.118,2.119] and one- and two-dimensional Gaussian random surfaces [2.120–2.124]. We note that for the fabrication of two-level structures one can use binary masks to expose the plate, but to generate more general profiles it is preferable to expose the plate directly to an interference pattern (periodic or random), without the use of masks.

To fabricate the kind of surfaces discussed in this chapter, one needs to find a suitable way to convert the computer-generated realization of a surface profile function into an exposure pattern. A linear response of the photoresist would also be required. To avoid nonlinearities caused by the intermediate step of producing a continuous-tone mask, it is desirable to employ a direct exposure method. One possibility is to fabricate the plates in a raster scan fashion, using an intensity-modulated beam focused on the plate. Such a system could also be used to fabricate multi-level two-dimensional surfaces. Although versatile, the method requires a dedicated and expensive instrument, and can also be quite slow. For instance, if a resolution of a few microns is required on the surface, the fabrication of a 2 by 2 inch sample could take a few days.

In this section, we describe a technique that has proved convenient and effective for the fabrication of one-dimensional surfaces like the ones described in this chapter. We start with a realization of a profile function $\zeta(x)$ with the desired statistical properties (pdf of slopes), which is then used to fabricate a slit of variable width in the manner shown in Fig. 2.49.

The mask has intensity transmittance

$$T(x, y) = \theta(y)\theta\big(\zeta(x) + \zeta_0 - y\big), \tag{2.10.1}$$

where ζ_0 is a positive constant that is larger than the largest negative value of $\zeta(x)$ and $\theta(y)$ is the Heaviside unit step function.

A good quality optical system is used to form an incoherent, demagnified image of the slit on the photoresist plate, as shown in Fig. 2.50. The illumination is provided by a

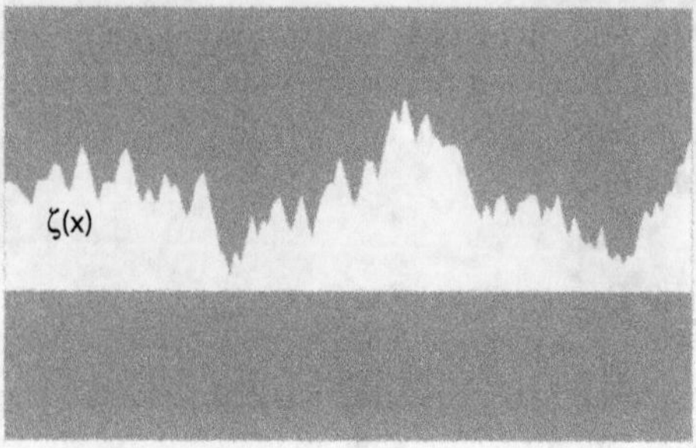

Figure 2.49. Mask generated with a realization of the surface profile function $\zeta(x)$.

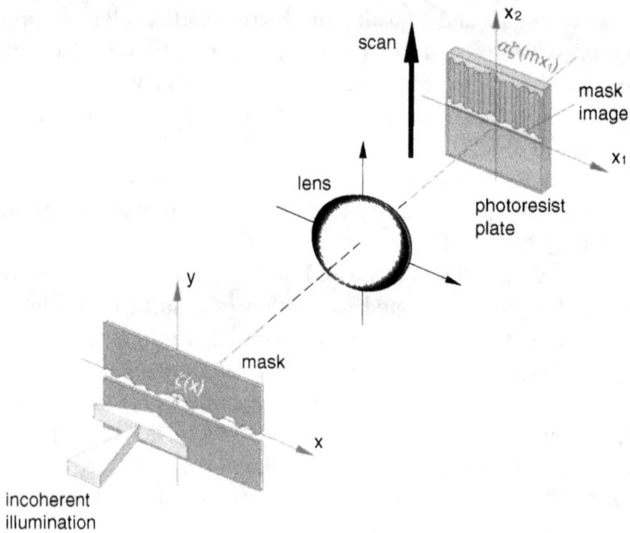

Figure 2.50. Schematic diagram of the experimental arrangement for the fabrication of surfaces with the desired surface profile function. The mask is imaged on the photoresist plate, which is then scanned along x_2.

rotating ground glass illuminated by a HeCd laser beam ($\lambda = 0.442$ μm). Assuming that the object is resolvable, we express the intensity image on the photoresist plate as (see e.g., [2.125])

$$\begin{aligned} I(x_1, x_2) &= I_0 m T(mx_1, mx_2) \\ &= I_0 m \theta(mx_2)\theta\big(\zeta(mx_1) + \zeta_0 - mx_2\big), \end{aligned} \tag{2.10.2}$$

where $m = 1/M$ represents the inverse of the magnification of the optical system, and I_0 is a constant.

The plate is then scanned along the x_2 direction. Its exposure (intensity times time) is then proportional to the integral of $I(x_1, x_2)$ along x_2. That is,

$$E(x_1) = K \int I(x_1, x_2)\, dx_2$$
$$= K I_0\big[\zeta(mx_1) + \zeta_0\big], \tag{2.10.3}$$

where K is a constant that depends on the speed of the scan. Then, assuming a linear response of the photoresist, the resulting heights can be written in the generic form

$$z(x_1) = z_0 + \alpha\zeta(mx_1), \tag{2.10.4}$$

where α is a constant that determines the vertical scale of the resulting profile, which can be adjusted through I_0 and the speed of the scan. The lateral scale can be adjusted through the magnification.

Expression (2.10.4) shows that the surface profile function results from a linear transformation of the numerically generated realization of the surface profile function. Based on this, one can find a relationship between the pdf of slopes of the developed plate and the $f(\gamma)$ employed to generate the surface profile function $\zeta(x)$:

$$p_{z'}(x) = \frac{1}{m\alpha} f\left(\frac{x}{m\alpha}\right). \tag{2.10.5}$$

2.11. Experimental Results

In this section, we present experimental results obtained with one-dimensional samples fabricated with the procedure described in Section 2.10. The results demonstrate the possibility of implementing the techniques discussed in the chapter and illustrate some of the issues discussed in Section 2.9 concerning averages. For simplicity we chose to work with uncoated samples in a transmission geometry, but the surfaces can be aluminized if one desires to work in reflection. The similarities and differences between the two geometries are discussed in Section 2.8.

We show results for a one-dimensional sample designed to produce a scattering pattern that is uniform in a predefined angular region, with no scattering outside that range. As discussed in Section 2.10, the angular width of the scattering pattern produced by the developed photoresist plate depends not only on the mask properties, but also on the exposure and the response of the photoresist. Thus, with this technique it is difficult to determine *a priori* the angular range of the plate that will be fabricated but, after only a few trials, a sample that produces uniform scattering in the $-10°$ and $10°$ range was produced.

To evaluate the scattering properties of the samples, we have used the setups illustrated in Fig. 2.51. They comprise, basically, an illumination system and a CCD detector array placed in the far-field of the scattering surface. Far-field conditions are attained through the

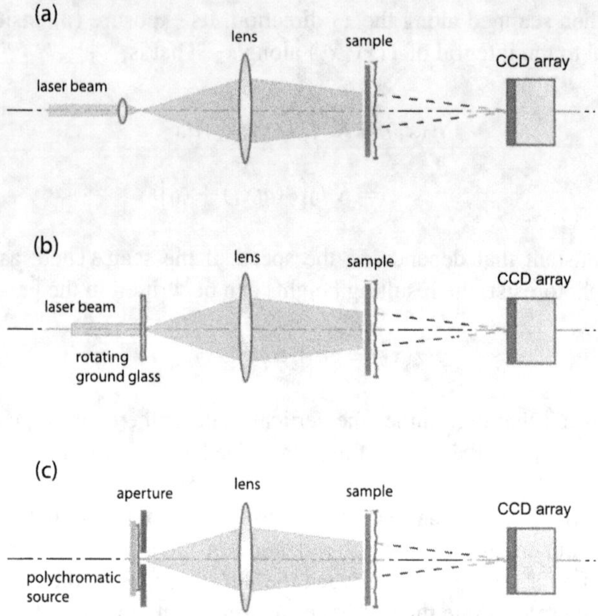

Figure 2.51. Schematic diagrams of the three different optical systems employed to observe the scattering pattern of the fabricated samples. (a) Coherent illumination, (b) partially coherent quasi-monochromatic illumination and, (c) polychromatic illumination.

use of a converging spherical beam whose focal plane coincides with that of the detector array. In the arrangement of Fig. 2.51(a) the sample is illuminated coherently. The scattering pattern obtained under these conditions is shown in Fig. 2.52(a). Since the profile function impressed on the photoresist plate has variations in only one direction, the diffuser should produce scattering in only one plane, which in this case is the horizontal one. One can see that although the scattering pattern presents speckle, its extent is fairly well defined in this horizontal direction. It should be mentioned that the contrast of the intensity fluctuations in the recorded image has already been reduced by integration in the detector pixels. The small amount of scattering observed in the orthogonal (vertical) direction is produced by imperfections in the photoresist plate and the optical system.

It is well known that speckle can be reduced by reducing the coherence of the illumination. With laser light, this can be achieved by passing the beam through a moving diffuser. The pattern shown in Fig. 2.52(b) was recorded with the arrangement shown in Fig. 2.51(b), which contains a ground glass diffuser that was rotated during the exposure. The illumination can then be considered to arise from a pseudo thermal partially coherent source. In this case, the size of the source is determined by the width of the beam, and explains the increased width of the pattern in the vertical direction. One observes that the speckle fluctuations have practically disappeared, and that the resulting pattern has the desired properties of being uniform within the design region. A similar result is obtained

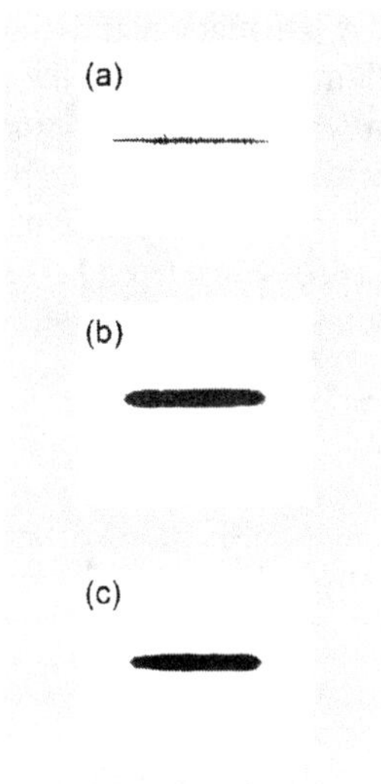

Figure 2.52. Scattering patterns produced by a uniform one-dimensional diffuser for the three cases of illumination illustrated in Fig. 2.51. (a) Coherent illumination, (b) illumination provided by a partially coherent quasi-monochromatic source and, (c) illumination provided by a small polychromatic source. The pictures are shown in reverse contrast and represent a region that is about 30°-wide.

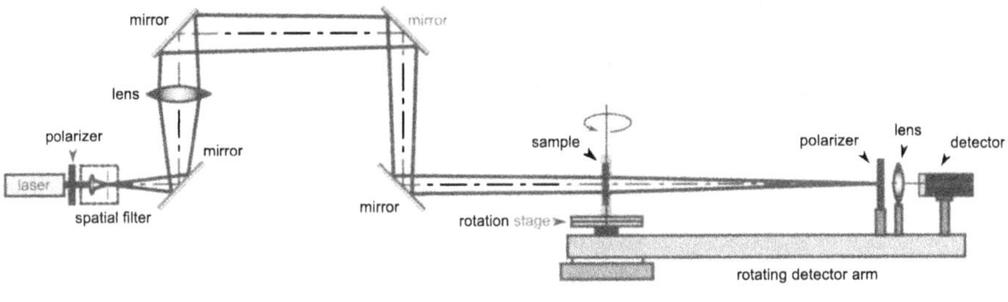

Figure 2.53. Schematic diagram of a scatterometer.

when one uses a small polychromatic source. The result, which was obtained with the arrangement of Fig. 2.51(c), is shown in Fig. 2.52(c).

The patterns of Figs. 2.52(b) and 2.52(c) show that, when illuminated by partially coherent illumination (spatial and/or temporal) the diffuser can be used to produce uniform illumination within a specified angular region.

The arrangements shown in Fig. 2.51 are useful for the characterization of samples that produce only small-angle scattering. For reflective surfaces, or surfaces that scatter light at large angles like the Lambertian ones, it is convenient to use a different setup. The light scattering properties of such surfaces can be determined using purposely built instruments called scatterometers. A schematic diagram of a typical instrument is shown in Fig. 2.53. The use of convergent illumination ensures the far-field condition on the detector, which is placed on a rotating arm whose axis of rotation coincides with the plane of the scattering surface. The speckle fluctuations are reduced by using a detecting aperture much larger than the speckle size. Polarization optical elements can be used to make polarization measurements but, in most of the cases considered here, the surfaces have properties that are independent of the wavelength and the polarization.

References

[2.1] A.E. Danese, Advanced Calculus, vol. I, Allyn and Bacon, Boston, 1965, p. 123.

[2.2] A.A. Maradudin, Interaction of surface polaritons and plasmons with surface roughness, in: V.M. Agranovich, D.L. Mills (Eds.), Surface Polaritons, North-Holland, Amsterdam, 1982, pp. 405–510.

[2.3] A.G. Voronovich, The Kirchhoff and related approximations, in: A.A. Maradudin (Ed.), Light Scattering and Nanoscale Surface Roughness, Springer, New York, 2006, pp. 35–60.

[2.4] J.W. Goodman, Statistical Optics, Wiley, New York, 1985 (Section 3.6.1).

[2.5] D. Middleton, An Introduction to Statistical Communication Theory, McGraw–Hill, New York, 1960, p. 370.

[2.6] W.H. Press, S.A. Teukolsky, W.T. Vetterling, B.P. Flannery, Numerical Recipes in Fortran, second ed., Cambridge Univ. Press, New York, 1992, pp. 281–282.

[2.7] M. Abramowitz, I.A. Stegun, Handbook of Mathematical Functions, US Department of Commerce, Washington, DC, 1972 (entries 9.1.3, 9.1.10, and 9.1.11).

[2.8] J.P. Toporkov, R.T. Marchand, G.S. Brown, On the discretization of the integral equation describing scattering by rough conducting surfaces, IEEE Trans. Antennas Propagat. 46 (1998) 150–161.

[2.9] T.A. Leskova, A.A. Maradudin, E.R. Méndez, J. Muñoz-Lopez, The design and photofabrication of random achromatic optical diffusers for uniform illumination, in: P. Kuchment (Ed.), Waves in Periodic and Random Media, American Mathematical Society, Providence, RI, 2003, pp. 117–140.

[2.10] E.R. Méndez, G. Martinez-Niconoff, A.A. Maradudin, T.A. Leskova, Design and synthesis of random uniform diffusers, Proc. SPIE 3426 (1998) 2–13.

[2.11] A.A. Maradudin, I. Simonsen, T.A. Leskova, E.R. Méndez, Random surfaces that suppress single scattering, Opt. Lett. 24 (1999) 1257–1259.

[2.12] A.A. Maradudin, T.A. Leskova, E.R. Méndez, A.V. Schegrov, The design and fabrication of one-dimensional random surfaces with specified scattering properties, Phys. Solid State 41 (1999) 835–841.

[2.13] I. Simonsen, T.A. Leskova, A.A. Maradudin, E.R. Méndez, Design of one-dimensional Lambertian diffusers of light, Waves Random Media 11 (2001) 419–533.

[2.14] J.C. Stover, Optical Scattering, SPIE Press, Bellingham, WA, 1995.

[2.15] F. Grum, G.W. Luckey, Optical sphere paint and a working standard reflectance, Appl. Opt. 1 (1968) 2239–2295.

[2.16] H.P. Baltes, Progress in inverse optical problems, in: H.P. Baltes (Ed.), Inverse Scattering Problems in Optics, Springer, New York, 1980, pp. 1–13.

[2.17] A.A. Maradudin, T. Michel, A.R. McGurn, E.R. Méndez, Enhanced backscattering of light from a random grating, Ann. Phys. (N.Y.) 203 (1990) 255–307.

[2.18] E. Burstein, A. Hartstein, J. Schoenwald, A.A. Maradudin, D.L. Mills, R.F. Wallis, Surface polaritons—electromagnetic waves at interfaces, in: E. Burstein, F. de Martini (Eds.), Polaritons, Pergamon, New York, 1974, pp. 89–110.

[2.19] E. Kröger, E. Kretschmann, Surface plasmon and polariton dispersion at rough boundaries, Phys. Stat. Solidi (b) 76 (1976) 515–523.

[2.20] A.A. Maradudin, W. Zierau, Effects of surface roughness on the surface-polariton dispersion relation, Phys. Rev. B 14 (1976) 484–499.

[2.21] F. Toigo, A. Marvin, V. Celli, N.R. Hill, Optical properties of rough surfaces: General theory and the small roughness limit, Phys. Rev. B 15 (1977) 5618–5626.

[2.22] D. Sornette, L. Macon, J. Coste, Transfer matrix theory of leaky guided waves, J. Phys. (France) 49 (1988) 1683.

[2.23] J.-P. Desideri, L. Macon, D. Sornette, Observation of critical modes in quasiperiodic systems, Phys. Rev. Lett. 63 (1989) 390–393.

[2.24] V.D. Freilikher, S.A. Gredeskul, Localization of waves in media with one-dimensional disorder, in: E. Wolf (Ed.), Progress in Optics, vol. XXX, North-Holland, Amsterdam, 1992, pp. 139–203.

[2.25] F. Pincemin, J.-J. Greffet, Propagation and localization of a surface plasmon polariton on a finite grating, J. Opt. Soc. Am. B 13 (1996) 1499–1509.

[2.26] J.A. Sánchez-Gil, A.A. Maradudin, Competition between Anderson localization and leakage of surface-plasmon polaritons on randomly rough periodic metal surfaces, Phys. Rev. B 56 (1997) 1103–1106.

[2.27] A.A. Maradudin, I. Simonsen, T.A. Leskova, E.R. Méndez, Localization of surface plasmon polaritons on a random surface, Physica B 296 (2001) 85–97.

[2.28] Lord Rayleigh, The Theory of Sound, vol. II, second ed., Macmillan, London, 1876, pp. 89–96, 297–311.

[2.29] R. Petit, M. Cadilhac, Sur la diffraction d'une onde plane par un réseau infiniment conducteur, C. R. Acad. Sci. B 262 (1966) 468.

[2.30] R.F. Millar, Singularities of two-dimensional exterior solutions of the Helmholtz equation, Proc. Camb. Philos. Soc. 69 (1971) 175–188.

[2.31] R.F. Millar, On the Rayleigh assumption in scattering by a periodic surface. II, Proc. Camb. Philos. Soc. 69 (1971) 217–225.

[2.32] N.R. Hill, V. Celli, Limits of convergence of the Rayleigh method for surface scattering, Phys. Rev. B 17 (1978) 2478–2481.

[2.33] C.S. West, K.A. O'Donnell, Observations of backscattering enhancement from polaritons on a rough metal surface, J. Opt. Soc. Am. A 12 (1995) 390–397.

[2.34] H. Ogura, Z.L. Wang, Y. Sasakura, V. Freilikher, Localization of surface plasmon waves on the surface of a random rough metallic grating with a narrow-band spectrum, Opt. Commun. 134 (1997) 1–6.

[2.35] R.M. Axline, A.K. Fung, Numerical computation of scattering from a perfectly conducting random surface, IEEE Trans. Antennas Propagat. AP-26 (1978) 482–488;
R.M. Axline, A.K. Fung, IEEE Trans. Antennas Propagat. AP-28 (1980) 949.

[2.36] N. García, E. Stoll, Monte Carlo calculation for electromagnetic-wave scattering from random rough surfaces, Phys. Rev. Lett. 52 (1984) 1798–1801.

[2.37] E.I. Thorsos, The validity of the Kirchhoff approximation for rough surface scattering using a Gaussian roughness spectrum, J. Acoust. Soc. Am. 83 (1988) 78–92.

[2.38] Ref. [2.17], Appendix A.

[2.39] V. Freilikher, E. Kanzieper, A.A. Maradudin, Coherent scattering enhancement in systems bounded by rough surfaces, Phys. Rep. 288 (1997) 127–204 (Appendix A).

[2.40] R. Kubo, Generalized cumulant expansion method, J. Phys. Soc. Jpn. 17 (1962) 1100–1120.

[2.41] S.M. Rytov, Yu.A. Kravtsov, V.I. Tatarskii, Principles of Statistical Radiophysics, vol. 3, Springer, New York, 1989, p. 29.

[2.42] A. Stuart, J. Keith Ord, Kendall's Advanced Theory of Statistics, vol. 1, fifth ed., Charles Griffin, London, 1987, p. 84ff.

[2.43] P.B. Johnson, R.W. Christy, Optical constants of the noble metals, Phys. Rev. B 6 (1972) 4370–4379.

[2.44] A.R. McGurn, A.A. Maradudin, V. Celli, Localization effects in the scattering of light from a randomly rough grating, Phys. Rev. B 31 (1985) 4866–4871.

[2.45] D.L. Mills, A.A. Maradudin, Attenuation of Rayleigh surface waves by surface roughness, Ann. Phys. (N.Y.) 100 (1976) 262–309.

[2.46] J.A. Sánchez-Gil, A.A. Maradudin, Near-field and far-field scattering of surface plasmon polaritons by one-dimensional surface defects, Phys. Rev. B 60 (1999) 8359–8367.

[2.47] J.A. Sánchez-Gil, Surface defect scattering of surface plasmon polaritons: Mirrors and light emitters, Appl. Phys. Lett. 73 (1998) 3509–3511.

[2.48] J.A. Sánchez-Gil, A.A. Maradudin, Resonant scattering of surface-plasmon polariton pulses by nanoscale metal defects, Opt. Lett. 28 (2003) 2255–2257.

[2.49] J.A. Sánchez-Gil, A.A. Maradudin, Dynamic near-field calculations of surface-plasmon polariton pulses resonantly scattered at sub-micron metal defects, Opt. Express 12 (2004) 883–894.

[2.50] J.A. Sánchez-Gil, A.A. Maradudin, Surface-plasmon polariton scattering from a finite array of nanogrooves/ridges: Efficient mirrors, Appl. Phys. Lett. 86 (2005) 251106.

[2.51] A.Yu. Nikitin, F. López-Tejeira, L. Martín-Moreno, Scattering of surface plasmon polaritons by one-dimensional inhomogeneities, Phys. Rev. B 75 (2007) 035129.

[2.52] A.Yu. Nikitin, L. Martín-Moreno, Scattering coefficients of surface plasmon polaritons impinging at oblique incidence onto one-dimensional surface relief defects, Phys. Rev. B 75 (2007) 81405.

[2.53] See e.g. R. Petit, Electromagnetic Theory of Gratings, Springer, Berlin, 1980.

[2.54] A. Marechal, G.W. Stroke, Sur l'origine des effets de polarisation et de diffraction dans les réseaux optiques, C. R. Acad. Sci. 249 (1959) 2042–2044.

[2.55] A. Hessel, J. Chmoys, D.Y. Tseng, Bragg-angle blazing of diffraction gratings, J. Opt. Soc. Am. 65 (1975) 380–384.

[2.56] D. Maystre, M. Cadilhac, J. Chandezon, Gratings: A phenomenological approach and its applications, perfect blazing in a non-zero deviation mounting, Optica Acta 28 (1981) 457–470.

[2.57] D. Maystre, M. Saillard, Enhanced backscattering and blazing effect from gratings, quasi-gratings and randomly rough surfaces, Waves Random Media 4 (1984) 467–485.

[2.58] M. Ciftan, T.A. Leskova, A.A. Maradudin, The fabrication of one-dimensional random surfaces that display enhanced backscattering for only one specified angle of incidence, Proc. SPIE 4447 (2001) 17–23.

[2.59] C.S. West, K.A. O'Donnell, Backscattering enhancement from plasmon polaritons on a deterministic metal surface, Opt. Commun. 123 (1996) 109–114.

[2.60] R. Goody, Cross correlating spectrometer, J. Opt. Soc. Am. 48 (1968) 900–908.

[2.61] H.O. Edwards, J.P. Daikin, Gas sensors using correlation spectroscopy compatible with fibreoptic operation, Sensors Actuators B 11 (1993) 9–19.

[2.62] F.W. Taylor, J.J. Houghton, G.D. Peskett, C.D. Rogers, E.J. Williamson, Radiometer for remote sounding of the upper atmosphere, Appl. Opt. 11 (1972) 135–141.

[2.63] M.B. Sinclair, M.A. Butler, A.J. Ricco, S.D. Senturia, Synthetic spectra: A tool for correlation spectroscopy, Appl. Opt. 36 (1997) 3342–3348.

[2.64] M.B. Sinclair, M.A. Butler, S.H. Kravitz, W.J. Zubrzycki, A.J. Ricco, Synthetic infrared spectra, Opt. Lett. 22 (1997) 1036–1038.

[2.65] G. Zhou, F.E.H. Tay, F.S. Chau, Design of the diffractive optical elements for synthetic spectra, Opt. Express 11 (2003) 1392–1399.

[2.66] R. Belikov, O. Solgaard, Optical wavelength filtering by diffraction from a surface relief, Opt. Lett. 28 (2003) 447–449.

[2.67] J.W. Goodman, Introduction to Fourier Optics, McGraw–Hill, New York, 1968, pp. 57–74.

[2.68] R.W. Gerchberg, W.O. Saxton, A practical algorithm for the determination of phase from image and diffraction plane pictures, Optik 35 (1972) 237–246.

[2.69] J.R. Fienup, Phase retrieval algorithms: A comparison, Appl. Opt. 21 (1982) 2758–2769.

[2.70] F. Wyrowski, O. Bryngdahl, Iterative Fourier-transform algorithm applied to computer holography, J. Opt. Soc. A 5 (1988) 1058–1065.

[2.71] T. Peter, F. Wyrowski, O. Bryngdahl, Comparison of iterative methods to calculate quantized digital holograms, in: F. Wyrowski (Ed.), Workshop on Digital Holography, Proc. SPIE 1718 (1992) 55–62.

[2.72] M.A. Wolf, Numerical Methods for Unconstrained Optimization: An Introduction, Van Nostrand Reinhold Company, Ltd., New York, 1978, pp. 161–167 (Chapter 6).

[2.73] A. Mendoza-Suárez, E.R. Méndez, Light scattering by a reentrant fractal surface, Appl. Opt. 36 (1997) 3521–3531.

[2.74] I. Simonsen, private communication, 2007.

[2.75] Ref. [2.7], p. 364, Formula 9.2.3.

[2.76] T.M. Apostol, Mathematical Analysis, Addison–Wesley, Reading, MA, 1964, pp. 169–176.
[2.77] Ref. [2.76], p. 315.
[2.78] A. Mendoza-Suárez, Métodos rigurosos para el esparcimiento de luz por superficies rugosas y medios estratificados con perfiles arbitrarios, Ph.D. Dissertation, Centro de Investigación Científica y de Educación Superior de Ensenada, 1996.
[2.79] http://vpl.ipac.caltech.edu/spectra/.
[2.80] See, for example, J. Mathews, R.L. Walker, Mathematical Methods of Physics, second ed., Addison–Wesley, Reading, MA, 1970, p. 90.
[2.81] See, for example, L. Tsang, J.A. Kong, R.T. Shin, Theory of Microwave Remote Sensing, Wiley, New York, 1985.
[2.82] A.A. Maradudin, R.E. Luna, E.R. Méndez, The Brewster effect for a one-dimensional random surface, Waves Random Media 3 (1993) 51–60.
[2.83] C. Baylard, J.-J. Greffet, A.A. Maradudin, Coherent reflection factor of a random rough surface: Application to the study of the shift of the Brewster angle, J. Opt. Soc. Am. A 10 (1993) 2637–2647.
[2.84] M.E. Veysoglu, H.A. Yuch, R.T. Shin, J.A. Kong, Polarimetric passive remote sensing of periodic surfaces, J. Electromagn. Waves Appl. 5 (1991) 267–280.
[2.85] H. Sai, H. Yugami, Y. Akiyama, Y. Kanamori, K. Hane, Spectral control of thermal emission by periodic microstructured surfaces in the near-infrared region, J. Opt. Soc. Am. A 18 (2001) 1471–1476.
[2.86] J.T. Johnson, R.T. Shin, J.A. Kong, L. Tsang, K. Pak, A numerical study of ocean polarimetric thermal emission, IEEE Trans. Geosci. Remote Sensing 37 (1999) 8–20.
[2.87] K.S. Chen, T.-D. Wu, L. Tsang, Q. Li, J. Shi, A.K. Fung, Emission of rough surfaces calculated by the integral equation method with comparison to three-dimensional moment method simulations, IEEE Trans. Geosci. Remote Sensing 41 (2003) 90–101.
[2.88] J. LeGall, M. Olivier, J.-J. Greffet, Experimental and theoretical study of reflection and coherent thermal expansion by a SiC grating supporting a surface-phonon polariton, Phys. Rev. B 55 (1997) 10105–10114.
[2.89] A. Heinzel, V. Boerner, A. Gombert, B. Bläsi, V. Wittwer, J. Luther, Radiation filters and emitters for the NIR based on periodically structured metal surfaces, J. Mod. Opt. 47 (2000) 2399–2419.
[2.90] L. Mandel, E. Wolf, Optical Coherence and Quantum Optics, Cambridge University Press, New York, 1995 (Section 4.3.2).
[2.91] R.E. Kinzly, Partially coherent imaging in a microdensitometer, J. Opt. Soc. Am. 62 (1972) 386–394.
[2.92] G.O. Reynolds, A.W. Smith, Experimental demonstration of coherence effects and linearity in microdensitometry, Appl. Opt. 12 (1953) 1259–1270.
[2.93] D. Nyyssonen, Linewidth measurement with an optical microscope: The effect of operating conditions on the image profile, Appl. Opt. 16 (1977) 2223–2230.
[2.94] D. Nyyssonen, Spatial coherence: The key to accurate optical metrology, in: W.H. Carter (Ed.), Applications of Optical Coherence, SPIE 194 (1979) 34–44.
[2.95] W.G. Oldham, S. Subramanian, A.R. Neureuther, Optical requirements for projection lithography, Solid State Electr. 24 (1981) 975–980.
[2.96] J. Wu, Propagation of a Gaussian–Schell beam through turbulent media, J. Mod. Opt. 37 (1990) 671–684.
[2.97] J. Wu, A.D. Boardman, Coherence length of a Gaussian–Schell model beam and atmospheric turbulence, J. Mod. Opt. 38 (1991) 1355–1363.
[2.98] M. Salem, T. Shirai, A. Dogariu, E. Wolf, Long-distance propagation of partially coherent beams through atmospheric turbulence, Opt. Commun. 216 (2003) 261–265.
[2.99] A.C. Schell, The Multiple Plate Antenna, Ph.D. Dissertation, Massachusetts Institute of Technology, 1961 (Section 7.5).
[2.100] Ref. [2.90], Section 5.4.3.
[2.101] J.D. Farina, L.M. Narducci, E. Collett, Generation of highly directional beams from a globally incoherent source, Opt. Commun. 32 (1980) 203–208.
[2.102] F. Scudieri, M. Bertolotti, R. Bartolino, Light scattered by a liquid crystal: A new quasi-thermal source, Appl. Opt. 13 (1974) 181–185.
[2.103] W.H. Carter, M. Bertolotti, An analysis of the far-field coherence and radiant intensity of light scattered from liquid crystals, J. Opt. Soc. Am. 68 (1978) 329–333.

[2.104] D. Courjon, J. Bulabois, Modifications of the coherence properties of a light beam: Applications in optical processing, in: W.H. Carter (Ed.), Applications of Optical Coherence, SPIE 194 (1979) 129–134.
[2.105] D. Courjon, J. Bulabois, W.H. Carter, Use of a holographic filter to modify the coherence of a light field, J. Opt. Soc. 71 (1981) 469–473.
[2.106] Y. Ohtsuka, Y. Imai, Partial coherence controlled by a progressive ultrasonic wave, J. Opt. Soc Am. 69 (1979) 684–689.
[2.107] Y. Imai, Y. Ohtsuka, Optical coherence modulation by ultrasonic waves. 1: Dependence of partial coherence on ultrasonic parameters, Appl. Opt. 19 (1980) 542–547.
[2.108] J. Turunen, E. Tervonen, A.T. Friberg, Acousto-optic control and modulation of optical coherence by electronically synthesized holographic gratings, J. Appl. Phys. 67 (1990) 49–59.
[2.109] E. Tervonen, A.T. Friberg, J. Turunen, Gaussian Schell-model beams generated with synthetic acousto-optic holograms, J. Opt. Soc. Am. A 9 (1992) 796–803.
[2.110] T.E. Horton, J.H. McDermit, Design of a specular aspheric surface to uniformly radiate a flat surface using a nonuniform collimated radiation source, J. Heat Transfer Trans. ASME C 94 (1972) 453–458.
[2.111] P.W. Rhodes, D.L. Sealy, Refractive optical systems for irradiance redistribution of collimated radiation: Their design and analysis, Appl. Opt. 19 (1980) 3545–3553.
[2.112] W.-H. Lee, Method for converting a Gaussian laser beam into a uniform beam, Opt. Commun. 36 (1981) 469–471.
[2.113] W.B. Veldkamp, Laser beam profile shaping with interlaced binary diffraction gratings, Appl. Opt. 21 (1982) 3209–3212.
[2.114] C.C. Aleksoff, K.K. Ellis, B.D. Neagle, Holographic conversion of a Gaussian beam to a near-field uniform beam, Opt. Eng. 3 (1991) 537–543.
[2.115] L.A. Romero, F.M. Dickey, Lossless laser beam shaping, J. Opt. Soc. Am. A 13 (1996) 751–760.
[2.116] J.M. Stone, Radiation and Optics, McGraw–Hill, New York, 1963 (Section 12-6).
[2.117] See, for example, C.-W. Tsai, Y.-C. Chang, G.Sh. Shmavonyan, Y.-S. Su, C.-F. Lin, Extremely broadband superluminescent didodes/semiconductor optical amplifiers in optical communications band, SPIE 4989 (2003) 69–77.
[2.118] M.C. Hutley, Interference (holographic) diffraction gratings, J. Phys. E: Sci. Instrum. 9 (1976) 513–520.
[2.119] M.C. Hutley, Diffraction Gratings, Academic Press, New York, 1982.
[2.120] P.F. Gray, A method of forming optical diffusers of simple known statistical properties, Optica Acta 25 (1978) 765–775.
[2.121] E.R. Méndez, K.A. O'Donnell, Observation of depolarization and backscattering enhancement in light scattering from Gaussian random surfaces, Opt. Commun. 61 (1987) 91–95.
[2.122] M.J. Kim, J.C. Dainty, A.T. Friberg, A.J. Sant, Experimental study of enhanced backscattering from one- and two-dimensional random rough surfaces, J. Opt. Soc. Am. A 7 (1990) 569–577.
[2.123] K.A. O'Donnell, M.E. Knotts, Polarization-dependence of scattering from one-dimensional rough surfaces, J. Opt. Soc. Am. A 9 (1992) 585–596.
[2.124] R.E. Luna, E.R. Mendez, J.Q. Lu, Z.H. Gu, Enhanced backscattering due to total internal reflection at a dielectric–air interface, J. Mod. Opt. 42 (1995) 257–269.
[2.125] J.W. Goodman, Introduction to Fourier Optics, second ed., McGraw–Hill, 1996.

Chapter 3

TWO-DIMENSIONAL SURFACES

In this chapter we turn to a description of approaches to the design of a two-dimensional randomly rough surface that produces a prescribed angular or spatial dependence of the scattered field when it is illuminated by a scalar plane wave or by a scalar beam. In these approaches it will be assumed that the randomly rough surface is impenetrable. In particular, it will be assumed that the total field in the vacuum region above the surface satisfies the Dirichlet boundary condition on the surface, namely that it vanishes on the surface. This assumption reflects the difficulties encountered at the present time in applying a vector theory of the scattering of an electromagnetic field from a two-dimensional randomly rough surface of a penetrable medium to the design of a surface that produces not only a specified angular dependence of the intensity of the scattered field, but also a specified angular dependence of its polarization. It is hoped that the treatment of the simpler problem addressed here will stimulate efforts to solve the latter more difficult problem.

In Section 3.1 we describe three approaches to the design of a two-dimensional randomly rough surface which when illuminated by a scalar plane wave produces a scattered field with a specified *angular* dependence of its mean intensity. In proceeding from the simplest and most restricted approaches, in terms of the kinds of scattering patterns they can produce, to the most flexible and general approach, we follow the historical development of the subject, which also reflects the growth of our own understanding of this class of inverse scattering problems.

In Section 3.2 we consider the problem of designing a two-dimensional randomly rough surface that produces a scattered field with a prescribed *spatial* dependence of its mean intensity. Specifically, we seek a surface in the x_1x_2 plane that produces a specified dependence of the mean intensity of the scattered field along the x_3 axis. It is then shown that such a surface can be used to extend the depth of focus of imaging systems.

Finally, one of the approaches developed in Section 3.1 is applied in Section 3.3 to the design of a two-dimensional random surface that scatters a scalar Gaussian beam incident normally on it in such a way that the two-point amplitude correlation function of the scattered field has a prescribed spatial dependence in the plane of the surface. An alternative approach to the solution of this problem is also presented in this section, and has the attractive feature of generating surfaces that can be fabricated by well-known experimental techniques. The scattered field generated by both of these approaches, which is partially coherent, is shown to produce the same far-field intensity distribution as a completely coherent laser source.

These examples illustrate the versatility of the approaches described for designing surfaces with different types of specified scattering properties, and indicate some of the kinds of applications in which specially designed surfaces can be used.

3.1. The Design of Two-Dimensional Randomly Rough Surfaces that Produce a Scattered Field with a Specified Angular Dependence of its Mean Intensity

A convenient starting point for the generation of two-dimensional randomly rough surfaces with specified scattering properties is the expression for the mean differential reflection coefficient in the case of the scattering of a scalar plane wave from the two-dimensional randomly rough surface of an impenetrable medium. The physical system we consider, therefore, consists of vacuum in the region $x_3 > \zeta(\mathbf{x}_\parallel)$, where $\mathbf{x}_\parallel = (x_1, x_2, 0)$ is a position vector in the plane $x_3 = 0$, and the scattering medium in the region $x_3 < \zeta(\mathbf{x}_\parallel)$. The surface profile function $\zeta(\mathbf{x}_\parallel)$ is assumed to be a single-valued function of $\mathbf{x}_\parallel$ that is differentiable with respect to x_1 and x_2, and constitutes a random process, but not necessarily a stationary one. The surface $x_3 = \zeta(\mathbf{x}_\parallel)$ is illuminated from the vacuum by a scalar plane wave.

3.1.1. The Scattering Problem

The problem, therefore, that concerns us is that of finding the solution of the scalar wave equation

$$\nabla^2 \Psi(\mathbf{x}; t) = \frac{1}{c^2}\frac{\partial^2}{\partial t^2}\Psi(\mathbf{x}; t) \tag{3.1.1}$$

in the vacuum region $x_3 > \zeta(\mathbf{x}_\parallel)$. In Eq. (3.1.1) $\mathbf{x} = (x_1, x_2, x_3)$, and c is the speed of light in vacuum. We assume that the field $\Psi(\mathbf{x}; t)$ is real and monochromatic of angular frequency ω, so that it has the form

$$\begin{aligned}\Psi(\mathbf{x}; t) &= \mathrm{Re}\{\psi(\mathbf{x}|\omega)\exp(-i\omega t)\} \\ &= \frac{1}{2}[\psi(\mathbf{x}|\omega)\exp(-i\omega t) + \psi^*(\mathbf{x}|\omega)\exp(i\omega t)],\end{aligned} \tag{3.1.2}$$

where Re denotes the real part. Substitution of Eq. (3.1.2) into Eq. (3.1.1) shows that the complex amplitude of the wave $\psi(\mathbf{x}|\omega)$ satisfies the Helmholtz equation

$$[\nabla^2 + (\omega/c)^2]\psi(\mathbf{x}|\omega) = 0. \tag{3.1.3}$$

If the operations on $\Psi(\mathbf{x}; t)$ are linear, one can drop the symbol Re in Eq. (3.1.2) and work directly with the complex function $\psi(\mathbf{x}|\omega)\exp(-i\omega t)$. The real part of the resulting expression is then understood to represent the physical quantity of interest [3.2]. However, in working with expressions that involve nonlinear operations, such as multiplying two fields, one must in general take the real parts first and operate with these alone.

We assume that the field $\psi(\mathbf{x}|\omega)$ satisfies either (a) the Dirichlet boundary condition

$$\psi(\mathbf{x}|\omega)\Big|_{x_3=\zeta(\mathbf{x}_\|)} = 0, \tag{3.1.4}$$

or (b) the Neumann boundary condition

$$\frac{\partial}{\partial n}\psi(\mathbf{x}|\omega)\Bigg|_{x_3=\zeta(\mathbf{x}_\|)} = 0 \tag{3.1.5}$$

on the rough surface $x_3 = \zeta(\mathbf{x}_\|)$. In Eq. (3.1.5) $\partial/\partial n$ is the derivative along the normal to this surface at each point, directed into the vacuum,

$$\frac{\partial}{\partial n} = \frac{1}{[1+(\nabla\zeta(\mathbf{x}_\|))^2]^{\frac{1}{2}}}\left(-\frac{\partial\zeta(\mathbf{x}_\|)}{\partial x_1}\frac{\partial}{\partial x_1} - \frac{\partial\zeta(\mathbf{x}_\|)}{\partial x_2}\frac{\partial}{\partial x_2} + \frac{\partial}{\partial x_3}\right). \tag{3.1.6}$$

In addition, we require that $\psi(\mathbf{x}|\omega)$ be the sum of an incoming incident wave and outgoing scattered waves at $x_3 = \infty$.

We introduce the free-space scalar Green's function $g_0(\mathbf{x}|\mathbf{x}')$ that satisfies the equation

$$\left(\nabla^2 + (\omega/c)^2\right)g_0(\mathbf{x}|\mathbf{x}') = -4\pi\delta(\mathbf{x}-\mathbf{x}'), \tag{3.1.7}$$

subject to an outgoing wave boundary condition at infinity. It has the representations

$$g_0(\mathbf{x}|\mathbf{x}') = \frac{\exp[i(\omega/c)|\mathbf{x}-\mathbf{x}'|]}{|\mathbf{x}-\mathbf{x}'|} \tag{3.1.8a}$$

$$= \int \frac{d^2k_\|}{(2\pi)^2}\frac{2\pi i}{\alpha_0(k_\|)}\exp\left[i\mathbf{k}_\|\cdot(\mathbf{x}_\|-\mathbf{x}'_\|) + i\alpha_0(k_\|)|x_3-x'_3|\right], \tag{3.1.8b}$$

where

$$\alpha_0(k_\|) = \left[(\omega/c)^2 - k_\|^2\right]^{\frac{1}{2}} \quad k_\| < \omega/c \tag{3.1.9a}$$

$$= i\left[k_\|^2 - (\omega/c)^2\right]^{\frac{1}{2}} \quad k_\| > \omega/c. \tag{3.1.9b}$$

3.1.1.1. The Incident and Scattered Fields We now apply Green's second integral identity [3.1]

$$\int_\Omega d^3x\,\left(u\nabla^2 v - v\nabla^2 u\right) = \int_\Sigma dS\left(u\frac{\partial v}{\partial \nu} - v\frac{\partial u}{\partial \nu}\right), \tag{3.1.10}$$

where $u(\mathbf{x})$ and $v(\mathbf{x})$ are arbitrary scalar functions of $\mathbf{x}$ defined in a volume Ω that is bounded by a closed surface Σ. The derivative $\partial/\partial\nu$ is taken along the normal to the surface Σ at each point, directed away from the volume Ω. We assume that the volume Ω is the region $x_3 > \zeta(\mathbf{x}_\parallel)$, while the surface Σ is the union of the rough surface $x_3 = \zeta(\mathbf{x}_\parallel)$, which we denote by S, and a hemispherical cap of infinite radius in the upper half space, which we denote by $S^{(+\infty)}$. Then, if we set $u(\mathbf{x}) = \psi(\mathbf{x}|\omega)$ and $v(\mathbf{x}) = g_0(\mathbf{x}|\mathbf{x}')$ in Eq. (3.1.10), and use Eqs. (3.1.3) and (3.1.7), we obtain

$$\begin{aligned} -4\pi\theta\big(x_3' - \zeta(\mathbf{x}_\parallel')\big)\psi(\mathbf{x}'|\omega) &= -\int_S dS\left[\psi(\mathbf{x}|\omega)\frac{\partial}{\partial n}g_0(\mathbf{x}|\mathbf{x}') - g_0(\mathbf{x}|\mathbf{x}')\frac{\partial}{\partial n}\psi(\mathbf{x}|\omega)\right] \\ &\quad + \int_{S^{(+\infty)}} dS\left[\psi(\mathbf{x}|\omega)\frac{\partial}{\partial \nu}g_0(\mathbf{x}|\mathbf{x}') - g_0(\mathbf{x}|\mathbf{x}')\frac{\partial}{\partial \nu}\psi(\mathbf{x}|\omega)\right], \end{aligned} \tag{3.1.11}$$

where $\theta(x)$ is the Heaviside unit step function. Because the scattered field will satisfy a radiation condition at infinity, its contribution to the integral over the hemispherical cap of infinite radius $S^{(+\infty)}$ on the right-hand side of Eq. (3.1.11) vanishes. The latter therefore yields $-4\pi\psi(\mathbf{x}'|\omega)_{\text{inc}}$, where $\psi(\mathbf{x}|\omega)_{\text{inc}}$ is the incident field. The incident field $\psi(\mathbf{x}|\omega)_{\text{inc}}$ satisfies the Helmholtz equation

$$\big(\nabla^2 + (\omega/c)^2\big)\psi(\mathbf{x}|\omega)_{\text{inc}} = 0. \tag{3.1.12}$$

We will represent it in the form of an incoming plane wave,

$$\psi(\mathbf{x}|\omega)_{\text{inc}} = \exp\big[i\mathbf{k}_\parallel \cdot \mathbf{x}_\parallel - i\alpha_0(k_\parallel)x_3\big]. \tag{3.1.13}$$

We can then rewrite Eq. (3.1.11) as

$$\begin{aligned} \theta\big(x_3 - \zeta(\mathbf{x}_\parallel)\big)\psi(\mathbf{x}|\omega) &= \psi(\mathbf{x}|\omega)_{\text{inc}} \\ &\quad + \frac{1}{4\pi}\int_S dS'\left[\left(\frac{\partial}{\partial n'}g_0(\mathbf{x}|\mathbf{x}')\right)\psi(\mathbf{x}'|\omega) - g_0(\mathbf{x}|\mathbf{x}')\frac{\partial}{\partial n'}\psi(\mathbf{x}'|\omega)\right]. \end{aligned} \tag{3.1.14}$$

In writing this equation we have used the symmetry of $g_0(\mathbf{x}|\mathbf{x}')$ in $\mathbf{x}$ and $\mathbf{x}'$.

We have assumed that the surface profile function $\zeta(\mathbf{x}_\parallel)$ is a single-valued function of $\mathbf{x}_\parallel$. We can therefore replace integration over the surface S by integration over the plane $x_3 = 0$, with the aid of the relation

$$dS = \big[1 + \big(\nabla\zeta(\mathbf{x}_\parallel)\big)^2\big]^{\frac{1}{2}}\, d^2x_\parallel. \tag{3.1.15}$$

Thus, Eq. (3.1.14) can be rewritten as

$$\theta\big(x_3-\zeta(\mathbf{x}_\parallel)\big)\psi(\mathbf{x}|\omega)=\psi(\mathbf{x}|\omega)_{\mathrm{inc}}+\frac{1}{4\pi}\int d^2x'_\parallel\,\big\{\big[\partial g_0(\mathbf{x}|\mathbf{x}')/\partial N'\big]_{x'_3=\zeta(\mathbf{x}'_\parallel)}H(\mathbf{x}'_\parallel|\omega)$$
$$-\big[g_0(\mathbf{x}|\mathbf{x}')\big]_{x'_3=\zeta(\mathbf{x}'_\parallel)}L(\mathbf{x}'_\parallel|\omega)\big\}, \tag{3.1.16}$$

where we have introduced the un-normalized normal derivative $\partial/\partial N$,

$$\frac{\partial}{\partial N}=-\frac{\partial\zeta(\mathbf{x}_\parallel)}{\partial x_1}\frac{\partial}{\partial x_1}-\frac{\partial\zeta(\mathbf{x}_\parallel)}{\partial x_2}\frac{\partial}{\partial x_2}+\frac{\partial}{\partial x_3}, \tag{3.1.17}$$

and have defined the source functions $H(\mathbf{x}_\parallel\omega)$ and $L(\mathbf{x}_\parallel|\omega)$ by

$$H(\mathbf{x}_\parallel|\omega)=\psi(\mathbf{x}|\omega)\big|_{x_3=\zeta(\mathbf{x}_\parallel)}, \tag{3.1.18}$$

$$L(\mathbf{x}_\parallel|\omega)=\big[\partial\psi(\mathbf{x}|\omega)/\partial N\big]\big|_{x_3=\zeta(\mathbf{x}_\parallel)}. \tag{3.1.19}$$

When $\psi(\mathbf{x}|\omega)$ satisfies the Dirichlet boundary condition, Eq. (3.1.4), then in view of Eq. (3.1.18), Eq. (3.1.16) becomes

$$\theta\big(x_3-\zeta(\mathbf{x}_\parallel)\big)\psi(\mathbf{x}|\omega)=\psi(\mathbf{x}|\omega)_{\mathrm{inc}}$$
$$-\frac{1}{4\pi}\int d^2x'_\parallel\,\big[g_0(\mathbf{x}|\mathbf{x}')\big]_{x'_3=\zeta(\mathbf{x}'_\parallel)}L(\mathbf{x}'_\parallel|\omega). \tag{3.1.20}$$

The scattered field is given by the second term on the right-hand side of this equation

$$\psi(\mathbf{x}|\omega)_{\mathrm{sc}}=-\frac{1}{4\pi}\int d^2x'_\parallel\,\big[g_0(\mathbf{x}|\mathbf{x}')\big]_{x'_3=\zeta(\mathbf{x}'_\parallel)}L(\mathbf{x}'_\parallel|\omega). \tag{3.1.21}$$

In the far field, $x_3\gg\zeta(\mathbf{x}_\parallel)_{\max}$, the use of Eq. (3.1.8b) enables us to rewrite Eq. (3.1.21) in the form

$$\psi(\mathbf{x}|\omega)_{\mathrm{sc}}=\int\frac{d^2q_\parallel}{(2\pi)^2}R_D(\mathbf{q}_\parallel|\mathbf{k}_\parallel)\exp\big[i\mathbf{q}_\parallel\cdot\mathbf{x}_\parallel+i\alpha_0(q_\parallel)x_3\big], \tag{3.1.22}$$

where the scattering amplitude $R_D(\mathbf{q}_\parallel|\mathbf{k}_\parallel)$ is

$$R_D(\mathbf{q}_\parallel|\mathbf{k}_\parallel)=-\frac{i}{2\alpha_0(q_\parallel)}\int d^2x_\parallel\,L(\mathbf{x}_\parallel|\omega)\exp\big[-i\mathbf{q}_\parallel\cdot\mathbf{x}_\parallel-i\alpha_0(q_\parallel)\zeta(\mathbf{x}_\parallel)\big]. \tag{3.1.23}$$

The dependence of $R_D(\mathbf{q}_\parallel|\mathbf{k}_\parallel)$ on $\mathbf{k}_\parallel$ is through the dependence of the source function $L(\mathbf{x}_\parallel|\omega)$ on the incident field (3.1.13).

When $\psi(\mathbf{x}|\omega)$ satisfies the Neumann boundary condition, Eq. (3.1.5), then in view of Eqs. (3.1.6), (3.1.15) and (3.1.19), Eq. (3.1.16) becomes

$$\theta\big(x_3 - \zeta(\mathbf{x}_\parallel)\big)\psi(\mathbf{x}|\omega) = \psi(\mathbf{x}|\omega)_{\text{inc}} + \frac{1}{4\pi}\int d^2x'_\parallel \big[\partial g_0(\mathbf{x}|\mathbf{x}')/\partial N'\big]_{x'_3=\zeta(\mathbf{x}'_\parallel)} H(\mathbf{x}'_\parallel|\omega). \quad (3.1.24)$$

Thus, the scattered field is now given by

$$\psi(\mathbf{x}|\omega)_{\text{sc}} = \frac{1}{4\pi}\int d^2x'_\parallel \big[\partial g_0(\mathbf{x}|\mathbf{x}')/\partial N'\big]_{x'_3=\zeta(\mathbf{x}'_\parallel)} H(\mathbf{x}'_\parallel|\omega). \quad (3.1.25)$$

In the far-field region, $x_3 \gg \zeta(\mathbf{x}_\parallel)_{\max}$, the use of Eq. (3.1.8b) enables us to rewrite Eq. (3.1.25) in the form

$$\psi(\mathbf{x}|\omega)_{\text{sc}} = \int \frac{d^2q_\parallel}{(2\pi)^2} R_N(\mathbf{q}_\parallel|\mathbf{k}_\parallel)\exp\big[i\mathbf{q}_\parallel\cdot\mathbf{x}_\parallel + i\alpha_0(q_\parallel)x_3\big], \quad (3.1.26)$$

where the scattering amplitude $R_N(\mathbf{q}_\parallel|\mathbf{k}_\parallel)$ is

$$R_N(\mathbf{q}_\parallel|\mathbf{k}_\parallel) = \frac{-1}{2\alpha_0(q_\parallel)}\int d^2x_\parallel\, H(\mathbf{x}_\parallel|\omega)\big[\mathbf{q}_\parallel\cdot\nabla\mathbf{x}_\parallel - \alpha_0(q_\parallel)\big] \times \exp\big[-i\mathbf{q}_\parallel\cdot\mathbf{x}_\parallel - i\alpha_0(q_\parallel)\zeta(\mathbf{x}_\parallel)\big]. \quad (3.1.27)$$

3.1.1.2. The Mean Differential Reflection Coefficient The differential reflection coefficient $(\partial R/\partial\Omega_s)$ is defined such that $(\partial R/\partial\Omega_s)d\Omega_s$ is the fraction of the total time-averaged incident flux that is scattered into the element of solid angle $d\Omega_s$ about a given scattering direction, in the limit as $d\Omega_s$ tends to zero. To calculate the energy flux vector of the real scalar field $\Psi(\mathbf{x};t)$ we use the approach presented by Born and Wolf [3.3]. We return to Eq. (3.1.1) and multiply it by $\partial\Psi(\mathbf{x};t)/\partial t$ to obtain

$$\frac{\partial\Psi(\mathbf{x};t)}{\partial t}\left[\nabla^2\Psi(\mathbf{x};t) - \frac{1}{c^2}\frac{\partial^2\Psi(x;t)}{\partial t^2}\right] = 0. \quad (3.1.28)$$

If we use the identities

$$\frac{\partial\Psi(\mathbf{x};t)}{\partial t}\nabla^2\Psi(\mathbf{x};t) = \nabla\cdot\left(\frac{\partial\Psi(\mathbf{x};t)}{\partial t}\nabla\Psi(\mathbf{x};t)\right) - \frac{1}{2}\frac{\partial}{\partial t}\big(\nabla\Psi(\mathbf{x};t)\big)^2 \quad (3.1.29)$$

and

$$\frac{\partial\Psi(\mathbf{x};t)}{\partial t}\frac{\partial^2\Psi(\mathbf{x};t)}{\partial t^2} = \frac{1}{2}\frac{\partial}{\partial t}\left(\frac{\partial\Psi(\mathbf{x};t)}{\partial t}\right)^2, \quad (3.1.30)$$

Eq. (3.1.28) can be rewritten in the form

$$\nabla \cdot \mathbf{S}(\mathbf{x}; t) + \frac{\partial W(\mathbf{x}; t)}{\partial t} = 0, \tag{3.1.31}$$

where

$$\mathbf{S}(\mathbf{x}; t) = -a \frac{\partial \Psi(\mathbf{x}; t)}{\partial t} \nabla \Psi(\mathbf{x}; t), \tag{3.1.32}$$

$$W(\mathbf{x}; t) = \frac{1}{2} a \left[\left(\nabla \Psi(\mathbf{x}; t) \right)^2 + \frac{1}{c^2} \left(\frac{\partial \Psi(\mathbf{x}; t)}{\partial t} \right)^2 \right]. \tag{3.1.33}$$

In Eqs. (3.1.32) and (3.1.33) a is an arbitrary constant which we assume to be positive. Equation (3.1.31) has the form of an equation of continuity, and this suggests that the vector $\mathbf{S}(\mathbf{x}; t)$ may be regarded as representing the energy flux vector of the field $\Psi(\mathbf{x}; t)$, while the scalar quantity $W(\mathbf{x}; t)$ represents its energy density.

In the case that the field $\Psi(\mathbf{x}; t)$ is monochromatic of frequency ω, and has the form given by Eq. (3.1.2), the expressions for $\mathbf{S}(\mathbf{x}; t)$ and $W(\mathbf{x}; t)$ become

$$\mathbf{S}(\mathbf{x}; t) = -\frac{a}{4}(i\omega)\left[\psi^* \nabla \psi - \psi \nabla \psi^* + \psi^* \nabla \psi^* \exp(i2\omega t) - \psi \nabla \psi \exp(-i2\omega t)\right], \tag{3.1.34}$$

$$W(\mathbf{x}; t) = \frac{a}{8} \Big\{ 2\nabla \psi \cdot \nabla \psi^* + (\nabla \psi)^2 \exp(-i2\omega t) + \left(\nabla \psi^*\right)^2 \exp(i2\omega t) - \left(\frac{\omega}{c}\right)^2 \left[-2\psi^* \psi + \psi^2 \exp(-i2\omega t) + \psi^{*2} \exp(i2\omega t)\right] \Big\}. \tag{3.1.35}$$

If we now evaluate the time averages of these expressions, which we will denote by an over-bar, over a time interval $-T' \leqslant t \leqslant T'$, where T' is large compared with the period $T = 2\pi/\omega$ of the oscillations of the field, the contributions from the time-dependent terms on the right-hand sides of Eqs. (3.1.34) and (3.1.35) are negligible, so that

$$\bar{\mathbf{S}}(\mathbf{x}; t) = -\frac{1}{4}(i\omega) a \left[\psi^* \nabla \psi - \psi \nabla \psi^*\right] \tag{3.1.36a}$$

$$= \frac{1}{2} \omega a \operatorname{Im} \psi^* \nabla \psi, \tag{3.1.36b}$$

$$\bar{W}(\mathbf{x}; t) = \frac{a}{4}\left[\nabla \psi^* \cdot \nabla \psi + (\omega/c)^2 \psi^* \psi\right]. \tag{3.1.37}$$

In calculating the incident or scattered energy flux we are interested only in the part of the flux approaching or leaving the surface, respectively. Thus, we are concerned with the component of the time-averaged energy flux vector that is parallel to the x_3 axis. From Eq. (3.1.36b) we see that it is given by

$$\bar{S}_3 = \frac{1}{2}\omega a \,\mathrm{Im}\, \psi^*(\mathbf{x}|\omega)\frac{\partial \psi(\mathbf{x}|\omega)}{\partial x_3}. \tag{3.1.38}$$

The magnitude of the total time-averaged flux incident on the region $-L_1/2 < x_1 < L_1/2$, $-L_2/2 < x_2 < L_2/2$ of the plane $x_3 = 0$ covered by the random surface is therefore

$$P_{\mathrm{inc}} = -\frac{1}{2}\omega a \,\mathrm{Im} \int_{-\frac{L_1}{2}}^{\frac{L_1}{2}} dx_1 \int_{-\frac{L_2}{2}}^{\frac{L_2}{2}} dx_2\, \psi^*(\mathbf{x}|\omega)_{\mathrm{inc}} \frac{\partial \psi(\mathbf{x}|\omega)_{\mathrm{inc}}}{\partial x_3} \tag{3.1.39a}$$

$$= L_1 L_2 \frac{1}{2}\omega a \alpha_0(k_\parallel), \tag{3.1.39b}$$

where the minus sign on the right-hand side of Eqs. (3.1.39a) compensates for the fact that the 3-component of the incident flux vector is negative.

The magnitude of the total time-averaged flux scattered from the region $-L_1/2 < x_1 < L_1/2$, $-L_2/2 < x_2 < L_2/2$ of the plane $x_3 = 0$ is

$$P_{\mathrm{sc}} = \frac{1}{2} a\omega \,\mathrm{Im} \int_{-\frac{L_1}{2}}^{\frac{L_1}{2}} dx_1 \int_{-\frac{L_2}{2}}^{\frac{L_2}{2}} dx_2\, \psi^*(\mathbf{x}|\omega)_{\mathrm{sc}} \frac{\partial \psi(\mathbf{x}|\omega)_{\mathrm{sc}}}{\partial x_3}$$

$$= \frac{1}{2} a\omega \,\mathrm{Im} \int_{-\frac{L_2}{2}}^{\frac{L_1}{2}} dx_1 \int_{-\frac{L_2}{2}}^{\frac{L_2}{2}} dx_2 \int \frac{d^2 q_\parallel}{(2\pi)^2} R^*(\mathbf{q}_\parallel|\mathbf{k}_\parallel) \exp\bigl[-i\mathbf{q}_\parallel \cdot \mathbf{x}_\parallel - i\alpha_0^*(q_\parallel)x_3\bigr]$$

$$\times \int \frac{d^2 q'_\parallel}{(2\pi)^2} i\alpha_0(q'_\parallel) R(\mathbf{q}'_\parallel|\mathbf{k}_\parallel) \exp\bigl[i\mathbf{q}'_\parallel \cdot \mathbf{x}_\parallel + i\alpha_0(q'_\parallel)x_3\bigr]$$

$$= \frac{1}{2} a\omega \,\mathrm{Im} \int \frac{d^2 q_\parallel}{(2\pi)^2} i\alpha_0(q_\parallel) \bigl|R(\mathbf{q}_\parallel|\mathbf{k}_\parallel)\bigr|^2 \exp\bigl[-2\,\mathrm{Im}\,\alpha_0(q_\parallel)x_3\bigr]$$

$$= \frac{1}{2} a\omega \int_{q_\parallel < \frac{\omega}{c}} \frac{d^2 q_\parallel}{(2\pi)^2} \alpha_0(q_\parallel) \bigl|R(\mathbf{q}_\parallel|\mathbf{k}_\parallel)\bigr|^2. \tag{3.1.40}$$

Here the scattering amplitude $R(\mathbf{q}_\parallel|\mathbf{k}_\parallel)$ is either $R_D(\mathbf{q}_\parallel|\mathbf{k}_\parallel)$ or $R_N(\mathbf{q}_\parallel|\mathbf{k}_\parallel)$. If we now introduce the polar and azimuthal angles of incidence (θ_0, ϕ_0) and scattering (θ_s, ϕ_s), respec-

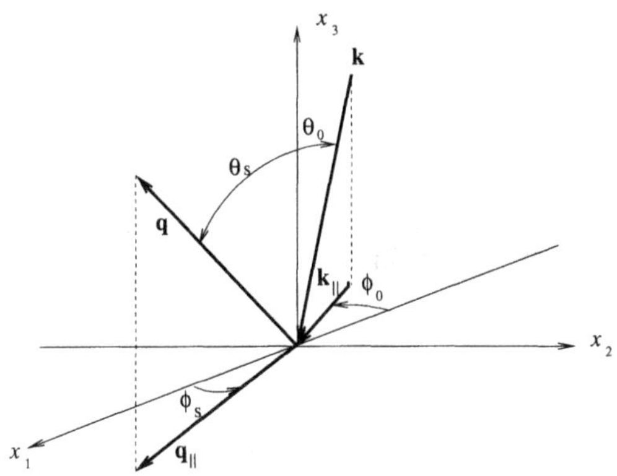

Figure 3.1. The scattering geometry assumed in this chapter.

tively, by (Fig. 3.1)

$$\mathbf{k}_{\parallel} = (\omega/c)\sin\theta_0(\cos\phi_0, \sin\phi_0, 0), \tag{3.1.41a}$$

$$\mathbf{q}_{\parallel} = (\omega/c)\sin\theta_s(\cos\phi_s, \sin\phi_s, 0), \tag{3.1.41b}$$

we find that

$$\alpha_0(k_{\parallel}) = (\omega/c)\cos\theta_0, \qquad \alpha_0(q_{\parallel}) = (\omega/c)\cos\theta_s, \tag{3.1.42}$$

$$d^2 q_{\parallel} = (\omega/c)^2 \cos\theta_s \, d\Omega_s, \tag{3.1.43}$$

where $d\Omega_s = \sin\theta_s \, d\theta_s \, d\phi_s$ is the element of solid angle. Thus, we can write the total incident flux as

$$P_{\text{inc}} = S\frac{1}{2}\omega a\frac{\omega}{c}\cos\theta_0, \tag{3.1.44}$$

where $S = L_1 L_2$ is the area of the $x_1 x_2$ plane covered by the rough surface, and we can write the total scattered flux as

$$P_{\text{sc}} = \int d\Omega_s \, P_{\text{sc}}(\Omega_s), \tag{3.1.45}$$

where

$$P_{\text{sc}}(\Omega_s) = \frac{1}{2}a\omega\frac{\omega}{c}\left(\frac{\omega}{2\pi c}\right)^2 \cos^2\theta_s \left|R(\mathbf{q}_{\parallel}|\mathbf{k}_{\parallel})\right|^2. \tag{3.1.46}$$

By definition the differential reflection coefficient is given by

$$\left(\frac{\partial R}{\partial \Omega_s}\right) = \frac{P_{sc}(\Omega_s)}{P_{inc}} = \frac{1}{S}\left(\frac{\omega}{2\pi c}\right)^2 \frac{\cos^2\theta_s}{\cos\theta_0}\left|R(\mathbf{q}_\parallel|\mathbf{k}_\parallel)\right|^2. \tag{3.1.47}$$

Since the surface defined by the profile function $\zeta(\mathbf{x}_\parallel)$ is a randomly rough surface, it is the mean differential reflection coefficient that is of interest to us. It is given by

$$\left\langle\frac{\partial R}{\partial \Omega_s}\right\rangle = \frac{1}{S}\left(\frac{\omega}{2\pi c}\right)^2 \frac{\cos^2\theta_s}{\cos\theta_0}\left\langle\left|R(\mathbf{q}_\parallel|\mathbf{k}_\parallel)\right|^2\right\rangle, \tag{3.1.48}$$

where as before the angle brackets denote an average over the ensemble of realizations of the surface profile function. Equations (3.1.47) and (3.1.48) apply to scattering from both Dirichlet and Neumann surfaces.

3.1.1.3. The Equations Satisfied by the Source Functions The source functions $L(\mathbf{x}_\parallel|\omega)$ and $H(\mathbf{x}_\parallel|\omega)$ satisfy integral equations that can be obtained in the following way. In the case of the source function $L(\mathbf{x}_\parallel|\omega)$ we set $x_3 = \zeta(\mathbf{x}_\parallel) + \eta$ and $x_3 = \zeta(\mathbf{x}_\parallel) - \eta$ in Eq. (3.1.20), where η is a positive infinitesimal, and add the resulting equations, taking the boundary condition (3.1.4) into account. The result can be written

$$2H(\mathbf{x}_\parallel|\omega)_{inc} = \frac{1}{2\pi}\int d^2x'_\parallel \left[g_0(\mathbf{x}|\mathbf{x}')\right]_{\substack{x'_3=\zeta(\mathbf{x}'_\parallel)\\ x_3=\zeta(\mathbf{x}_\parallel)}} L(\mathbf{x}'_\parallel|\omega), \tag{3.1.49}$$

where

$$H(\mathbf{x}_\parallel|\omega)_{inc} = \psi(\mathbf{x}|\omega)_{inc}\big|_{x_3=\zeta(\mathbf{x}_\parallel)}. \tag{3.1.50}$$

In obtaining this equation we have used the result that [3.4]

$$\int d^2x'_\parallel \left[g_0(\mathbf{x}|\mathbf{x}')\right]_{\substack{x'_3=\zeta(\mathbf{x}'_\parallel)\\ x_3=\zeta(\mathbf{x}_\parallel)\pm\eta}} L(\mathbf{x}'_\parallel|\omega) = \int d^2x'_\parallel \left[g_0(\mathbf{x}|\mathbf{x}')\right]_{\substack{x'_3=\zeta(\mathbf{x}'_\parallel)\\ x_3=\zeta(\mathbf{x}_\parallel)}} L(\mathbf{x}'_\parallel|\omega), \tag{3.1.51}$$

i.e., that the singularity of $[g_0(\mathbf{x}|\mathbf{x}')]_{x'_3=\zeta(\mathbf{x}'_\parallel),\, x_3=\zeta(\mathbf{x}_\parallel)}$ at $\mathbf{x}_\parallel = \mathbf{x}'_\parallel$ is integrable.

A more convenient equation for $L(\mathbf{x}_\parallel|\omega)$ is obtained if we assume that $x_3 > \zeta(\mathbf{x}_\parallel)$ and apply the operator $\partial/\partial N$ to both sides of Eq. (3.1.20). In this way we obtain

$$\begin{aligned}\theta\big(x_3 - \zeta(\mathbf{x}_\parallel)\big)\big(\partial\psi(\mathbf{x}|\omega)/\partial N\big) &= \big(\partial\psi(\mathbf{x}|\omega)_{inc}/\partial N\big)\\ &\quad - \frac{1}{4\pi}\int d^2x'_\parallel \left[\partial g_0(\mathbf{x}|\mathbf{x}')/\partial N\right]_{x'_3=\zeta(\mathbf{x}'_\parallel)} L(\mathbf{x}'_\parallel|\omega).\end{aligned} \tag{3.1.52}$$

We next set $x_3 = \zeta(\mathbf{x}_\parallel) + \eta$ and $x_3 = \zeta(\mathbf{x}_\parallel) - \eta$ in this equation, where η is a positive infinitesimal, and add the resulting equations. The result is the equation

$$L(\mathbf{x}_\parallel|\omega) = 2L(\mathbf{x}_\parallel|\omega)_{\text{inc}} - \frac{1}{2\pi} P \int d^2x'_\parallel \left[\partial g_0(\mathbf{x}|\mathbf{x}')/\partial N\right]_{\substack{x'_3=\zeta(\mathbf{x}'_\parallel)\\ x_3=\zeta(\mathbf{x}_\parallel)}} L(\mathbf{x}'_\parallel|\omega), \tag{3.1.53}$$

where P denotes the Cauchy principal value, and

$$L(\mathbf{x}_\parallel|\omega)_{\text{inc}} = \left[\partial\psi(\mathbf{x}|\omega)_{\text{inc}}/\partial N\right]_{x_3=\zeta(\mathbf{x}_\parallel)}. \tag{3.1.54}$$

In writing Eq. (3.1.53) we have used the result that

$$\left[\partial g_0(\mathbf{x}|\mathbf{x}')/\partial N\right]_{\substack{x'_3=\zeta(\mathbf{x}'_\parallel)\\ x_3=\zeta(\mathbf{x}_\parallel)+\eta}} + \left[\partial g_0(\mathbf{x}|\mathbf{x}')/\partial N\right]_{\substack{x'_3=\zeta(\mathbf{x}'_\parallel)\\ x_3=\zeta(\mathbf{x}_\parallel)-\eta}}$$
$$= \begin{cases} 2[\partial g_0(\mathbf{x}|\mathbf{x}')/\partial N]_{\substack{x'_3=\zeta(\mathbf{x}'_\parallel)\\ x_3=\zeta(\mathbf{x}_\parallel)}} & \mathbf{x}_\parallel \neq \mathbf{x}'_\parallel, \\ 0 & \mathbf{x}_\parallel = \mathbf{x}'_\parallel\,. \end{cases} \tag{3.1.55}$$

To obtain the equation satisfied by the source function $H(\mathbf{x}_\parallel|\omega)$ we let x_3 take the values $\zeta(\mathbf{x}_\parallel) + \eta$ and $\zeta(\mathbf{x}_\parallel) - \eta$ in turn in Eq. (3.1.26) and add the resulting equations. In this way we obtain the equation

$$H(\mathbf{x}_\parallel|\omega) = 2H(\mathbf{x}_\parallel|\omega)_{\text{inc}} + \frac{1}{2\pi} P \int d^2x'_\parallel \left[\partial g_0(\mathbf{x}|\mathbf{x}')/\partial N'\right]_{\substack{x'_3=\zeta(\mathbf{x}'_\parallel)\\ x_3=\zeta(\mathbf{x}_\parallel)}} H(\mathbf{x}'_\parallel|\omega), \tag{3.1.56}$$

where $H(\mathbf{x}_\parallel|\omega)_{\text{inc}}$ has been defined in Eq. (3.1.50).

In obtaining Eq. (3.1.56) we used the result that

$$\left[\partial g_0(\mathbf{x}|\mathbf{x}')/\partial N'\right]_{\substack{x'_3=\zeta(\mathbf{x}'_\parallel)\\ x_3=\zeta(\mathbf{x}_\parallel)+\eta}} + \left[\partial g_0(\mathbf{x}|\mathbf{x}')/\partial N'\right]_{\substack{x'_3=\zeta(\mathbf{x}'_\parallel)\\ x_3=\zeta(\mathbf{x}_\parallel)-\eta}}$$
$$= \begin{cases} 2[\partial g_0(\mathbf{x}|\mathbf{x}')/\partial N']_{\substack{x'_3=\zeta(\mathbf{x}'_\parallel)\\ x_3=\zeta(\mathbf{x}_\parallel)}} & \mathbf{x}_\parallel \neq \mathbf{x}'_\parallel, \\ 0 & \mathbf{x}_\parallel = \mathbf{x}'_\parallel. \end{cases} \tag{3.1.57}$$

3.1.1.4. The Kirchhoff Approximation and its Geometrical Optics Limit We will return to Eqs. (3.1.53) and (3.1.56), especially Eq. (3.1.53), later in this chapter. However to proceed farther here, we will use the Kirchhoff approximation to each source function due to its simplicity. It can be obtained from the exact equations (3.1.53) and (3.1.56) in the following way. In this approximation the source term $L(\mathbf{x}_\parallel|\omega)$ is given by the inhomogeneous term on the right-hand side of Eq. (3.1.53),

$$L(\mathbf{x}_\parallel|\omega)_K = 2\left(-\frac{\partial\zeta(\mathbf{x}_\parallel)}{\partial x_1}\frac{\partial}{\partial x_1} - \frac{\partial\zeta(\mathbf{x}_\parallel)}{\partial x_2}\frac{\partial}{\partial x_2} + \frac{\partial}{\partial x_3}\right)\psi(\mathbf{x}_\parallel\omega)_{\text{inc}}\Big|_{x_3=\zeta(\mathbf{x}_\parallel)}$$

$$= -2i\left(k_1\frac{\partial\zeta(\mathbf{x}_\parallel)}{\partial x_1} + k_2\frac{\partial\zeta(\mathbf{x}_\parallel)}{\partial x_2} + \alpha_0(k)\right)\exp\left[i\mathbf{k}_\parallel\cdot\mathbf{x}_\parallel - i\alpha_0(k_\parallel)\zeta(\mathbf{x}_\parallel)\right]. \tag{3.1.58}$$

This approximation is equivalent to the assumption that the scattering takes place as reflection from the plane tangent to the surface at each point [3.5].

Substitution of Eq. (3.1.58) into Eq. (3.1.25) yields the result that the amplitude for scattering from a Dirichlet surface is

$$R_D(\mathbf{q}_\parallel|\mathbf{k}_\parallel) = -\frac{1}{\alpha_0(q_\parallel)}\int d^2x_\parallel\left(k_1\frac{\partial\zeta(\mathbf{x}_\parallel)}{\partial x_1} + k_2\frac{\partial\zeta(\mathbf{x}_\parallel)}{\partial x_2} + \alpha_0(k_\parallel)\right)$$
$$\times\exp\left[-i(\mathbf{q}_\parallel - \mathbf{k}_\parallel)\cdot\mathbf{x}_\parallel - i\big(\alpha_0(q_\parallel) + \alpha_0(k_\parallel)\big)\zeta(\mathbf{x}_\parallel)\right]. \tag{3.1.59}$$

We can rewrite Eq. (3.1.59) in such a way that derivatives of the surface profile function no longer appear in it. To accomplish this we begin by introducing the function $I(\gamma|\mathbf{Q}_\parallel)$ defined by

$$\exp\left[-i\gamma\zeta(\mathbf{x}_\parallel)\right] = \int\frac{d^2Q_\parallel}{(2\pi)^2}I(\gamma|\mathbf{Q}_\parallel)\exp(i\mathbf{Q}_\parallel\cdot\mathbf{x}_\parallel). \tag{3.1.60}$$

If we differentiate both sides of this equation with respect to x_α ($\alpha = 1, 2$), we obtain

$$\frac{\partial\zeta(\mathbf{x}_\parallel)}{\partial x_\alpha}\exp\left[-i\gamma\zeta(\mathbf{x}_\parallel)\right] = -\int\frac{d^2Q_\parallel}{(2\pi)^2}\frac{Q_\alpha}{\gamma}I(\gamma|\mathbf{Q}_\parallel)\exp(i\mathbf{Q}_\parallel\cdot\mathbf{x}_\parallel). \tag{3.1.61}$$

The use of Eqs. (3.1.60)–(3.1.61) in Eq. (3.1.59) yields the result

$$R_D(\mathbf{q}_\parallel|\mathbf{k}_\parallel)$$
$$= -\frac{(\omega/c)^2 + \alpha_0(q_\parallel)\alpha_0(k_\parallel) - \mathbf{q}_\parallel\cdot\mathbf{k}_\parallel}{\alpha_0(q_\parallel)[\alpha_0(q_\parallel) + \alpha_0(k_\parallel)]}I\big(\alpha_0(q_\parallel) + \alpha_0(k_\parallel)|\mathbf{q}_\parallel - \mathbf{k}_\parallel\big). \tag{3.1.62}$$

The inverse of Eq. (3.1.60),

$$I(\gamma|\mathbf{Q}_\parallel) = \int d^2x_\parallel\exp(-i\mathbf{Q}_\parallel\cdot\mathbf{x}_\parallel)\exp\left[-i\gamma\zeta(\mathbf{x}_\parallel)\right], \tag{3.1.63}$$

enables us to rewrite Eq. (3.1.62) finally as

$$R_D(\mathbf{q}_\parallel|\mathbf{k}_\parallel) = -\frac{(\omega/c)^2 + \alpha_0(q_\parallel)\alpha_0(k_\parallel) - \mathbf{q}_\parallel\cdot\mathbf{k}_\parallel}{\alpha_0(q_\parallel)[\alpha_0(q_\parallel)+\alpha_0(k_\parallel)]} \times \int d^2x_\parallel \exp\bigl[-i(\mathbf{q}_\parallel - \mathbf{k}_\parallel)\cdot\mathbf{x}_\parallel\bigr]\exp\bigl[-i\bigl(\alpha_0(q_\parallel)+\alpha_0(k_\parallel)\bigr)\zeta(\mathbf{x}_\parallel)\bigr]. \tag{3.1.64}$$

In the Kirchhoff approximation the source function $H(\mathbf{x}_\parallel|\omega)$ is given by the inhomogeneous term on the right-hand side of Eq. (3.1.56),

$$H(\mathbf{x}_\parallel|\omega)_K = 2\psi(\mathbf{x}|\omega)_{\mathrm{inc}}\big|_{x_3=\zeta(\mathbf{x}_\parallel)} = 2\exp\bigl[i\mathbf{k}_\parallel\cdot\mathbf{x}_\parallel - i\alpha_0(k_\parallel)\zeta(\mathbf{x}_\parallel)\bigr]. \tag{3.1.65}$$

This approximation is also equivalent to the assumption that scattering takes place as reflection from the plane tangent to the surface at each point [3.5]. Substitution of Eq. (3.1.65) into Eq. (3.1.27) yields the result that the amplitude for scattering from a Neumann surface is

$$R_N(\mathbf{q}_\parallel|\mathbf{k}_\parallel) = -\frac{1}{\alpha_0(q_\parallel)}\int d^2x_\parallel\left[q_1\frac{\partial\zeta(\mathbf{x}_\parallel)}{\partial x_1} + q_2\frac{\partial\zeta(\mathbf{x}_\parallel)}{\partial x_2} - \alpha_0(q_\parallel)\right] \times \exp\bigl[-i(\mathbf{q}_\parallel - \mathbf{k}_\parallel)\cdot\mathbf{x}_\parallel - i\bigl(\alpha_0(q_\parallel)+\alpha_0(k_\parallel)\bigr)\zeta(\mathbf{x}_\parallel)\bigr]. \tag{3.1.66}$$

The application of Eqs. (3.1.60)–(3.1.61) to Eq. (3.1.66) produces the result

$$R_N(\mathbf{q}_\parallel|\mathbf{k}_\parallel) = \frac{(\omega/c)^2 + \alpha_0(q_\parallel)\alpha_0(k_\parallel) - \mathbf{q}_\parallel\cdot\mathbf{k}_\parallel}{\alpha_0(q_\parallel)[\alpha_0(q_\parallel)+\alpha_0(k_\parallel)]} I\bigl(\alpha_0(q_\parallel)+\alpha_0(k_\parallel)|\mathbf{q}_\parallel - \mathbf{k}_\parallel\bigr). \tag{3.1.67}$$

The use of Eq. (3.1.63) gives us finally that

$$R_N(\mathbf{q}_\parallel|\mathbf{k}_\parallel) = \frac{(\omega/c)^2 + \alpha_0(q_\parallel)\alpha_0(k_\parallel) - \mathbf{q}_\parallel\cdot\mathbf{k}_\parallel}{\alpha_0(q_\parallel)[\alpha_0(q_\parallel)+\alpha_0(k_\parallel)]} \times \int d^2x_\parallel \exp\bigl[-i(\mathbf{q}_\parallel - \mathbf{k}_\parallel)\cdot\mathbf{x}_\parallel\bigr]\exp\bigl[-i\bigl(\alpha_0(q_\parallel)+\alpha_0(k_\parallel)\bigr)\zeta(\mathbf{x}_\parallel)\bigr]. \tag{3.1.68}$$

A comparison of Eqs. (3.1.64) and (3.1.68) reveals that the amplitude for the scattering of a scalar plane wave from a two-dimensional rough surface on which the Dirichlet boundary condition is satisfied differs only in sign from the amplitude for the scattering of the same scalar plane wave from a two-dimensional rough surface on which the Neumann boundary condition is satisfied. Since the mean differential reflection coefficient, Eq. (3.1.48) is indifferent to the sign of the scattering amplitude, it follows that it is given by

$$\left\langle\frac{\partial R}{\partial\Omega_s}\right\rangle = \frac{1}{S}\frac{1}{4\pi^2\cos\theta_0}\frac{[(\omega/c)^2+\alpha_0(q_\parallel)\alpha_0(k_\parallel)-\mathbf{q}_\parallel\cdot\mathbf{k}_\parallel]^2}{[\alpha_0(q_\parallel)+\alpha_0(k_\parallel)]^2}$$
$$\times\int d^2x_\parallel\int d^2x'_\parallel \exp\bigl[-i(\mathbf{q}_\parallel-\mathbf{k}_\parallel)\cdot(\mathbf{x}_\parallel-\mathbf{x}'_\parallel)\bigr]$$
$$\times\bigl\langle\exp\bigl[-i\bigl(\alpha_0(q_\parallel)+\alpha_0(k_\parallel)\bigr)\bigl(\zeta(\mathbf{x}_\parallel)-\zeta(\mathbf{x}'_\parallel)\bigr)\bigr]\bigr\rangle \tag{3.1.69a}$$
$$=\frac{1}{S}\left(\frac{\omega}{2\pi c}\right)^2\frac{[1+\cos\theta_s\cos\theta_0-\sin\theta_s\sin\theta_0\cos(\phi_s-\phi_0)]^2}{\cos\theta_0(\cos\theta_0+\cos\theta_s)^2}$$
$$\times\int d^2x_\parallel\int d^2x'_\parallel \exp\bigl[-i(\mathbf{q}_\parallel-\mathbf{k}_\parallel)\cdot(\mathbf{x}_\parallel-\mathbf{x}'_\parallel)\bigr]$$
$$\times\bigl\langle\exp\bigl[-i(\omega/c)(\cos\theta_s+\cos\theta_0)\bigl(\zeta(\mathbf{x}_\parallel)-\zeta(\mathbf{x}'_\parallel)\bigr)\bigr]\bigr\rangle. \tag{3.1.69b}$$

Because of the equality, within the Kirchhoff approximation, of the mean differential reflection coefficient for scattering from a Dirichlet and a Neumann surface, in what follows, for specificity, we will confine our attention to Dirichlet surfaces whenever the Kirchhoff approximation is used.

Our aim is to find a function $\zeta(\mathbf{x}_\parallel)$ that produces a prescribed form for $\langle\partial R/\partial\Omega_s\rangle$. As it stands, Eqs. (3.1.69) are too complicated to invert to obtain $\zeta(\mathbf{x}_\parallel)$ in terms of $\langle\partial R/\partial\Omega_s\rangle$. To simplify them we pass to the geometrical optics limit of the Kirchhoff approximation, which is obtained by making the change of variable $\mathbf{x}'_\parallel=\mathbf{x}_\parallel-\mathbf{u}_\parallel$, expanding the difference $\zeta(\mathbf{x}_\parallel)-\zeta(\mathbf{x}_\parallel-\mathbf{u}_\parallel)$ in powers of $\mathbf{u}_\parallel$, and keeping only the linear term. In this way we obtain the result

$$\left\langle\frac{\partial R}{\partial\Omega_s}\right\rangle = \frac{1}{S}\frac{1}{4\pi^2\cos\theta_0}\frac{[(\omega/c)^2+\alpha_0(q_\parallel)\alpha_0(k_\parallel)-\mathbf{q}_\parallel\cdot\mathbf{k}_\parallel]^2}{[\alpha_0(q_\parallel)+\alpha_0(k_\parallel)]^2}$$
$$\times\int d^2x_\parallel\int d^2u_\parallel \exp\bigl[-i(\mathbf{q}_\parallel-\mathbf{k}_\parallel)\cdot\mathbf{u}_\parallel\bigr]$$
$$\times\bigl\langle\exp\bigl[-i\bigl(\alpha_0(q_\parallel)+\alpha_0(k_\parallel)\bigr)\mathbf{u}_\parallel\cdot\nabla\zeta(\mathbf{x}_\parallel)\bigr]\bigr\rangle \tag{3.1.70a}$$
$$=\frac{1}{S}\left(\frac{\omega}{2\pi c}\right)^2\frac{[1+\cos\theta_s\cos\theta_0-\sin\theta_s\sin\theta_0\cos(\phi_s-\phi_0)]^2}{\cos\theta_0(\cos\theta_0+\cos\theta_s)^2}$$
$$\times\int d^2x_\parallel\int d^2u_\parallel \exp\bigl[-i(\mathbf{q}_\parallel-\mathbf{k}_\parallel)\cdot\mathbf{u}_\parallel\bigr]$$
$$\times\bigl\langle\exp\bigl[-i(\omega/c)(\cos\theta_s+\cos\theta_0)\mathbf{u}_\parallel\cdot\nabla\zeta(\mathbf{x}_\parallel)\bigr]\bigr\rangle. \tag{3.1.70b}$$

Equation (3.1.70b) simplifies greatly in the case of normal incidence, where $\theta_0 = 0°$ and $\mathbf{k}_\parallel = 0$:

$$\left\langle \frac{\partial R}{\partial \Omega_s} \right\rangle = \frac{1}{S} \left(\frac{\omega}{2\pi c} \right)^2 \int d^2 u_\parallel \exp(-i\mathbf{q}_\parallel \cdot \mathbf{u}_\parallel) \times \int d^2 x_\parallel \left\langle \exp\left[-i(\omega/c)(1 + \cos\theta_s)\mathbf{u}_\parallel \cdot \nabla\zeta(\mathbf{x}_\parallel)\right] \right\rangle. \tag{3.1.71}$$

We now turn to the application of the results obtained in this section to the design of two-dimensional randomly rough surfaces with several different types of scattering properties.

3.1.2. The Design of a Two-Dimensional Randomly Rough Surface that Acts as a Band-Limited Uniform Diffuser Within a Rectangular Domain of Scattering Angles

The earliest effort to design a two-dimensional randomly rough surface with prescribed scattering processes was devoted to the design of such a surface that produces a mean differential reflection coefficient that is constant within a rectangular domain of scattering angles, and produces no scattering outside this domain [3.6]. This is perhaps the simplest two-dimensional random surface that scatters an incident field in a prescribed fashion. The approach used in this work consisted of representing the surface profile function of the random surface as the superposition of two orthogonal sets of equally spaced identical trapezoidal grooves whose amplitudes were assumed to be independent, identically distributed random deviates drawn from a probability density function determined in such a way that the field scattered from the resulting surface produced a mean differential reflection coefficient that possessed the prescribed form. Such a surface was fabricated on photoresist [3.6], and was shown to produce an intensity distribution in transmission that resembled the specified one [3.6].

In this section we revisit the problem on the basis of an approach [3.7] that is simpler to implement, both theoretically and experimentally, than the one that was employed in Ref. [3.6]. Thus, we begin by representing the surface profile function $\zeta(\mathbf{x}_\parallel)$ in the form

$$\zeta(\mathbf{x}_\parallel) = \zeta_1(x_1) + \zeta_2(x_2), \tag{3.1.72}$$

where $\zeta_1(x_1)$ and $\zeta_2(x_2)$ are statistically independent random processes. We assume that this surface is illuminated at normal incidence by a scalar plane wave. In this case Eq. (3.1.71) becomes

$$\left\langle \frac{\partial R}{\partial \Omega_s} \right\rangle = \frac{1}{L_1 L_2} \left(\frac{\omega}{2\pi c} \right)^2 I_1(q_1) I_2(q_2), \tag{3.1.73}$$

where L_1 and L_2 are the lengths of the x_1 and x_2 axes covered by the random surface, and where $(j = 1, 2)$

$$I_j(q_j) = \int_{-\infty}^{\infty} dx_1 \int_{-\infty}^{\infty} du_j \exp(-iq_j u_j)\langle \exp[-ia_0 u_j \zeta_j'(x_j)]\rangle. \tag{3.1.74}$$

In writing Eq. (3.1.74) we have simplified the notation by setting

$$a_0(\theta_s) \equiv (\omega/c)(1 + \cos\theta_s). \tag{3.1.75}$$

We assume for $\zeta_j(x_j)$ the form

$$\zeta_j(x_j) = a_{n_j} x_j + b_j, \quad n_j b < x_j < (n_j + 1)b, \; n_j = -N_j, -N_j + 1, \ldots, N_j - 1, \tag{3.1.76}$$

where the $\{a_{n_j}\}$ are independent, identically distributed random deviates, b is a characteristic length, and the $\{b_{n_j}\}$ are obtained from the requirement that $\zeta_j(x_j)$ be a continuous function of x_j. This condition leads to the recurrence relation

$$b_{n_j+1} = b_{n_j} - (n_j + 1)(a_{n_j+1} - a_{n_j})b, \tag{3.1.77}$$

which determines all of the $\{b_{n_j}\}$ once the value of one of them is known. In applying this relation it is convenient to set $b_0 = 0$, and we do so. The independence of the $\{a_{n_j}\}$, and the fact that they are identically distributed, have the consequence that the probability density function of a_{n_j},

$$f_j(\gamma_j) = \langle \delta(\gamma_j - a_{n_j})\rangle \tag{3.1.78}$$

is independent of n_j. The angle brackets in Eq. (3.1.78) denote an average over the ensemble of realizations of the surface profile function $\zeta_j(x_j)$. In the present case this is equivalent to an average over the ensemble of realizations of a_{n_j}.

When the result given by Eq. (3.1.76) is substituted into Eq. (3.1.74), the latter becomes

$$I_j(q_j) = \int_{-\infty}^{\infty} du_j \exp(-iq_j u_j) \sum_{n_j=-N_j}^{N_j-1} \int_{n_j b}^{(n_j+1)b} dx_j \langle \exp[-ia_0 u_j a_{n_j}]\rangle. \tag{3.1.79}$$

The evaluation of this expression is carried out exactly as this was done in Section 2.1.1, with the result

$$I_j(q_j) = L_j \frac{2\pi}{a_0} f_j\left(-\frac{q_j}{a_0}\right), \tag{3.1.80}$$

where we have set $L_j = 2N_j b$. On combining Eqs. (3.1.73) and (3.1.80) we obtain for the mean differential reflection coefficient

$$\left\langle \frac{\partial R}{\partial \Omega_s} \right\rangle = \frac{1}{(1+\cos\theta_s)^2} f_1\left(-\frac{q_1}{a_0}\right) f_2\left(-\frac{q_2}{a_0}\right). \tag{3.1.81}$$

We seek a surface profile function $\zeta(\mathbf{x}_\parallel)$ that produces a mean differential reflection coefficient that is a constant within a rectangular domain of scattering angles, and vanishes outside this domain. It can be represented in the form

$$\left\langle \frac{\partial R}{\partial \Omega_s} \right\rangle = A_1\theta\big(q_{1m} - |q_1|\big) A_2\theta\big(q_{2m} - |q_2|\big), \tag{3.1.82}$$

where A_1 and A_2 are constants to be determined. Thus, on combining Eqs. (3.1.81) and (3.1.82), we find that we can define $f_1(\gamma_1)$ and $f_2(\gamma_2)$ through the relations

$$\frac{1}{1+\cos\theta_s} f_1\left(-\frac{q_1}{a_0}\right) = A_1\theta\big(q_{1m} - |q_1|\big) \tag{3.1.83a}$$

and

$$\frac{1}{1+\cos\theta_s} f_2\left(-\frac{q_2}{a_0}\right) = A_2\theta\big(q_{2m} - |q_2|\big). \tag{3.1.83b}$$

We now make the changes of variables

$$\gamma_1 = -\frac{q_1}{a_0} = -\frac{\sin\theta_s \cos\phi_s}{1+\cos\theta_s}, \tag{3.1.84a}$$

$$\gamma_2 = -\frac{q_2}{a_0} = -\frac{\sin\theta_s \sin\phi_s}{1+\cos\theta_s}, \tag{3.1.84b}$$

from which it follows that

$$\cos\theta_s = \frac{1-\gamma_\parallel^2}{1+\gamma_\parallel^2}, \qquad \sin\theta_s = \frac{2\gamma_\parallel}{(1+\gamma_\parallel^2)}, \qquad 1+\cos\theta_s = \frac{2}{(1+\gamma_\parallel^2)}, \tag{3.1.85}$$

where

$$\gamma_\parallel = \big(\gamma_1^2 + \gamma_2^2\big)^{\frac{1}{2}}. \tag{3.1.86}$$

In order for $f_1(\gamma_1)$ to be a function of γ_1 alone, and not of γ_1 and γ_2, we have to work in the limit of small scattering angles, where we can neglect $\gamma_\parallel^2$ compared with unity. From Eq. (3.1.85) we see that this is equivalent to replacing $1+\cos\theta_s$ by 2. This is not a very restrictive condition. If θ_s is smaller than 23°, $1+\gamma_{\|}^2$ differs from unity by only 3% at most. In this limit, Eqs. (3.1.83) yield the result that

$$f_j(\gamma_j) = 2A_j\theta(\gamma_{jm} - |\gamma_j|), \tag{3.1.87}$$

where

$$\gamma_{jm} = (cq_{jm}/2\omega) \cong \theta_{jm}/2, \tag{3.1.88}$$

which defines the angles θ_{jm}. The normalization of $f_j(\gamma_j)$

$$\int_{-\infty}^{\infty} d\gamma_j\, f_j(\gamma_j) = 2A_j \int_{-\gamma_{jm}}^{\gamma_{jm}} d\gamma_j = 4A_j\gamma_{jm} = 1, \tag{3.1.89}$$

yields the result that

$$A_j = \frac{1}{4\gamma_{jm}}. \tag{3.1.90}$$

It follows that

$$\left\langle \frac{\partial R}{\partial \Omega_s} \right\rangle = \frac{\theta(\gamma_{1m} - |\gamma_1|)}{4\gamma_{1m}} \frac{\theta(\gamma_{2m} - |\gamma_2|)}{4\gamma_{2m}}, \tag{3.1.91}$$

$$f_j(\gamma_j) = \frac{\theta(\gamma_{jm} - |\gamma_1|)}{2\gamma_{jm}}. \tag{3.1.92}$$

With the probability density functions $f_1(\gamma_1)$ and $f_2(\gamma_2)$ in hand, the rejection method [3.8] can be used to generate a long sequence of the $\{a_{n_j}\}$, from which the corresponding $\{b_{n_j}\}$ are obtained by use of Eq. (3.1.77). The partial surface profile functions $\zeta_1(x_1)$ and $\zeta_2(x_2)$ are then obtained from Eq. (3.1.76), and a realization of the surface profile function $\zeta(\mathbf{x}_{\|})$ is obtained from Eq. (3.1.72).

3.1.2.1. Solution of the Scattering Problem To determine how well the angular dependence of the mean differential reflection coefficient obtained from the field scattered from the random surface generated by the method just described agrees with the expression for $\langle\partial R/\partial\Omega_s\rangle$ used as the input for this method, a large number N_p of realizations of the surface profile function $\zeta(\mathbf{x}_{\|})$ is generated, and for each realization the scattering problem for a scalar plane wave incident normally on it is solved. The differential reflection coefficient $\partial R/\partial\Omega_s$ is calculated from each solution, and an arithmetic average of the N_p results for $\partial R/\partial\Omega_s$ yields the mean differential reflection coefficient $\langle\partial R/\partial\Omega_s\rangle$. A comparison of this result for $\langle\partial R/\partial\Omega_s\rangle$ with the expression used as the input to the determination of the surface reveals the quality of that surface.

3.1.2.1.1. A Rigorous Numerical Simulation Approach For this comparison to be meaningful, the scattering problem should be solved by an approach that is more rigorous than the geometrical optics limit of the Kirchhoff approximation, on which the determination of the surface is based. This can be done by solving Eq. (3.1.53) for the source function $L(\mathbf{x}_{\|}|\omega)$ numerically. For this purpose we first replace integration over the entire $x_1'x_2'$ plane by integration over the finite square region $-L/2 \leqslant x_1', x_2' \leqslant L/2$. The edge of this square region is divided into N equal segments of length $\Delta x = L/N$. The large square region is thus subdivided into N^2 square regions, each of area $(\Delta x)^2$. The centers of these small square regions are given by the N^2 vectors

$$\mathbf{x}_{\|}(m_1, m_2) = \left(-\frac{L}{2} + \left(m_1 - \frac{1}{2}\right)\Delta x, -\frac{L}{2} + \left(m_2 - \frac{1}{2}\right)\Delta x\right), \quad m_1, m_2 = 1, 2, 3, \ldots, N. \tag{3.1.93}$$

Integration with respect to $\mathbf{x}_{\|}'$ is then replaced by summation over these vectors according to

$$\int d^2x_{\|}' \, f(\mathbf{x}_{\|}') = (\Delta x)^2 \sum_{m_1'=1}^{N} \sum_{m_2'=1}^{N} f\big(\mathbf{x}_{\|}(m_1', m_2')\big). \tag{3.1.94}$$

Of course, more sophisticated numerical quadrature schemes than the one given by Eqs. (3.1.93) and (3.1.94) can be used to replace integration by summation. However, the one given by Eqs. (3.1.93) and (3.1.94) is simple to use, and quite accurate if Δx is made small enough. Then by evaluating $L(\mathbf{x}_{\|}|\omega)$ in Eq. (3.1.53) at the same set of vectors (3.1.93) used in evaluating the integral, we are led to the following matrix equation for $L(\mathbf{x}_{\|}|\omega)$:

$$\begin{aligned} L\big(\mathbf{x}_{\|}(m_1, m_2)|\omega\big) &= 2L\big(\mathbf{x}_{\|}(m_1, m_2)|\omega\big)_{\mathrm{inc}} \\ &\quad - \frac{(\Delta x)^2}{2\pi} \sum_{m_1'=1}^{N} \sum_{m_2'=1}^{N}{}' F\big(\mathbf{x}_{\|}(m_1, m_2)|\mathbf{x}_{\|}(m_1', m_2')\big) L\big(\mathbf{x}_{\|}(m_1', m_2')|\omega\big), \\ m_1, m_2 &= 1, 2, 3, \ldots, N, \end{aligned} \tag{3.1.95}$$

where the kernel $F(\mathbf{x}_{\|}|\mathbf{x}_{\|}')$ is given explicitly by

$$\begin{aligned} F(\mathbf{x}_{\|}|\mathbf{x}_{\|}') &= \big[\partial g_0(\mathbf{x}|\mathbf{x}')/\partial N\big]_{\substack{x_3'=\zeta(\mathbf{x}_{\|}') \\ x_3=\zeta(\mathbf{x}_{\|})}} \\ &= \big[-(\mathbf{x}_{\|} - \mathbf{x}_{\|}') \cdot \nabla\zeta(\mathbf{x}_{\|}) + \big(\zeta(\mathbf{x}_{\|}) - \zeta(\mathbf{x}_{\|}')\big)\big]\left(\frac{i(\omega/c)}{R} - \frac{1}{R^2}\right) \\ &\quad \times \frac{\exp[i(\omega/c)R]}{R} \end{aligned} \tag{3.1.96}$$

with

$$R = \left[(\mathbf{x}_\parallel - \mathbf{x}'_\parallel)^2 + \left(\zeta(\mathbf{x}_\parallel) - \zeta(\mathbf{x}'_\parallel) \right)^2 \right]^{\frac{1}{2}}. \tag{3.1.97}$$

The prime on the sums in Eq. (3.1.95) indicates that the terms with $(m'_1, m'_2) = (m_1, m_2)$ are to be omitted. This is the simplest way of evaluating the Cauchy principal values of the integral in Eq. (3.1.53).

When $L(\mathbf{x}_\parallel(m_1, m_2)|\omega)$ has been obtained from the solution of Eq. (3.1.95), the scattering amplitude, Eq. (3.1.25), can be expressed as

$$R(\mathbf{q}_\parallel|\mathbf{k}_\parallel) = \frac{-i}{2(\omega/c)\cos\theta_s} r(\mathbf{q}_\parallel|\mathbf{k}_\parallel), \tag{3.1.98}$$

where

$$r(\mathbf{q}_\parallel|\mathbf{k}_\parallel) = (\Delta x)^2 \sum_{m_1=1}^{N} \sum_{m_2=1}^{N} L\left(\mathbf{x}_\parallel(m_1, m_2)|\omega\right) \times \exp\left[-i\mathbf{q}_\parallel \cdot \mathbf{x}_\parallel(m_1, m_2) - i\alpha_0(q_\parallel)\zeta\left(\mathbf{x}_\parallel(m_1, m_2)\right) \right]. \tag{3.1.99}$$

The dependence of $R(\mathbf{q}_\parallel|\mathbf{k}_\parallel)$ and $r(\mathbf{q}_\parallel|\mathbf{k}_\parallel)$ on $\mathbf{k}_\parallel$ is due to the dependence of $L(\mathbf{x}_\parallel|\omega)$ on $\mathbf{k}_\parallel$ through the dependence of $L(\mathbf{x}_\parallel|\omega)_{\mathrm{inc}}$ on $\mathbf{k}_\parallel$. The mean differential reflection coefficient (3.1.48) is then given by

$$\left\langle \frac{\partial R}{\partial \Omega_s} \right\rangle = \frac{1}{16\pi^2(N\Delta x)^2} \frac{\langle |r(\mathbf{q}_\parallel|\mathbf{k}_\parallel)|^2 \rangle}{\cos\theta_0}, \tag{3.1.100}$$

where we have used the result that $S = L^2 = (N\Delta x)^2$.

In Fig. 3.2 we present a plot of $\langle \partial R/\partial \Omega_s \rangle$ calculated by means of Eqs. (3.1.95)–(3.1.100), for the scattering of a scalar plane wave incident normally on a two-dimensional randomly rough Dirichlet surface that has been designed to act as a band-limited uniform diffuser within a rectangular domain of scattering angles. The wavelength of the incident field was $\lambda = 632.8$ nm. The roughness parameters employed in this calculation were $b = 5\lambda = 3.164$ μm, $\theta_{1m} = 10°$, and $\theta_{2m} = 20°$. The results for $N_p = 30\,000$ realizations of the surface profile function were averaged to obtain this result. We see that the angular dependence of the mean differential reflection is very nearly constant within the range $|\theta_{1s}| < 10°$, $|\theta_{2s}| < 20°$, where $\theta_{js} = cq_j/\omega$, and is essentially zero outside this range.

3.1.2.1.2. The Kirchhoff Approximation However, such rigorous numerical simulation calculations of the scattering of a scalar plane wave from a two-dimensional randomly rough Dirichlet surface are computationally intensive [3.9–3.11]. Calculations of the mean differential reflection coefficient are readily carried out on the basis of the Kirchhoff approximation, without passing to the geometrical optics limit. In this approximation the

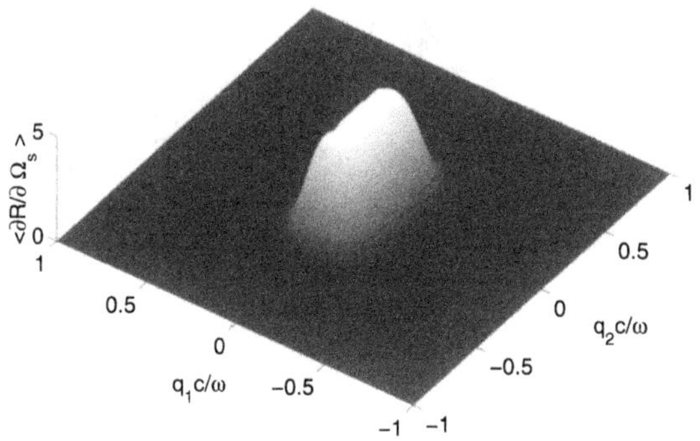

Figure 3.2. The mean differential reflection coefficient estimated from $N_p = 30\,000$ realizations of the surface profile function for the scattering of a scalar plane wave of wavelength $\lambda = 632.8$ nm incident normally on a two-dimensional randomly rough Dirichlet surface designed to act as a band-limited uniform diffuser within a rectangular region of scattering angles. The calculations were carried out on the basis of the rigorous approach defined by Eqs. (3.1.95)–(3.1.100). The values of the parameters employed in these calculations were $b = 5\lambda$, $\theta_{1m} = 10°$, and $\theta_{2m} = 20°$.

scattering amplitude in the case of normal incidence is obtained from Eq. (3.1.64) in the form

$$\begin{aligned} R(\mathbf{q}_\parallel|\mathbf{0}) &= -\frac{1}{\cos\theta_s}\int d^2x_\parallel \exp\left[-i\mathbf{q}_\parallel\cdot\mathbf{x}_\parallel - ia_0\zeta(\mathbf{x}_\parallel)\right] \\ &= -\frac{b^2}{\cos\theta_s} r_1(q_2) r_2(q_2), \end{aligned} \tag{3.1.101}$$

where

$$\begin{aligned} r_j(q_j) &= \frac{1}{b}\int\limits_{-\infty}^{\infty} dx_j \exp\left[-iq_jx_j - ia_0\zeta_j(x_j)\right] \\ &= \frac{1}{b}\sum_{n_j=-N_j}^{N_j-1}\int\limits_{n_jb}^{(n_j+1)b} dx_j \exp[-iq_jx_j - ia_0a_{n_j}x_j - ia_0b_{n_j}] \\ &= \sum_{n_j=-N_j}^{N_j-1} \exp(-ia_0b_{n_j}) \exp\left[-i(q_j + a_0a_{n_j})\left(n_j + \frac{1}{2}\right)b\right] \\ &\quad \times \operatorname{sinc}(q_j + a_0a_{n_j})\frac{b}{2}. \end{aligned} \tag{3.1.102}$$

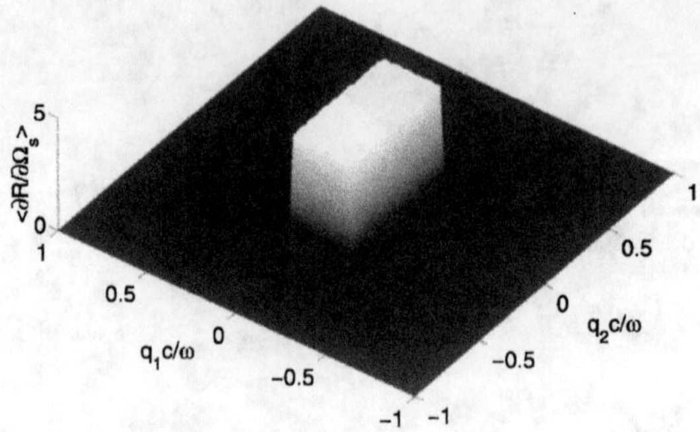

Figure 3.3. The mean differential reflection coefficient $\langle \partial R/\partial \Omega_s \rangle$ estimated from $N_p = 20\,000$ realizations of the surface profile function for the case of normal incidence and a wavelength $\lambda = 632.8$ nm. The calculations were carried out on the basis of the Kirchhoff approximation, Eqs. (3.1.102)–(3.1.103), for a two-dimensional Dirichlet surface that acts as a band-limited uniform diffuser within a rectangular region of scattering angles. The values of the parameters employed in this calculation were $b = 200$ μm, $\theta_{1m} = 10°$, and $\theta_{2m} = 20°$.

Since $\zeta_1(x_1)$ and $\zeta_2(x_2)$ are independent random processes, the substitution of Eq. (3.1.101) into Eq. (3.1.48) (with $\theta_0 = 0$ and $\mathbf{k}_\parallel = 0$) yields the mean differential reflection coefficient in the form

$$\left\langle \frac{\partial R}{\partial \Omega_s} \right\rangle = \frac{b^2}{4N_1N_2}\left(\frac{\omega}{2\pi c}\right)^2 \langle |r_1(q_1)|^2 \rangle \langle |r_2(q_2)|^2 \rangle, \tag{3.1.103}$$

where we have used the result that $S = 4N_1N_2b^2$.

3.1.2.2. Results In Fig. 3.3 we present a plot of $\langle \partial R/\partial \Omega_s \rangle$ calculated on the basis of Eqs. (3.1.102) and (3.1.103). The wavelength of the incident field was $\lambda = 632.8$ nm. The roughness parameters employed were $b = 200$ μm, $\theta_{1m} = 10°$, and $\theta_{2m} = 20°$. The results for 20 000 realizations of the surface profile function were averaged to obtain this result. It is seen that the angular dependence of the mean differential reflection coefficient is very close to the desired result. There is almost no scattered intensity outside the range $-10° < \theta_{1s} < 10°$, $-20° < \theta_{2s} < 20°$, where $\theta_{js} = cq_j/\omega$, and inside this range the mean differential reflection coefficient is nearly constant.

It is because the two-dimensional random surface considered here is a superposition of two one-dimensional random surfaces whose generators are perpendicular to each other, Eq. (3.1.72), that this is a particularly simple case to treat.

3.1.2.3. Experimental Results Surfaces with rectangular symmetry in their scattering properties can be fabricated with a simple modification of the fabrication technique pre-

Figure 3.4. Photograph of an exposed photoresist plate. A one-dimensional surface profile has been written on the left, and a two-dimensional surface profile with rectangular symmetry has been written on the right.

sented in Section 2.10. First, a one-dimensional surface profile function is impressed on the photoresist plate in the manner described there. Then, the photoresist plate is rotated by 90° and the operation is repeated changing, if desired, the exposure time to obtain a different angular range of the scattering pattern. In Fig. 3.4 we show a photograph of a plate that has, on the left, a one-dimensional pattern, and, on the right, a pattern with rectangular symmetry recorded in the manner just described.

In Fig. 3.5, we present the scattering distributions produced by a fabricated diffuser designed to scatter light uniformly in a rectangular domain. The conditions of illumination are those illustrated in Fig. 2.51. The pattern obtained with coherent illumination, shown in Fig. 3.5(a), is significant only on a fairly well-defined region and contains speckle fluctuations that, in this case, have rectangular symmetry. The use of a rotating ground glass in the illumination results in a fairly uniform distribution of intensity in a rectangular region, as shown in Fig. 3.5(b). Similarly, the use of polychromatic illumination eliminates the speckle and produces a uniform intensity distribution in the same region. This is shown in Fig. 3.5(c).

It is also illustrative to consider the results obtained with a binary diffractive optical element designed to scatter light uniformly within a square region. The scattering patterns obtained under the three different conditions of illumination are shown in Fig. 3.6. The coherent pattern, shown in Fig. 3.6(a) contains speckle in a well-defined square region, with a small amount of scattered light outside it. The use of the rotating ground glass eliminates the speckle and produces the pattern shown in Fig. 3.6(b). The result is similar to the one obtained with the refractive diffuser, which is shown in Fig. 3.5(b). However, as shown in Fig. 3.6(c), the use of white light illumination with the diffractive optical element does not produce acceptable results. In Fig. 3.7, we present a comparison of the white light scattering patterns obtained with the refractive (a) and diffractive (b) diffusers as three-dimensional images. These results illustrate one of the advantages of refractive diffusers over the diffractive ones.

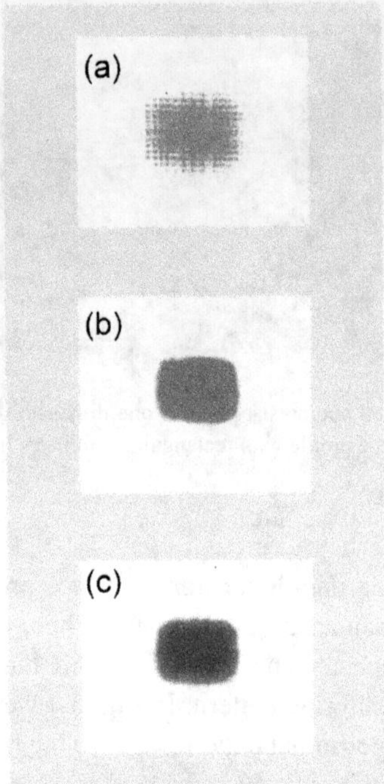

Figure 3.5. Scattering patterns produced by a uniform rectangular optical diffuser for the three cases of illumination illustrated in Fig. 2.51. (a) Coherent illumination, (b) illumination provided by a partially coherent quasi-monochromatic source and, (c) illumination provided by a small polychromatic source. The pictures are shown in reverse contrast and represent a region that is about 30°-wide.

3.1.3. The Design of Two-Dimensional Randomly Rough Surfaces that Produce a Scattered Field with Circular Symmetry

The approach described in Section 3.1.1 is effective for designing only two-dimensional randomly rough surfaces that produce a scattered field that leads to a mean differential reflection coefficient that is constant within a rectangular domain of scattering angles and vanishes outside this domain. For purposes of applications it is desirable to be able to design two-dimensional randomly rough surfaces that produce a more general specified angular dependence of the mean differential reflection coefficient in more general regions of scattering angles. Thus, in this section we present a method for designing a two-dimensional randomly rough surface which, when illuminated at normal incidence by a scalar plane wave, scatters it with a prescribed circularly symmetric distribution of intensity.

3.1.3.1. The Random Surface and its Statistical Properties The starting point for this approach is Eq. (3.1.71). The idea underlying this approach is that if the surface we seek to design is to scatter in a circularly symmetric fashion a scalar plane wave incident nor-

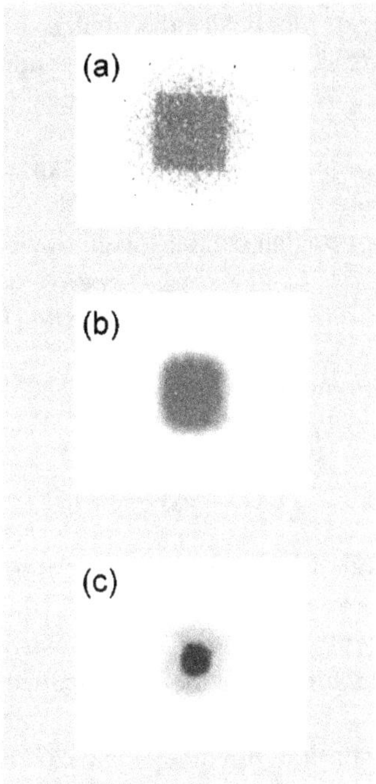

Figure 3.6. Scattering patterns produced by a uniform rectangular diffractive optical diffuser for the three cases of illumination illustrated in Fig. 2.51. (a) Coherent illumination, (b) illumination provided by a partially coherent quasi-monochromatic source and, (c) illumination provided by a small polychromatic source. The pictures are shown in reverse contrast and represent a region that is about 30°-wide.

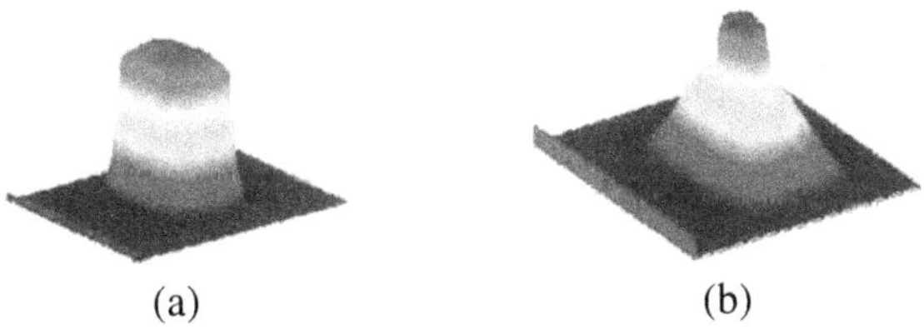

Figure 3.7. Three-dimensional representations of the scattering patterns produced by a uniform rectangular optical diffuser under the white light illumination illustrated in Fig. 2.51(c). (a) Refractive diffuser and, (b) diffractive diffuser. The x_3 axis represents the scattered intensity and the region on the x_1x_2 plane is about 30°-wide.

mally on it, the surface itself, although randomly rough, should have circular symmetry. Consequently, we assume that the surface profile function $\zeta(\mathbf{x}_\parallel)$ is a function of the radial coordinate $r = |\mathbf{x}_\parallel|$ alone, $\zeta(\mathbf{x}_\parallel) = H(r)$, and we choose $H(r)$ to have the form

$$H(r) = a_n r + b_n, \quad nb < r < (n+1)b, \ n = 0, 1, 2, \ldots, N_s - 1, \tag{3.1.104}$$

where the $\{a_n\}$ are independent, identically distributed, random deviates, b is a characteristic length, and the $\{b_n\}$ are determined from the condition that $H(r)$ is a continuous function of r. The continuity of $H(r)$ at $r=(n+1)b$ yields the condition

$$a_n(n+1)b+b_n=a_{n+1}(n+1)b+b_{n+1}. \tag{3.1.105}$$

This is a recurrence relation for b_n that can be solved to yield

$$b_n=b_0+(a_0+a_1+\cdots+a_{n-1}-na_n)b, \quad n\geqslant 1. \tag{3.1.106}$$

We can assume that $b_0=0$ with no loss of generality, and do so.

We seek the probability density function of a_n,

$$f(\gamma)=\langle\delta(\gamma-a_n)\rangle, \tag{3.1.107}$$

where the angle brackets denote an average over the ensemble of realizations of a_n, such that the surface profile function defined by Eqs. (3.1.104) and (3.1.106) yields a mean differential reflection coefficient of a prescribed circularly symmetric form. Due to the independence of the $\{a_n\}$ and the fact that they are identically distributed, the pdf $f(\gamma)$ is independent of n.

For the form of the surface profile function we have chosen, its gradient is given by

$$\nabla\zeta(\mathbf{x}_\parallel)=\hat{\mathbf{x}}_\parallel a_n, \quad nb<r<(n+1)b,\ n=0,1,\ldots,N_s-1. \tag{3.1.108}$$

The double integral in Eq. (3.1.71) therefore becomes

$$\begin{aligned}
&\int d^2u_\parallel \exp(-i\mathbf{q}_\parallel\cdot\mathbf{u}_\parallel)\int d^2x_\parallel \langle\exp[-ia_0\mathbf{u}_\parallel\cdot\nabla\zeta(\mathbf{x}_\parallel)]\rangle\\
&\quad=\int d^2u_\parallel \exp(-i\mathbf{q}_\parallel\cdot\mathbf{u}_\parallel)\\
&\qquad\times\sum_{n=0}^{N_s-1}\int_{nb}^{(n+1)b} dx_\parallel\, x_\parallel\int_{-\pi}^{\pi} d\phi_x\,\langle\exp[-ia_0a_nu_\parallel\cos(\phi_u-\phi_x)]\rangle,
\end{aligned} \tag{3.1.109}$$

where ϕ_x and ϕ_u are the azimuthal angles of the vectors $\mathbf{x}_\parallel$ and $\mathbf{u}_\parallel$, respectively. Equation (3.1.109) then becomes

$$\begin{aligned}
&\int d^2u_\parallel \exp(-i\mathbf{q}_\parallel\cdot\mathbf{u}_\parallel)\sum_{n=0}^{N_s-1}\pi[(n+1)^2b^2-n^2b^2]\langle J_0(a_0a_nu_\parallel)\rangle\\
&\quad=\int d^2u_\parallel \exp(-i\mathbf{q}_\parallel\cdot\mathbf{u}_\parallel)\sum_{n=0}^{N_s-1}\pi[(n+1)^2b^2-n^2b^2]\int_{-\infty}^{\infty} d\gamma\, f(\gamma)J_0(a_0\gamma u_\parallel)
\end{aligned}$$

$$= S \int_{-\infty}^{\infty} du_\parallel \, u_\parallel \exp\big[-iq_\parallel u_\parallel \cos(\phi_q - \phi_u)\big] \int_{-\infty}^{\infty} d\gamma \, f(\gamma) J_0(a_0 \gamma u_\parallel)$$

$$= 2\pi S \int_{-\infty}^{\infty} d\gamma \, f(\gamma) \int_{-\infty}^{\infty} du_\parallel \, u_\parallel J_0(q_\parallel u_\parallel) J_0(a_0 \gamma u_\parallel)$$

$$= 2\pi S \int_{-\infty}^{\infty} d\gamma \, f(\gamma) \frac{1}{q_\parallel} \delta(q_\parallel - a_0 \gamma)$$

$$= S \frac{2\pi}{a_0} \frac{1}{q_\parallel} f\left(\frac{q_\parallel}{a_0}\right), \tag{3.1.110}$$

where $S = \pi(N_s b)^2$. On combining Eqs. (3.1.71) and (3.1.110) we find that the mean differential reflection coefficient is given by

$$\left\langle \frac{\partial R}{\partial \Omega_s} \right\rangle = \frac{2\pi}{a_0} \left(\frac{\omega}{2\pi c}\right)^2 \frac{1}{q_\parallel} f\left(\frac{q_\parallel}{a_0}\right). \tag{3.1.111}$$

We see from Eq. (3.1.111) that the mean differential reflection coefficient is a function of the wave vector $\mathbf{q}_\parallel$ only though its magnitude $q_\parallel$. Equivalently, it is a function of the polar scattering angle θ_s alone, and is independent of the azimuthal scattering angle ϕ_s. Thus, the mean differential reflection coefficient has an angular dependence that is circularly symmetric.

If we make the change of variable

$$\gamma = \frac{q_\parallel}{a_0} = \tan \frac{\theta_s}{2}, \tag{3.1.112}$$

we find that $f(\gamma)$ is given by

$$f(\gamma) = 8\pi \frac{\gamma}{(1+\gamma^2)^2} \left\langle \frac{\partial R}{\partial \Omega_s} \right\rangle (2\tan^{-1}\gamma), \tag{3.1.113}$$

where $\langle \partial R/\partial \Omega_s \rangle (2\tan^{-1}\gamma)$ is the form that $\langle \partial R/\partial \Omega_s \rangle (\theta_s)$ takes when θ_s is replaced by γ through the use of Eq. (3.1.112).

For a particular choice of $\langle \partial R/\partial \Omega_s \rangle (\theta_s)$, from the result given by Eqs. (3.1.113) a long sequence of $\{a_n\}$ is obtained by the rejection method [3.8], and a realization of the surface profile function $H(r)$ is constructed on the basis of Eqs. (3.1.104) and (3.1.106).

3.1.3.2. Solution of the Scattering Problem To assess how well the surface designed in this fashion produces a scattered field that reproduces the mean differential reflection coefficient that served as the input to the determination of that surface, the problem of the scattering of a normally incident scalar plane wave from the surface must be solved, and the mean differential reflection coefficient determined from the solution. There are two approaches to the solution of the scattering problem that can be used.

3.1.3.2.1. A Rigorous Numerical Simulation Approach A rigorous numerical approach to the solution of the scattering problem exploits the circular symmetry of the problem. Our starting point is the representation given by Eq. (3.1.8b) for the Green's function $g_0(\mathbf{x}|\mathbf{x}')$. It can be rewritten more explicitly as

$$g_0(\mathbf{x}|\mathbf{x}') = \frac{i}{2\pi}\int_0^\infty dq \frac{q}{\alpha_0(q)} \int_{-\pi}^{\pi} d\phi_s \exp\big[iqr\cos(\phi_s - \phi_x)\big] \times \exp\big[-iqr'\cos(\phi_s - \phi_{x'})\big]\exp\big[i\alpha_0(q)|x_3 - x_3'|\big], \quad (3.1.114)$$

where ϕ_s, ϕ_x, and $\phi_{x'}$ are the azimuthal angles of the two-dimensional vectors $\mathbf{q}_\|, \mathbf{x}_\|$, and $\mathbf{x}'_\|$, respectively, while $r = |\mathbf{x}_\||$ and $r' = |\mathbf{x}'_\||$. With the use of the expansion

$$\exp(iz\cos\phi) = \sum_{\ell=-\infty}^{\infty} i^\ell \exp(i\ell\phi) J_\ell(z), \quad (3.1.115)$$

where $J_\ell(z)$ is a Bessel function of the first kind of order ℓ, the angular integration in Eq. (3.1.114) can be carried out, with the result that

$$g_0(\mathbf{x}|\mathbf{x}') = \sum_{\ell=-\infty}^{\infty} \exp\big[-i\ell(\phi_x - \phi_{x'})\big] G_\ell(r, x_3|r', x_3'), \quad (3.1.116)$$

where

$$G_\ell(r, x_3|r', x_3') = i\int_0^\infty dq \frac{q}{\alpha_0(q)} J_\ell(qr) J_\ell(qr')\exp\big[i\alpha_0(q)|x_3 - x_3'|\big] \quad (3.1.117a)$$

$$= G_{-\ell}(r_1, x_3|r_1', x_3'). \quad (3.1.117b)$$

We also expand the source function $L(\mathbf{x}_\||\omega)$ as

$$L(\mathbf{x}_\||\omega) = \sum_{m=-\infty}^{\infty} i^m \exp(im\phi_x)\ell_m(r|\omega). \quad (3.1.118)$$

Substitution of Eqs. (3.1.116) and (3.1.118) into the right-hand side of Eq. (3.1.25) yields the scattering amplitude $R(\mathbf{q}_\parallel|\mathbf{k}_\parallel)$ in the form

$$R(\mathbf{q}_\parallel|\mathbf{k}_\parallel) = -\frac{i\pi}{(\omega/c)\cos\theta_s} r(\theta_s, \phi_s), \tag{3.1.119}$$

where

$$r(\theta_s, \phi_s) = \sum_{m=-\infty}^{\infty} \exp(im\phi_s) \int_0^R dr \exp\bigl[-i(\omega/c)H(r)\cos\theta_s\bigr] \times J_m\bigl((\omega r/c)\sin\theta_s\bigr) r\ell_m(r|\omega). \tag{3.1.120}$$

Here R is the radius of the circle in the plane $x_3 = 0$ that is covered by the randomly rough surface. On combining Eqs. (3.1.48) and (3.1.119), we obtain the mean differential reflection coefficient in the form

$$\left\langle \frac{\partial R}{\partial \Omega_s} \right\rangle = \frac{1}{4\pi R^2} \frac{\langle |r(\theta_s, \phi_s)|^2 \rangle}{\cos\theta_0}. \tag{3.1.121}$$

The coefficient function $\ell_m(r|\omega)$ entering the expansion (3.1.118) satisfies an integral equation that is obtained by substituting Eqs. (3.1.116) and (3.1.118) into the right-hand side of Eq. (3.1.49), and introducing the expansion

$$\begin{aligned}\psi(\mathbf{x}|\omega)_{\mathrm{inc}}\big|_{x_3=H(r)} &= \exp\bigl[i\mathbf{k}_\parallel \cdot \mathbf{x}_\parallel - i\alpha_0(k_\parallel)H(r)\bigr] \\ &= \exp\bigl[ik_\parallel r\cos(\phi_0 - \phi_x)\bigr]\exp\bigl[-i\alpha_0(k_\parallel)H(r)\bigr] \\ &= \sum_{\ell=-\infty}^{\infty} i^\ell \exp\bigl[i\ell(\phi_0 - \phi_x)\bigr] J_\ell(k_\parallel r)\exp\bigl[-i\alpha_0(k_\parallel)H(r)\bigr]\end{aligned} \tag{3.1.122}$$

into the left-hand side. The result is

$$\begin{aligned}&2J_m\bigl((\omega r/c)\sin\theta_0\bigr)\exp\bigl[-im\phi_0 - i(\omega/c)H(r)\cos\theta_0\bigr] \\ &\quad = \int_{-\infty}^{\infty} dr'\, G_m\bigl(r, H(r)|r', H(r')\bigr) r'\ell_m(r'|\omega), \\ &m = 0, \pm 1, \pm 2, \ldots.\end{aligned} \tag{3.1.123}$$

This equation simplifies in the case of interest here, namely that of normal incidence, $\theta_0 = 0°$. By the use of the result that $J_m(0) = \delta_{m0}$, where δ_{m0} is the Kronecker symbol, we obtain the system of equations

$$2\delta_{m0}\exp\big[-i(\omega/c)H(r)\big]=\int_0^\infty dr'\,G_m\big(r,H(r)|r',H(r')\big)r'\ell_m(r'|\omega),$$

$$m=0,\pm1,\pm2,\ldots. \tag{3.1.124}$$

This system separates into two systems of equations,

$$2\exp\big[-i(\omega/c)H(r)\big]=\int_0^\infty dr'\,G_0\big(r,H(r)|r',H(r')\big)r'\ell_0(r'|\omega),\quad m=0, \tag{3.1.125a}$$

$$0=\int_0^\infty dr'\,G_m\big(r,H(r)|r',H(r')\big)r'\ell_m(r'|\omega),\quad m\neq 0. \tag{3.1.125b}$$

The homogeneous nature of Eq. (3.1.125b) has the consequence that $\ell_m(r|\omega)\equiv 0$ for $m\neq 0$. The scattering amplitude $r(\theta_s,\phi_s)$, Eq. (3.1.120), now becomes

$$r(\theta_s)=\int_0^R dr\,\exp\big[-i(\omega/c)H(r)\cos\theta_s\big]J_0\big((\omega r/c)\sin\theta_s\big)r\ell_0(r|\omega). \tag{3.1.126}$$

The mean differential reflection coefficient in the present case is obtained by replacing $r(\theta_s,\phi_s)$ by $r(\theta_s)$ in Eq. (3.1.121).

The integral equations (3.1.123) and (3.1.125a) have to be solved numerically. They have the generic form

$$g_\ell(r)=\int_0^\infty dr'\,G_\ell(r|r')f_\ell(r'), \tag{3.1.127}$$

a Fredholm equation of the first kind, where

$$G_\ell(r|r')=G_\ell\big(r,H(r)|r',H(r')\big). \tag{3.1.128}$$

To solve this equation numerically we begin by replacing the infinite upper limit by a finite one,

$$g_\ell(r)=\int_0^{R_0} dr'\,G_\ell(r|r')f_\ell(r'). \tag{3.1.129}$$

We next introduce the set of equally spaced points $\{r_n\}$ defined by

$$r_n = \left(n - \frac{1}{2}\right)\Delta r, \qquad \Delta r = R_0/N, \ n = 1, 2, \ldots, N. \tag{3.1.130}$$

Equation (3.1.129) can then be written exactly as

$$g_\ell(r) = \sum_{n=1}^{N} \int_{r_n - \frac{1}{2}\Delta r}^{r_n + \frac{1}{2}\Delta r} dr'\, G_\ell(r|r') f_\ell(r'). \tag{3.1.131}$$

We now assume that the unknown function $f_\ell(r)$ is a slowly varying function of r in each of the intervals $(r_n - \frac{1}{2}\Delta r, r_n + \frac{1}{2}\Delta r)$. We therefore evaluate it at the midpoint of each interval, and remove it from the integral:

$$\begin{aligned} g_\ell(r) &\cong \sum_{n=1}^{N} \left\{ \int_{r_n - \frac{1}{2}\Delta r}^{r_n + \frac{1}{2}\Delta r} dr'\, G_\ell(r|r') \right\} f_\ell(r_n) \\ &= \sum_{n=1}^{N} \left\{ \int_{-\frac{1}{2}\Delta r}^{\frac{1}{2}\Delta r} du\, G_\ell(r|r_n + u) \right\} f_\ell(r_n). \end{aligned} \tag{3.1.132}$$

Finally, we set r equal to r_m $(m = 1, 2, \ldots, N)$ and obtain a matrix equation satisfied by $f(r_n)$:

$$g_\ell(r_m) = \sum_{n=1}^{N} K_\ell(r_m|r_n) f_\ell(r_n), \qquad m = 1, 2, \ldots, N, \tag{3.1.133}$$

where

$$K_\ell(r_m|r_n) = \int_{-\frac{1}{2}\Delta r}^{\frac{1}{2}\Delta r} du\, G_\ell(r_m|r_n + u). \tag{3.1.134}$$

The off-diagonal matrix element $K_\ell(r_m|r_n)$ $(m \neq n)$ to lowest order in Δr is given by

$$K_\ell(r_m|r_n) = \Delta r G_\ell(r_m|r_n), \qquad m \neq n. \tag{3.1.135}$$

The evaluation of the diagonal element $K_\ell(r_m|r_n)$ requires some care because it is shown in Appendix C that $G_\ell(r|r')$ has a logarithmic singularity as $r' \to r$:

$$G_\ell(r|r') = -\frac{1}{\pi r}\ell n \frac{\{1+[H'(r)]^2\}^{\frac{1}{2}}|r-r'|}{8r} + \frac{1}{\pi r}C_\ell(\omega r/c)$$
$$+ \text{terms that vanish as } r' \to r. \tag{3.1.136}$$

An explicit expression for $C_\ell(\omega r/c)$ is given in Appendix C. Consequently, the contribution to $K_\ell(r_m|r_m)$ of the two lowest orders in Δr is

$$K_\ell(r_m|r_m) = -\frac{\Delta r}{\pi r_m}\left\{\ell n \frac{\{1+[H'(r_m)]^2\}^{\frac{1}{2}}\Delta r}{16 e r_m} - C_\ell(\omega r_m/c)\right\} + o(\Delta r). \tag{3.1.137}$$

The convergence of the solution of Eq. (3.1.133) is tested by increasing the upper limit R_0 of the integral in Eq. (3.1.129), and also the number of abscissas used in replacing integration by summation in Eq. (3.1.131), until the solution no longer depends on R_0.

3.1.3.2.2. The Kirchhoff Approximation A computationally simpler approach to the solution of the scattering problem is provided by the Kirchhoff approximation. In this approximation the scattering amplitude $R(\mathbf{q}_\parallel|\mathbf{0})$ is obtained from Eq. (3.1.64) in the form

$$R(\mathbf{q}_\parallel|\mathbf{0}) = -\frac{1}{\cos\theta_s} r(\mathbf{q}_\parallel), \tag{3.1.138}$$

where

$$\begin{aligned} r(\mathbf{q}_\parallel) &= \int d^2x_\parallel \exp(-i\mathbf{q}_\parallel \cdot \mathbf{x}_\parallel)\exp[-ia_0 H(r)] \\ &= \sum_{n=0}^{N_s-1} \int_{nb}^{(n+1)b} dr\, r \exp[-ia_0(a_n r + b_n)] \int_{-\pi}^{\pi} d\phi_x \exp[-iq_\parallel r\cos(\phi_s - \phi_x)] \\ &= 2\pi \sum_{n=0}^{N_s-1} \exp(-ia_0 b_n) \int_{nb}^{(n+1)b} dr\, r J_0(q_\parallel r)\exp(-a_0 a_n r), \end{aligned} \tag{3.1.139}$$

where we recall that $a_0 = (\omega/c)(1+\cos\theta_s)$ and $q_\parallel = (\omega/c)\sin\theta_s$. The integrals in this expression have to be evaluated numerically. The mean differential reflection coefficient in this approximation is then given by

$$\left\langle \frac{\partial R}{\partial \Omega_s} \right\rangle = \frac{1}{S}\left(\frac{\omega}{2\pi c}\right)^2 \langle |r(\mathbf{q}_\parallel)|^2 \rangle, \tag{3.1.140}$$

where the area S is $S = \pi(N_s b)^2$.

3.1.3.3. Results To illustrate the approach developed in this section we apply it to the design of a two-dimensional circularly symmetric surface that acts as a band-limited uniform diffuser, and a surface that acts as a Lambertian diffuser. We consider these two design problems in turn.

3.1.3.3.1. A Band-Limited Uniform Diffuser In designing a surface that acts as a band-limited uniform diffuser within the circular region $0 \leqslant \theta_s < \theta_m$ and $0 \leqslant \phi_s \leqslant 2\pi$. The mean differential reflection coefficient that we seek to produce is therefore

$$\left\langle \frac{\partial R}{\partial \Omega_s} \right\rangle = A\theta(\theta_m - \theta_s)\theta(\theta_s), \tag{3.1.141}$$

where A is an amplitude to be determined. Since $\tan(\theta_s/2)$ is a monotonically increasing function of θ_s for $0 \leqslant \theta_s \leqslant \pi/2$, we can rewrite Eq. (3.1.141) as

$$\left\langle \frac{\partial R}{\partial \Omega_s} \right\rangle = A\theta\big(\tan(\theta_m/2) - \tan(\theta_s/2)\big)\theta\big(\tan(\theta_s/2)\big). \tag{3.1.142}$$

The change of variable (3.1.112) transforms Eq. (3.1.142) into

$$\left\langle \frac{\partial R}{\partial \Omega_s} \right\rangle = A\theta(\gamma_m - \gamma)\theta(\gamma), \tag{3.1.143}$$

where $\gamma_m = \tan(\theta_m/2)$. It follows from Eq. (3.1.113) that the probability density function $f(\gamma)$ in this case is given by

$$f(\gamma) = 8\pi A \frac{\gamma}{(1+\gamma^2)^2}\theta(\gamma_m - \gamma)\theta(\gamma). \tag{3.1.144}$$

The normalization condition for $f(\gamma)$ is

$$\begin{aligned} \int_{-\infty}^{\infty} d\gamma\, f(\gamma) = 1 &= 8\pi A \int_0^{\gamma_m} d\gamma \frac{\gamma}{(1+\gamma^2)^2} \\ &= 4\pi A \frac{\gamma_m^2}{1+\gamma_m^2}. \end{aligned} \tag{3.1.145}$$

We find, therefore, that

$$A = \frac{1}{4\pi}\frac{1+\gamma_m^2}{\gamma_m^2}, \tag{3.1.146}$$

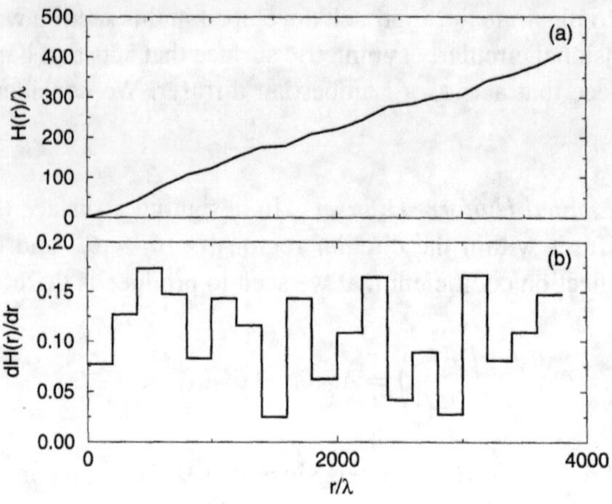

Figure 3.8. A numerically generated segment of one realization of the surface profile function $H(r)$ (a) and its derivative (b) for a two-dimensional randomly rough surface that acts as a band-limited uniform diffuser within a circular region of scattering angles. The values of the parameters employed in generating this surface were $\theta_m = 20°$ and $b = 200\lambda$.

so that

$$\left\langle \frac{\partial R}{\partial \Omega_s} \right\rangle = \frac{1}{4\pi} \frac{1+\gamma_m^2}{\gamma_m^2} \theta(\theta_m - \theta_s)\theta(\theta_s), \tag{3.1.147}$$

$$f(\gamma) = 2\frac{1+\gamma_m^2}{\gamma_m^2} \frac{\gamma}{(1+\gamma^2)^2} \theta(\gamma_m - \gamma)\theta(\gamma). \tag{3.1.148}$$

From the results given by Eq. (3.1.148) a long sequence of $\{a_n\}$ is obtained by the rejection method [3.8], and the corresponding sequence of $\{b_n\}$ is obtained from Eq. (3.1.106). A realization of the surface profile function $H(r)$ is then calculated by the use of Eq. (3.1.104). A large number of realizations of $H(r)$ is generated in this way, and for each realization the corresponding function $|r(\theta_s)|^2$ is calculated on the basis of Eq. (3.1.126). The arithmetic average of the N_p functions $|r(\theta_s)|^2$ obtained in this way yields the average indicated in Eq. (3.1.121).

In Fig. 3.8 we present a segment of one realization of the surface profile function $H(r)$ and its derivative calculated by the approach described here for the circular band-limited uniform diffuser defined by Eq. (3.1.147). The parameters used in generating these functions were $\theta_m = 20°$, and $b = 200\lambda$. Due to the form of $f(\gamma)$ given by Eq. (3.1.148), each slope a_n is positive, with the result that the surface profile function $H(r)$ increases monotonically with increasing r.

In Fig. 3.9 we present a plot of the mean differential reflection coefficient $\langle \partial R/\partial \Omega_s \rangle$ as a function of θ_s for the circular band-limited uniform diffuser defined by Eq. (3.1.147) obtained by the approach presented in this section. The parameters employed in obtain-

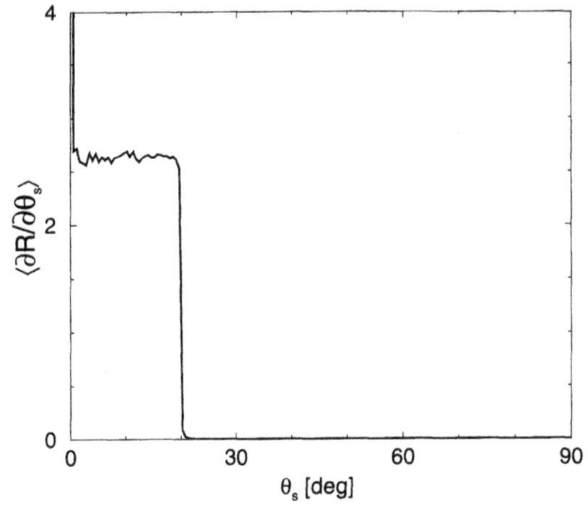

Figure 3.9. The mean differential reflection coefficient estimated from $N_p = 80\,000$ realizations of the surface profile function for the scattering of a scalar plane wave incident normally on a two-dimensional circularly symmetric randomly rough Dirichlet surface designed to act as a band-limited uniform diffuser within a circular region of scattering angles. The values of the parameters employed in this calculation were $\theta_m = 20°$, $b = 20\lambda$, and $R = Nb = 400\lambda$.

ing the result plotted in this figure were $\theta_m = 20°$, $b = 20\lambda$, and $R = Nb = 400\lambda$. The results obtained for $N_p = 80\,000$ realizations of the surface profile function were used in calculating the average indicated in Eq. (3.1.121). We see from this figure that the random surface defined by the $f(\gamma)$ given by Eq. (3.1.148) indeed acts as a band-limited uniform diffuser. There is no scattered intensity for scattering angles greater than 20°, and the cutoff is very sharp. For values of θ_s in the interval $0 \leqslant \theta_s \leqslant 20°$ the scattered intensity is very closely a constant. However, in the immediate vicinity of the specular direction $\theta_s = 0$ a peak in the mean differential reflection coefficient is present. We will return to this peak below.

3.1.3.3.2. A Lambertian Diffuser If we wish to design a surface that acts as a Lambertian diffuser, i.e., a surface that gives rise to a mean differential reflection coefficient that is proportional to the cosine of the polar scattering angle, we have that

$$\left\langle \frac{\partial R}{\partial \Omega_s} \right\rangle = A \cos\theta_s \quad 0 \leqslant \theta_s \leqslant \pi/2 \tag{3.1.149a}$$

$$= A \frac{1-\gamma^2}{1+\gamma^2} \theta(1-\gamma)\theta(\gamma), \tag{3.1.149b}$$

where the second form is obtained by the use of Eq. (3.1.112). It follows from Eq. (3.1.113)

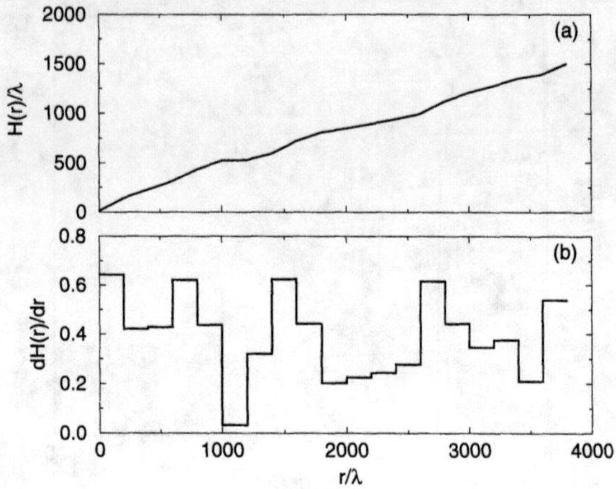

Figure 3.10. A numerically generated segment of one realization of the surface profile function $H(r)$ (a) and its derivative (b) for a two-dimensional randomly rough surface that acts as a Lambertian diffuser. The value of the parameter b assumed in generating this surface was $b = 200\lambda$.

that the probability density function $f(\gamma)$ in this case is given by

$$f(\gamma) = 8\pi A \frac{\gamma(1-\gamma^2)}{(1+\gamma^2)^3}\theta(1-\gamma)\theta(\gamma). \tag{3.1.150}$$

The normalization of $f(\gamma)$,

$$8\pi A \int_0^1 d\gamma \frac{\gamma(1-\gamma^2)}{(1+\gamma^2)^3} = \pi A = 1, \tag{3.1.151}$$

yields the result that

$$A = \frac{1}{\pi}, \tag{3.1.152}$$

from which it follows that

$$\left\langle \frac{\partial R}{\partial \Omega_s} \right\rangle = \frac{1}{\pi}\cos\theta_s, \quad 0 \leqslant \theta_s \leqslant \pi/2, \tag{3.1.153}$$

$$f(\gamma) = 8\frac{\gamma(1-\gamma^2)}{(1+\gamma^2)^3}\theta(1-\gamma)\theta(\gamma). \tag{3.1.154}$$

In Fig. 3.10 we present a plot of a segment of a single realization of the surface profile function $H(r)$ and its derivative calculated by the approach described here, for the Lam-

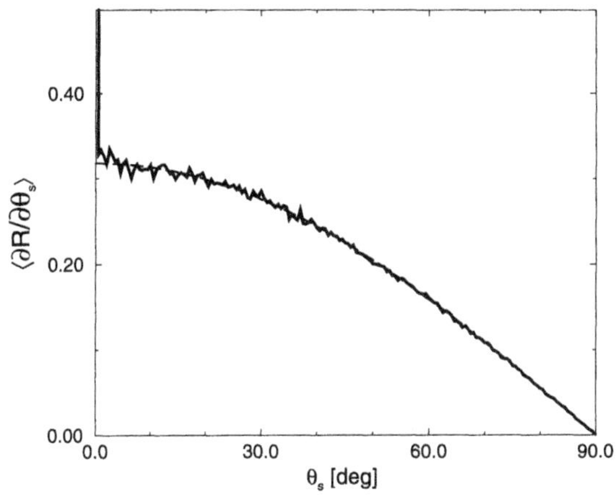

Figure 3.11. The mean differential reflection calculation estimated from $N_p = 80\,000$ realizations of the surface profile function for the scattering of a scalar plane wave incident normally on a two-dimensional circularly symmetric randomly rough Dirichlet surface designed to act as a Lambertian diffuser. The values of the parameters employed in this calculation were $b = 20\lambda$ and $R = Nb = 400\lambda$.

bertian diffuser defined by Eq. (3.1.153). The parameter b used in calculating these functions was $b = 200\lambda$. Again, because the probability density function of a_n, $f(\gamma)$, given by Eq. (3.1.154) is nonzero only for positive values of γ, each slope a_n is positive, so that the surface profile function $H(r)$ increases monotonically with increasing r.

Similarly, in Fig. 3.11 we have plotted the mean differential reflection coefficient $\langle \partial R/\partial \Omega_s \rangle$ as a function of θ_s for the Lambertian diffuser defined by Eq. (3.1.153), obtained by the method presented in this section. The parameters used in obtaining the result plotted in this figure were $b = 20\lambda$, and $R = Nb = 400\lambda$. The results obtained from $N_p = 80\,000$ realizations of the surface profile function were used in obtaining the average indicated in Eq. (3.1.121). The random surface defined by the $f(\gamma)$ given by Eq. (3.1.154) is seen from this figure to act as a Lambertian diffuser: the mean differential coefficient follows the cosine law Eq. (3.1.153) very closely, apart from the presence of a peak in the immediate vicinity of the specular direction $\theta_s = 0$.

Thus, the results presented in Figs. 3.9 and 3.11 show that although the surfaces assumed in obtaining these figures were designed on the basis of the geometrical optics limit of the Kirchhoff approximation, apart from the narrow peak in the specular direction in the mean differential reflection coefficients they produce, they nevertheless retain the scattering properties they were designed to possess when multiple-scattering processes of all orders are taken into account in calculations of scattering from them. We now turn to the discussion of these peaks.

3.1.3.4. Discussion If the coefficients $\{a_n\}$ are obtained by a straightforward application of the rejection method to the pdfs given by Eqs. (3.1.148) and (3.1.154), a fraction of the surfaces that are generated possess regions in which several consecutive a_n are very

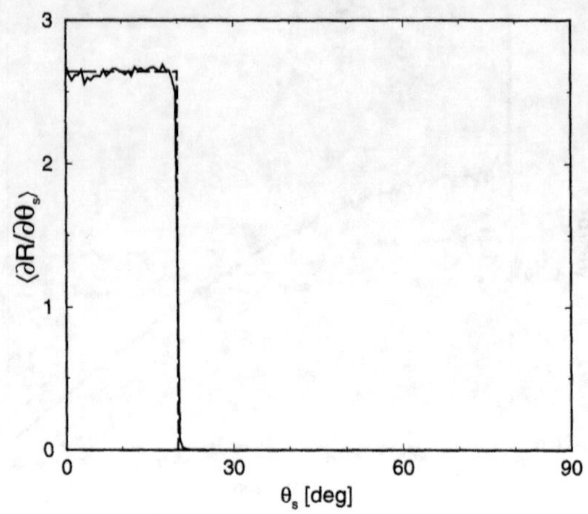

Figure 3.12. The mean differential reflection coefficient for scattering from a circular band-limited uniform diffuser, estimated from $N_p = 8000$ realizations of the surface profile function for the case of normal incidence. The values of the parameters employed in this calculation were $\theta_m = 20°$, $b = 200\lambda$, and $N_s = 200$. The dashed curve is a plot of the function defined by Eq. (3.1.147).

small ($< 10^{-2}$). The presence of such practically horizontal regions on the surface gives rise to a peak in the mean differential reflection coefficient in the immediate vicinity of the specular direction $\theta_s = 0°$. To eliminate this peak, as a particular realization of the surface profile function $H(r)$ was being generated, the corresponding differential reflection coefficient at $\theta_s = 0°$ was also calculated. If its value exceeded $2\max\langle\partial R/\partial\Omega_s\rangle$, that surface was discarded. Although this procedure alters the statistics of the resulting surface from the statistics defined by the pdf $f(\gamma)$, the altered statistics produce surfaces with the desired scattering properties, and can be used in fabricating surfaces with these properties.

The circular symmetry of the surface, which gives rise to the rings of specular points on the surface that produce constructive interference of the scattered waves in the specular direction, also contributes to the peak in the mean differential reflection coefficient in the immediate vicinity of $\theta_s = 0$. The angular width of this peak is smaller the larger $R = Nb$ is, i.e., the larger the portion of the plane $x_3 = 0$ covered by the random surface.

To show that the peak in the specular direction can be eliminated by the approaches described, we present in Figs. 3.12 and 3.13 plots of the mean differential reflection coefficients for scattering from a circular band-limited uniform diffuser and from a Lambertian diffuser, respectively, calculated in the manner just described. These calculations were carried out on the basis of the Kirchhoff approximation, Eqs. (3.1.139)–(3.1.140), since its use permits the use of larger surfaces than can be used in the rigorous computational approach.

In Fig. 3.12, we present a plot of $\langle\partial R/\partial\Omega_s\rangle$ as a function of θ_s for the circular band-limited uniform diffuser defined by Eq. (3.1.147). The values of the parameters assumed in obtaining the result plotted in this figure were the same as those assumed in obtaining the surface profile function plotted in Fig. 3.4 and the corresponding mean differential

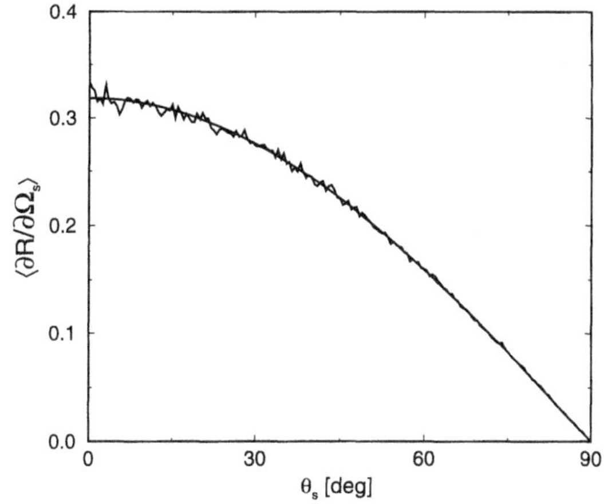

Figure 3.13. The mean differential reflection coefficient for scattering from a two-dimensional Lambertian diffuser, estimated from $N_p = 15\,000$ realizations of the surface profile function for the case of normal incidence. The values of the parameters assumed in this calculation were $b = 200\lambda$ and $N_s = 200$. The dashed curve is a plot of the function defined by Eq. (3.1.153).

reflection coefficient plotted in Fig. 3.9. The results obtained for $N_p = 8000$ realizations of the surface profile function were used in calculating the average indicated in Eq. (3.1.140). We see from this figure that the random surface defined by the pdf (3.1.148) indeed acts as a band-limited uniform diffuser: there is no scattered intensity for scattering angles larger than 20°, and the cutoff is very sharp. For values of θ_s in the interval $0 \leqslant \theta_s < 20°$ the scattered intensity is very closely a constant.

A plot of $\langle \partial R/\partial \Omega_s \rangle$ as a function of θ_s for the Lambertian diffuser defined by Eq. (3.1.153) is presented in Fig. 3.13. The values of the parameters used in obtaining the result plotted in this figure were the same as those assumed in obtaining the surface profile function plotted in Fig. 3.10 and the corresponding mean differential reflection coefficient plotted in Fig. 3.11. The results obtained for $N_p = 15\,000$ realizations of the surface profile function were used in obtaining the average appearing in Eq. (3.1.140). The random surface defined by the pdf (3.1.154) is seen to act as a Lambertian diffuser: the mean differential reflection coefficient follows the cosine law given by Eq. (3.1.153) very closely.

The results presented in Figs. 3.12 and 3.13 display no peaks in the specular direction.

The peak in the mean differential reflection coefficient in the immediate vicinity of the specular direction $\theta_s = 0$ can be also eliminated in the following way. The differential reflection coefficient $(\partial R/\partial \Omega_s)$ is calculated for a single realization of the surface profile function $H(r)$. An angular average of this function is then carried out. This, in fact, is what the detector does in a measurement of the intensity of the field scattered from a single realization of a rough surface. If the angular range over which the average is carried out is broader than that under the peak in the specular direction, the peak is eliminated from the average. In Figs. 3.14 and 3.15 we have plotted the angle-averaged differential reflection

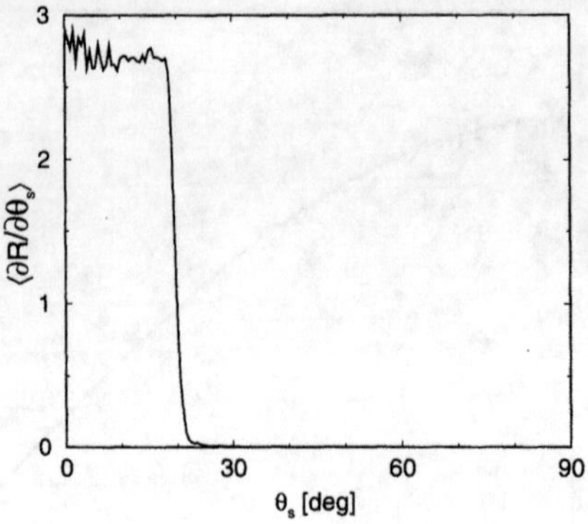

Figure 3.14. The angle-averaged differential reflection coefficient for the surface designed to act as a band-limited uniform diffuser within a circular domain of scattering angles. The values of the parameters employed in this calculation were the ones used in obtaining Fig. 3.9.

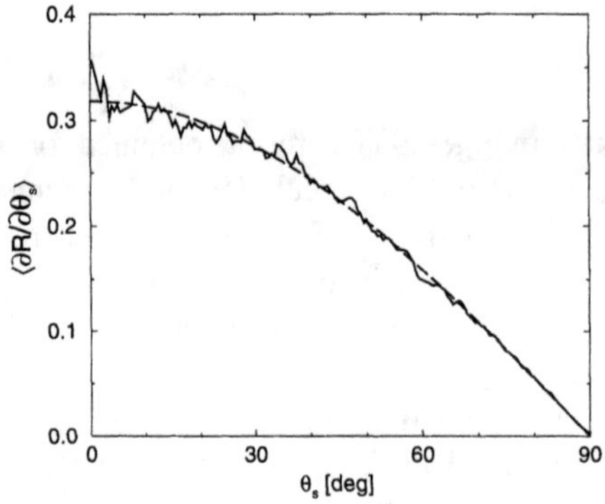

Figure 3.15. The angle-averaged differential reflection coefficient for the surface designed to act as a Lambertian diffuser. The values of the parameters employed in this calculation were those used in obtaining Fig. 3.11.

coefficients $\langle \partial R/\partial \Omega_s \rangle$ calculated for the surface designed to act as a band-limited uniform diffuser within a circular domain of scattering angles, and for the surface designed to act as a Lambertian diffuser, respectively. It is seen that for both surfaces the peak in the specular direction has been removed by the averaging. Since the angular width of the peak in the specular direction is smaller the larger the portion of the plane $x_3 = 0$ covered by the random surface is, the angular averaging is more effective when the Kirchhoff approximation is used because in this case we can use considerably longer surfaces than when using the rigorous approach.

Circularly symmetric randomly rough surfaces of the type considered here can be fabricated on photoresist by the approach described in Section 3.4.

3.1.4. The Design of Two-Dimensional Randomly Rough Surfaces, Formed from Triangular Facets, that Scatter Light in a Prescribed Fashion

In Section 3.1.2 a method was presented for designing a two-dimensional rough surface that when illuminated at normal incidence by a scalar plane wave scatters it with a constant intensity within a rectangular domain of scattering angles, and produces no scattering outside this domain. In Section 3.1.3 a method was presented for designing a surface that scatters the plane wave and produces a circularly symmetric distribution of the scattered intensity. However, the approach presented in Section 3.1.2 cannot design a surface that scatters a plane wave to produce a scattered field that has a circularly symmetric distribution of its intensity. Similarly, the approach presented in Section 3.1.3 cannot design a surface that scatters a plane wave to produce a scattered field that has a constant intensity within a rectangular domain of scattering angles, and zero intensity outside this domain. Moreover, the direct application of the latter approach yields an undesired peak in the retroreflection direction in the angular dependence of the intensity of the scattered field that requires some modification of the original approach to eliminate it.

In this section we present a method for designing a two-dimensional randomly rough Dirichlet surface which, when illuminated by a scalar plane wave, produces a scattered field with an arbitrary angular dependence of its intensity within an arbitrary domain of scattering angles. For greater generality we also assume non-normal incidence of the incident field.

Our starting point, therefore, is Eq. (3.1.70a), which we write in the form

$$\left\langle \frac{\partial R}{\partial \Omega_s} \right\rangle = \frac{(\omega/c)}{4\pi^2 S} \frac{[F(\mathbf{q}_\parallel, \mathbf{k}_\parallel)]^2}{\alpha_0(k_\parallel)} \int d^2x_\parallel \int d^2u_\parallel \exp[-i(\mathbf{q}_\parallel - \mathbf{k}_\parallel) \cdot \mathbf{u}_\parallel] \times \left\langle \exp[-ia\mathbf{u}_\parallel \cdot \nabla\zeta(\mathbf{x}_\parallel)] \right\rangle, \tag{3.1.155}$$

where

$$F(\mathbf{q}_\parallel, \mathbf{k}_\parallel) = \frac{[(\omega/c)^2 + \alpha_0(q_\parallel)\alpha_0(k_\parallel) - \mathbf{q}_\parallel \cdot \mathbf{k}_\parallel]}{[\alpha_0(q_\parallel) + \alpha_0(k_\parallel)]} \tag{3.1.156}$$

and

$$a(\theta_0, \theta_s) = \alpha_0(k_\parallel) + \alpha_0(q_\parallel) = (\omega/c)(\cos\theta_0 + \cos\theta_s). \tag{3.1.157}$$

3.1.4.1. The Random Surface and its Statistical Properties We evaluate the double integral in Eq. (3.1.155) in the following way. We begin by covering the x_1x_2 plane by equilateral triangles of edge b (Fig. 3.16). The vertices of these triangles are given by the vectors $\{\mathbf{x}_\parallel(m, n)\}$ that are defined by

$$\mathbf{x}_\parallel(m, n) = m\mathbf{a}_1 + n\mathbf{a}_2, \tag{3.1.158}$$

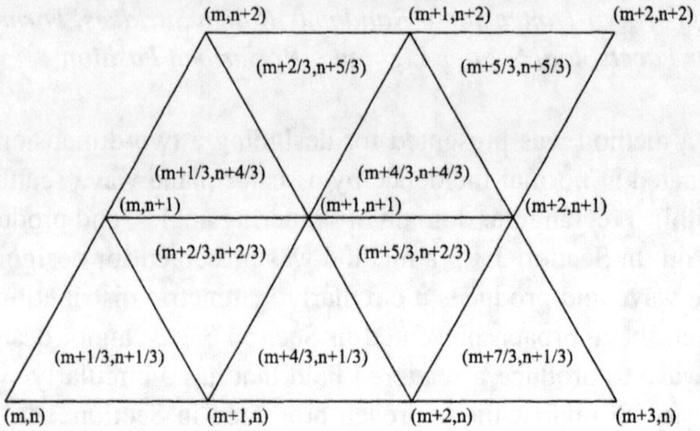

Figure 3.16. A segment of the x_1x_2 plane showing the equilateral triangles above which the triangular facets are placed that generate the two-dimensional randomly rough surface defined by Eqs. (3.1.160) and (3.1.161).

where $m, n, 0, \pm 1, \pm 2, \ldots$, and the basis vectors $\mathbf{a}_1$ and $\mathbf{a}_2$ are

$$\mathbf{a}_1 = (b, 0), \qquad \mathbf{a}_2 = \left(\frac{b}{2}, \frac{\sqrt{3}b}{2}\right). \tag{3.1.159}$$

Each triangle is labeled by the coordinates of its center of gravity. These are given by the mean values of the coordinates of its three vertices. Therefore, the triangle defined by the vertices (m, n), $(m+1, n)$, and $(m, n+1)$ is the $(m+\frac{1}{3}, n+\frac{1}{3})$ triangle. Similarly, the triangle defined by the vertices $(m+1, n)$, $(m+1, n+1)$, and $(m, n+1)$ is the $(m+\frac{2}{3}, n+\frac{2}{3})$ triangle. As m and n take the values $0, \pm 1, \pm 2, \ldots$, the $(m+\frac{1}{3}, n+\frac{1}{3})$ triangles generated constitute the subset of triangles with horizontal bases and peaks at the top, while the $(m+\frac{2}{3}, n+\frac{2}{3})$ triangles generated constitute the subset of triangles with peaks at the bottom and horizontal tops. Together these two sets of triangles cover the x_1x_2 plane.

For $\mathbf{x}_\parallel$ contained within the triangle $(m+\frac{1}{3}, n+\frac{1}{3})$ the surface profile function is assumed to be given by

$$\zeta(\mathbf{x}_\parallel) = b^{(0)}_{m+\frac{1}{3},n+\frac{1}{3}} + a^{(1)}_{m+\frac{1}{3},n+\frac{1}{3}} x_1 + a^{(2)}_{m+\frac{1}{3},n+\frac{1}{3}} x_2. \tag{3.1.160}$$

For $\mathbf{x}_\parallel$ contained within the triangle $(m+\frac{2}{3}, n+\frac{2}{3})$ the surface profile function is assumed to be given by

$$\zeta(\mathbf{x}_\parallel) = b^{(0)}_{m+\frac{2}{3},n+\frac{2}{3}} + a^{(1)}_{m+\frac{2}{3},n+\frac{2}{3}} x_1 + a^{(2)}_{m+\frac{2}{3},n+\frac{2}{3}} x_2. \tag{3.1.161}$$

The coefficients $a^{(1,2)}_{m+\frac{1}{3},n+\frac{1}{3}}$ and $a^{(1,2)}_{m+\frac{2}{3},n+\frac{2}{3}}$ are assumed to be independent identically distributed random deviates. Therefore, the joint probability density functions of the two coefficients associated with a given triangle,

$$\begin{aligned}\left\langle\delta\left(s_1 - a^{(1)}_{m+\frac{1}{3},n+\frac{1}{3}}\right)\delta\left(s_2 - a^{(2)}_{m+\frac{1}{3},n+\frac{1}{3}}\right)\right\rangle &= \left\langle\delta\left(s_1 - a^{(1)}_{m+\frac{2}{3},n+\frac{2}{3}}\right)\delta\left(s_2 - a^{(2)}_{m+\frac{2}{3},n+\frac{2}{3}}\right)\right\rangle \\ &= f(s_1, s_2), \end{aligned} \tag{3.1.162}$$

are independent of the coordinates labeling the triangles.

The double integral in Eq. (3.1.155) can now be written in the form

$$\begin{aligned} &\int d^2x_\parallel \int d^2u_\parallel \exp\left[-i(\mathbf{q}_\parallel - \mathbf{k}_\parallel)\cdot\mathbf{u}_\parallel\right]\left\langle\exp\left[-ia\mathbf{u}_\parallel\cdot\nabla\zeta(\mathbf{x}_\parallel)\right]\right\rangle \\ &= \int d^2u_\parallel \exp\left[-i(q_1-k_1)u_1 - i(q_2-k_2)u_2\right] \\ &\quad\times\sum_{m,n}\left\{\int\limits_{(m+\frac{1}{3},n+\frac{1}{3})} d^2x_\parallel \left\langle\exp\left[-iau_1a^{(1)}_{m+\frac{1}{3},n+\frac{1}{3}} - iau_2a^{(2)}_{m+\frac{1}{3},n+\frac{1}{3}}\right]\right\rangle\right. \\ &\quad+\left.\int\limits_{(m+\frac{2}{3},n+\frac{2}{3})} d^2x_\parallel \left\langle\exp\left[-iau_1a^{(1)}_{m+\frac{2}{3},n+\frac{2}{3}} - iau_2a^{(2)}_{m+\frac{2}{3},n+\frac{2}{3}}\right]\right\rangle\right\}, \end{aligned} \tag{3.1.163}$$

where the notation $\int_{(m+\frac{1}{3},n+\frac{1}{3})} d^2x_\parallel$ indicates that the integration is carried out over the area of the triangle $(m+\frac{1}{3}, n+\frac{1}{3})$. With the use of Eqs. (3.1.162) the averages in Eq. (3.1.163) can be evaluated, with the result that the double integral becomes

$$\begin{aligned} &\int d^2u_\parallel \exp\left\{\left[-i(q_1-k_1)u_1 + (q_2-k_2)u_2\right]\right\} \\ &\quad\times\sum_{m,n}\left\{\int\limits_{(m+\frac{1}{3},n+\frac{1}{3})} d^2x_\parallel \int d^2s_\parallel f(\mathbf{s}_\parallel)\exp[-iau_1s_1 - iau_2s_2]\right. \\ &\quad+\left.\int\limits_{(m+\frac{2}{3},n+\frac{2}{3})} d^2x_\parallel \int d^2s_\parallel\, f(\mathbf{s}_\parallel)\exp[-iau_1s_1 - iau_2s_2]\right\}. \end{aligned} \tag{3.1.164}$$

The integrand in each integral over $\mathbf{x}_\parallel$ is independent of $\mathbf{x}_\parallel$. Consequently each integral yields just the area of the corresponding triangle. Each sum over m and n thus gives half the area of the plane $x_3 = 0$ covered by the randomly rough surface. Therefore, the expression given by (3.1.164) becomes

$$S\int d^2u_\parallel \int d^2s_\parallel\, f(\mathbf{s}_\parallel)\exp\left[-i(q_1-k_1)u_1 - i(q_2-k_2)u_2\right]\exp\left[-ia(u_1s_1 + u_2s_2)\right]$$

$$= S \int d^2 s_\parallel \, f(\mathbf{s}_\parallel) 2\pi \delta(q_1 - k_1 + a s_1) 2\pi \delta(q_2 - k_2 + a s_2)$$

$$= S \frac{(2\pi)^2}{a^2} f\left(\frac{k_1 - q_1}{a}, \frac{k_2 - q_2}{a}\right). \tag{3.1.165}$$

With this result the mean differential reflection coefficient, Eq. (3.1.155), takes the form

$$\left\langle \frac{\partial R}{\partial \Omega_s} \right\rangle = \frac{\omega}{c} \frac{[F(\mathbf{q}_\parallel, \mathbf{k}_\parallel)]^2}{\alpha_0(k_\parallel)[\alpha_0(q_\parallel) + \alpha_0(k_\parallel)]^2} f\left(\frac{k_1 - q_1}{a}, \frac{k_2 - q_2}{a}\right). \tag{3.1.166}$$

We can invert Eq. (3.1.166) to obtain

$$f\left(\frac{k_1 - q_1}{a}, \frac{k_2 - q_2}{a}\right) = \frac{c}{\omega} \frac{\alpha_0(k_\parallel)[\alpha_0(q_\parallel) + \alpha_0(k_\parallel)]^4}{[(\omega/c)^2 + \alpha_0(q_\parallel)\alpha_0(k_\parallel) - \mathbf{q}_\parallel \cdot \mathbf{k}_\parallel]^2} \left\langle \frac{\partial R}{\partial \Omega_s} \right\rangle (q_1, q_2). \tag{3.1.167}$$

In writing Eq. (3.1.167) we have indicated explicitly that the mean differential reflection coefficient is a function of the two-dimensional wave vector $\mathbf{q}_\parallel$. It is also a function of $\mathbf{k}_\parallel$, but we have not indicated this explicitly.

We now make the changes of variables

$$\frac{k_1 - q_1}{a} = s_1, \qquad \frac{k_2 - q_2}{a} = s_2. \tag{3.1.168}$$

To invert these equations to obtain $\mathbf{q}_\parallel$ in terms of $\mathbf{k}_\parallel$ and $\mathbf{s}_\parallel$ we begin by writing them with the aid of Eqs. (3.1.41) as

$$\sin\theta_s \cos\phi_s = \sin\theta_0 \cos\phi_0 - s_1(\cos\theta_0 + \cos\theta_s), \tag{3.1.169a}$$

$$\sin\theta_s \sin\phi_s = \sin\theta_0 \sin\phi_0 - s_2(\cos\theta_0 + \cos\theta_s). \tag{3.1.169b}$$

On squaring each equation and adding the results we obtain a quadratic equation for $\cos\theta_s$

$$\left(1 + s_\parallel^2\right)\cos^2\theta_s - 2(a_1 s_1 + a_2 s_2)\cos\theta_s + a_1^2 + a_2^2 - 1 = 0, \tag{3.1.170}$$

where

$$a_1 = \sin\theta_0 \cos\phi_0 - s_1 \cos\theta_0, \tag{3.1.171a}$$

$$a_2 = \sin\theta_0 \sin\phi_0 - s_2 \cos\theta_0. \tag{3.1.171b}$$

The solution of Eq. (3.1.170) is

$$\cos\theta_s = \frac{1}{1 + s_\parallel^2}\left[A \sin\theta_0 - s_\parallel^2 \cos\theta_0 \pm (A \sin\theta_0 + \cos\theta_0)\right], \tag{3.1.172}$$

where

$$A = s_1 \cos\phi_0 + s_2 \sin\phi_0. \tag{3.1.173}$$

The "+" sign has to be chosen on the right-hand side of Eq. (3.1.172) in order that the expression for $\cos\theta_s$ reduce to the result $\cos\theta_s = (1 - s_\parallel^2)/(1 + s_\parallel^2)$, rather than to the unphysical result $\cos\theta_s = -1$, when $\theta_0 = 0$. With a second use of Eqs. (3.1.41) we find that

$$\cos\theta_s = \frac{c}{\omega}\frac{1}{1+s_\parallel^2}\left[(1 - s_\parallel^2)\alpha_0(k_\parallel) + 2\mathbf{s}_\parallel \cdot \mathbf{k}_\parallel\right]. \tag{3.1.174}$$

From this result and Eqs. (3.1.169) we obtain finally

$$q_1 = k_1 - 2s_1\frac{\alpha_0(k_\parallel) + \mathbf{s}_\parallel \cdot \mathbf{k}_\parallel}{1+s_\parallel^2}, \tag{3.1.175a}$$

$$q_2 = k_2 - 2s_2\frac{\alpha_0(k_\parallel) + \mathbf{s}_\parallel \cdot \mathbf{k}_\parallel}{1+s_\parallel^2}. \tag{3.1.175b}$$

Equation (3.1.167) then becomes

$$f(s_1, s_2) = \frac{c}{\omega}\frac{\alpha_0(k_\parallel)[\alpha_0(k_\parallel) + \alpha_0(|\mathbf{k}_\parallel - B\mathbf{s}_\parallel|)]^4}{[\alpha_0^2(k_\parallel) + \alpha_0(k_\parallel)\alpha_0(|\mathbf{k}_\parallel - B\mathbf{s}_\parallel|) + B\mathbf{k}_\parallel \cdot \mathbf{s}_\parallel]^2} \times \left\langle\frac{\partial R}{\partial\Omega_s}\right\rangle(k_1 - Bs_1, k_2 - Bs_2), \tag{3.1.176}$$

where we have introduced the function

$$B = 2\frac{\alpha_0(k_\parallel) + \mathbf{k}_\parallel \cdot \mathbf{s}_\parallel}{1+s_\parallel^2}. \tag{3.1.177}$$

The dependence of $f(s_1, s_2)$ on $\mathbf{k}_\parallel$ means that if we wish to obtain the same function $\langle\partial R/\partial\Omega_s\rangle(q_1, q_2)$ for different angles of incidence, a new surface has to be generated for each set of values of (θ_0, ϕ_0).

For constructing the surface we will also need the (marginal) pdf of $a^{(1)}_{m+\frac{1}{3},n+\frac{1}{3}}$ and of $a^{(1)}_{m+\frac{2}{3},n+\frac{2}{3}}$,

$$f(s_1) = \left\langle\delta\left(s_1 - a^{(1)}_{m+\frac{1}{3},n+\frac{1}{3}}\right)\right\rangle = \left\langle\delta\left(s_1 - a^{(1)}_{m+\frac{2}{3},n+\frac{2}{3}}\right)\right\rangle = \int_{-\infty}^{\infty} ds_2\, f(s_1, s_2), \tag{3.1.178}$$

and the conditional pdf of $a^{(2)}_{m+\frac{1}{3},n+\frac{1}{3}}$ $(a^{(2)}_{m+\frac{2}{3},n+\frac{2}{3}})$ given $a^{(1)}_{m+\frac{1}{3},n+\frac{1}{3}}$ $(a^{(1)}_{m+\frac{2}{3},n+\frac{2}{3}})$,

$$f(s_2|s_1) = \frac{f(s_1, s_2)}{f(s_1)}. \tag{3.1.179}$$

The physical meaning of the coefficients $a^{(1,2)}_{m+\frac{1}{3},n+\frac{1}{3}}$ $(a^{(1,2)}_{m+\frac{2}{3},n+\frac{2}{3}})$ is that they are the derivatives of the surface profile function $\zeta(\mathbf{x}_\|)$ with respect to x_1 and x_2, respectively, when $\mathbf{x}_\|$ is within the triangle $(m+\frac{1}{3}, n+\frac{1}{3})$ $((m+\frac{2}{3}, n+\frac{2}{3}))$. That is to say, they are the derivatives of $\zeta(\mathbf{x}_\|)$ along the horizontal edges of these triangles and along the normals to these edges inside the triangles.

To construct the surface we will also need the joint pdf of the derivatives of $\zeta(\mathbf{x}_\|)$ along the other edges of the triangles and along the normals to these edges, and the corresponding marginal and conditional pdfs. Thus, let us consider the triangle $(m+\frac{1}{3}, n+\frac{1}{3})$. The unit vector along the left edge of this triangle is $(\frac{1}{2}, \frac{\sqrt{3}}{2})$. If we denote the derivative of $\zeta(\mathbf{x}_\|)$, within this triangle, along this vector by $\alpha^{(1)}_{m+\frac{1}{3},n+\frac{1}{3}}$, we obtain

$$\alpha^{(1)}_{m+\frac{1}{3},n+\frac{1}{3}} = \left(\frac{1}{2}, \frac{\sqrt{3}}{2}\right) \cdot \nabla\zeta(\mathbf{x}_\|) = \frac{1}{2}a^{(1)}_{m+\frac{1}{3},n+\frac{1}{3}} + \frac{\sqrt{3}}{2}a^{(2)}_{m+\frac{1}{3},n+\frac{1}{3}}. \tag{3.1.180a}$$

The unit vector along the normal to this edge, directed into the triangle, is $(\frac{\sqrt{3}}{2}, -\frac{1}{2})$. If we denote the derivative of $\zeta(\mathbf{x}_\|)$ within this triangle along this vector by $\alpha^{(2)}_{m+\frac{1}{3},n+\frac{1}{3}}$ we obtain

$$\alpha^{(2)}_{m+\frac{1}{3},n+\frac{1}{3}} = \left(\frac{\sqrt{3}}{2}, -\frac{1}{2}\right) \cdot \nabla\zeta(\mathbf{x}_\|) = \frac{\sqrt{3}}{2}a^{(1)}_{m+\frac{1}{3},n+\frac{1}{3}} - \frac{1}{2}a^{(2)}_{m+\frac{1}{3},n+\frac{1}{3}}. \tag{3.1.180b}$$

The inverses of Eqs. (3.1.180a) and (3.1.180b) are

$$a^{(1)}_{m+\frac{1}{3},n+\frac{1}{3}} = \frac{1}{2}\alpha^{(1)}_{m+\frac{1}{3},n+\frac{1}{3}} + \frac{\sqrt{3}}{2}\alpha^{(2)}_{m+\frac{1}{3},n+\frac{1}{3}}, \tag{3.1.181a}$$

$$a^{(2)}_{m+\frac{1}{3},n+\frac{1}{3}} = \frac{\sqrt{3}}{2}\alpha^{(1)}_{m+\frac{1}{3},n+\frac{1}{3}} - \frac{1}{2}\alpha^{(2)}_{m+\frac{1}{3},n+\frac{1}{3}}. \tag{3.1.181b}$$

Let us now consider the triangle $(m+\frac{2}{3}, n+\frac{2}{3})$. The unit vector along the left edge of this triangle is $(-\frac{1}{2}, \frac{\sqrt{3}}{2})$. We denote the derivative of $\zeta(\mathbf{x}_\|)$ within this triangle along the vector by $\beta^{(1)}_{m+\frac{1}{3},n+\frac{1}{3}}$ and obtain

$$\beta^{(1)}_{m+\frac{1}{3},n+\frac{1}{3}} = \left(-\frac{1}{2}, \frac{\sqrt{3}}{2}\right) \cdot \nabla\zeta(\mathbf{x}_\|) = -\frac{1}{2}a^{(1)}_{m+\frac{2}{3},n+\frac{2}{3}} + \frac{\sqrt{3}}{2}a^{(2)}_{m+\frac{2}{3},n+\frac{2}{3}}. \tag{3.1.182a}$$

The unit vector perpendicular to this vector, directed into the triangle is $(\frac{\sqrt{3}}{2}, \frac{1}{2})$. We denote the derivative of $\zeta(\mathbf{x}_\|)$ within this triangle along this vector by $\beta^{(2)}_{m+\frac{2}{3},n+\frac{2}{3}}$, and obtain

$$\beta^{(2)}_{m+\frac{2}{3},n+\frac{2}{3}} = \left(\frac{\sqrt{3}}{2}, \frac{1}{2}\right) \cdot \nabla\zeta(\mathbf{x}_\|) = \frac{\sqrt{3}}{2} a^{(1)}_{m+\frac{2}{3},n+\frac{2}{3}} + \frac{1}{2} a^{(2)}_{m+\frac{2}{3},n+\frac{2}{3}}. \tag{3.1.182b}$$

The inverses of Eqs. (3.1.182a) and (3.1.182b) are

$$a^{(1)}_{m+\frac{2}{3},n+\frac{2}{3}} = -\frac{1}{2}\beta^{(1)}_{m+\frac{2}{3},n+\frac{2}{3}} + \frac{\sqrt{3}}{2}\beta^{(2)}_{m+\frac{2}{3},n+\frac{2}{3}}, \tag{3.1.183a}$$

$$a^{(2)}_{m+\frac{2}{3},n+\frac{2}{3}} = \frac{\sqrt{3}}{2}\beta^{(1)}_{m+\frac{2}{3},n+\frac{2}{3}} + \frac{1}{2}\beta^{(2)}_{m+\frac{2}{3},n+\frac{2}{3}}. \tag{3.1.183b}$$

The joint pdf of $\alpha^{(1)}_{m+\frac{1}{3},n+\frac{1}{3}}$ and $\alpha^{(2)}_{m+\frac{1}{3},n+\frac{1}{3}}$ will be denoted by $\tilde{f}(s_1, s_2)$ and is given by

$$\begin{aligned}
\tilde{f}(s_1, s_2) &= \left\langle \delta\left(s_1 - \alpha^{(1)}_{m+\frac{1}{3},n+\frac{1}{3}}\right)\delta\left(s_2 - \alpha^{(2)}_{m+\frac{1}{3},n+\frac{1}{3}}\right)\right\rangle \\
&= \left\langle \delta\left(s_1 - \frac{1}{2}a^{(1)}_{m+\frac{1}{3},n+\frac{1}{3}} - \frac{\sqrt{3}}{2}a^{(2)}_{m+\frac{1}{3},n+\frac{1}{3}}\right)\right. \\
&\qquad \left. \times\, \delta\left(s_2 - \frac{\sqrt{3}}{2}a^{(1)}_{m+\frac{1}{3},n+\frac{1}{3}} + \frac{1}{2}a^{(2)}_{m+\frac{1}{3},n+\frac{1}{3}}\right)\right\rangle \\
&= \int_{-\infty}^{\infty} d\gamma_1 \int_{-\infty}^{\infty} d\gamma_2\, f(\gamma_1, \gamma_2)\delta\left(s_1 - \frac{1}{2}\gamma_1 - \frac{\sqrt{3}}{2}\gamma_2\right) \\
&\qquad \times\delta\left(s_2 - \frac{\sqrt{3}}{2}\gamma_1 + \frac{1}{2}\gamma_2\right).
\end{aligned} \tag{3.1.184}$$

We make the change of variables $\frac{1}{2}\gamma_1 + \frac{\sqrt{3}}{2}\gamma_2 = v_1$, $\frac{\sqrt{3}}{2}\gamma_1 - \frac{1}{2}\gamma_2 = v_2$, so that $\gamma_1 = \frac{1}{2}v_1 + \frac{\sqrt{3}}{2}v_2$, $\gamma_2 = \frac{\sqrt{3}}{2}v_1 - \frac{1}{2}v_2$. In this way we obtain the result that

$$\begin{aligned}
\tilde{f}(s_1, s_2) &= \int_{-\infty}^{\infty} dv_1 \int_{-\infty}^{\infty} dv_2\, f\left(\frac{1}{2}v_1 + \frac{\sqrt{3}}{2}v_2, \frac{\sqrt{3}}{2}v_1 - \frac{1}{2}v_2\right)\delta(v_1 - s_1)\delta(v_2 - s_2) \\
&= f\left(\frac{1}{2}s_1 + \frac{\sqrt{3}}{2}s_2, \frac{\sqrt{3}}{2}s_1 - \frac{1}{2}s_2\right).
\end{aligned} \tag{3.1.185}$$

The marginal pdf $\tilde{f}(s_1) = \langle \delta(s_1 - \alpha^{(1)}_{m+\frac{1}{3},n+\frac{1}{3}}) \rangle$ is then given by

$$\tilde{f}(s_1) = \int_{-\infty}^{\infty} ds_2\, f\left(\frac{1}{2}s_1 + \frac{\sqrt{3}}{2}s_2, \frac{\sqrt{3}}{2}s_1 - \frac{1}{2}s_2\right). \tag{3.1.186}$$

The joint pdf of $\beta^{(1)}_{m+\frac{2}{3},n+\frac{2}{3}}$ and $\beta^{(2)}_{m+\frac{2}{3},n+\frac{2}{3}}$ will be denoted by $\tilde{\tilde{f}}(s_1, s_2)$ and is given by

$$\begin{aligned}
\tilde{\tilde{f}}(s_1, s_2) &= \left\langle \delta\left(s_1 - \beta^{(1)}_{m+\frac{2}{3},n+\frac{2}{3}}\right)\delta\left(s_2 - \beta^{(2)}_{m+\frac{2}{3},n+\frac{2}{3}}\right)\right\rangle \\
&= \left\langle \delta\left(s_1 + \frac{1}{2}a^{(1)}_{m+\frac{2}{3},n+\frac{2}{3}} - \frac{\sqrt{3}}{2}a^{(2)}_{m+\frac{2}{3},n+\frac{2}{3}}\right)\right. \\
&\quad \left.\times \delta\left(s_2 - \frac{\sqrt{3}}{2}a^{(1)}_{m+\frac{2}{3},n+\frac{2}{3}} - \frac{1}{2}a^{(2)}_{m+\frac{2}{3},n+\frac{2}{3}}\right)\right\rangle \\
&= \int_{-\infty}^{\infty} d\gamma_1 \int_{-\infty}^{\infty} d\gamma_2\, f(\gamma_1, \gamma_2)\delta\left(s_1 + \frac{1}{2}\gamma_1 - \frac{\sqrt{3}}{2}\gamma_2\right) \\
&\quad \times \delta\left(s_2 - \frac{\sqrt{3}}{2}\gamma_1 - \frac{1}{2}\gamma_2\right).
\end{aligned} \tag{3.1.187}$$

We make the changes of variables $-\frac{1}{2}\gamma_1 + \frac{\sqrt{3}}{2}\gamma_2 = v_1$, $\frac{\sqrt{3}}{2}\gamma_1 + \frac{1}{2}\gamma_2 = v_2$, so that $\gamma = -\frac{1}{2}v_1 + \frac{\sqrt{3}}{2}v_2$, $\gamma_2 = \frac{\sqrt{3}}{2}v_1 + \frac{1}{2}v_2$. We thus obtain the result that

$$\begin{aligned}
\tilde{\tilde{f}}(s_1, s_2) &= \int_{-\infty}^{\infty} dv_1 \int_{-\infty}^{\infty} dv_2\, f\left(\frac{1}{2}v_1 + \frac{\sqrt{3}}{2}v_2, \frac{\sqrt{3}}{2}v_1 + \frac{1}{2}v_2\right)\delta(v_1 - s_1)\delta(v_2 - s_2) \\
&= f\left(-\frac{1}{2}s_1 + \frac{\sqrt{3}}{2}s_2, \frac{\sqrt{3}}{2}s_1 + \frac{1}{2}s_2\right).
\end{aligned} \tag{3.1.188}$$

The marginal pdf $\tilde{\tilde{f}}(s_1) = \langle \delta(s_1 - \beta^{(1)}_{m+\frac{2}{3},n+\frac{2}{3}}) \rangle$ is then given by

$$\tilde{\tilde{f}}(s_1) = \int_{-\infty}^{\infty} ds_2\, f\left(\frac{1}{2}s_1 + \frac{\sqrt{3}}{2}s_2, \frac{\sqrt{3}}{2}s_1 + \frac{1}{2}s_2\right). \tag{3.1.189}$$

We note that if the joint pdf $f(s_1, s_2)$ is a function of s_1 and s_2 only through the combination $s_\parallel = (s_1^2 + s_2^2)^{\frac{1}{2}}$, i.e., if the angular dependence of the mean differential reflection coefficient is circularly symmetric, then in this case

$$\tilde{f}(s_1, s_2) = \tilde{\tilde{f}}(s_1, s_2) = f(s_1, s_2), \tag{3.1.190}$$

and

$$\tilde{f}(s_1) = \tilde{\tilde{f}}(s_1) = f(s_1). \tag{3.1.191}$$

In concluding this section we note the following useful results. If we denote the heights of the surface at the three vertices of the triangle $(m + \frac{1}{3}, n + \frac{1}{3})$ by $h_{m,n}, h_{m+1,n}, h_{m,n+1}$, we find that

$$\begin{pmatrix} b^{(0)}_{m+\frac{1}{3},n+\frac{1}{3}} \\ a^{(1)}_{m+\frac{1}{3},n+\frac{1}{3}} \\ a^{(2)}_{m+\frac{1}{3},n+\frac{1}{3}} \end{pmatrix} = \begin{pmatrix} (m+n+1) & -m & -n \\ -\frac{1}{b} & \frac{1}{b} & 0 \\ -\frac{1}{\sqrt{3}b} & -\frac{1}{\sqrt{3}b} & \frac{2}{\sqrt{3}b} \end{pmatrix} \begin{pmatrix} h_{m,n} \\ h_{m+1,n} \\ h_{m,n+1} \end{pmatrix}. \tag{3.1.192}$$

In particular, we have the result

$$b^{(0)}_{m+\frac{1}{3},n+\frac{1}{3}} = (m+n+1)h_{m,n} - mh_{m+1,n} - nh_{m,n+1}. \tag{3.1.193}$$

Similarly, if we denote the heights of the surface at the three vertices of the triangle $(m + \frac{2}{3}, n + \frac{2}{3})$ by $h_{m+1,n}, h_{m+1,n}, h_{m,n+1}$, we find that

$$\begin{pmatrix} b^{(0)}_{m+\frac{2}{3},n+\frac{2}{3}} \\ a^{(1)}_{m+\frac{2}{3},n+\frac{2}{3}} \\ a^{(2)}_{m+\frac{2}{3},n+\frac{2}{3}} \end{pmatrix} = \begin{pmatrix} n+1 & -(m+n+1) & m+1 \\ 0 & \frac{1}{b} & -\frac{1}{b} \\ -\frac{2}{\sqrt{3}b} & \frac{1}{\sqrt{3}b} & \frac{1}{\sqrt{3}b} \end{pmatrix} \begin{pmatrix} h_{m+1,n} \\ h_{m+1,n+1} \\ h_{m,n+1} \end{pmatrix}. \tag{3.1.194}$$

In particular we obtain the result

$$b^{(0)}_{m+\frac{2}{3},n+\frac{2}{3}} = (n+1)h_{m+1,n} - (m+n+1)h_{m+1,n+1} + (m+1)h_{m,n+1}. \tag{3.1.195}$$

We now turn to a discussion of how the coefficients $b^{(0)}_{m+\frac{1}{3},n+\frac{1}{3}}, a^{(1,2)}_{m+\frac{1}{3},n+\frac{1}{3}}$, and $b^{(0)}_{m+\frac{2}{3},n+\frac{2}{3}}, a^{(1,2)}_{m+\frac{2}{3},n+\frac{2}{3}}$ are obtained in practice.

3.1.4.2. Construction of the Surface Let us consider the system of triangles depicted in Fig. 3.16.

To construct a realization of the surface we assume with no loss of generality that the height of the surface at the vertex (m, n) is zero, $h_{m,n} = 0$. The marginal pdf $f(s_1)$ is used with the rejection method [3.8] to obtain the coefficient $a^{(1)}_{m+\frac{1}{3},n+\frac{1}{3}}$, which yields the height $h_{m+1,n}$. The conditional pdf $f(s_2|s_1)$ is used with this value of $a^{(1)}_{m+\frac{1}{3},n+\frac{1}{3}}$ as s_1 to obtain the coefficient $a^{(2)}_{m+\frac{1}{3},n+\frac{1}{3}}$ and hence the height $h_{m,n+1}$. The coefficient $b^{(0)}_{m+\frac{1}{3},n+\frac{1}{3}}$ is determined from $h_{m,n}$, $h_{m+1,n}$, and $h_{m,n+1}$ according to Eq. (3.1.193). The surface above the triangle $(m+\frac{1}{3}, n+\frac{1}{3})$ is now specified.

The slope of the edge joining the vertices $(m+1, n)$ and $(m, n+1)$, namely $(h_{m,n+1} - h_{m+1,n})/b$ is $\beta^{(1)}_{m+\frac{2}{3},n+\frac{2}{3}}$. If this value of $\beta^{(1)}_{m+\frac{2}{3},n+\frac{2}{3}}$ is used as the value of s_1 in the marginal pdf $\tilde{\tilde{f}}(s_1)$ and in the joint pdf $\tilde{\tilde{f}}(s_1, s_2)$, the use of the conditional pdf $\tilde{\tilde{f}}(s_2|s_1)$ with the rejection method gives the values of the slope $\beta^{(2)}_{m+\frac{2}{3},n+\frac{2}{3}}$ in the direction normal to the edge joining $(m+1, n)$ and $(m, n+1)$, and hence the height $h_{m+1,n+1}$. The values of $a^{(1,2)}_{m+\frac{2}{3},n+\frac{2}{3}}$ are then obtained from these values of $\beta^{(1,2)}_{m+\frac{2}{3},n+\frac{2}{3}}$ by the use of Eq. (3.1.183). From the heights $h_{m+1,n}$, $h_{m+1,n+1}$, and $h_{m,n+1}$ the coefficient $b^{(0)}_{m+\frac{2}{3},n+\frac{2}{3}}$ is determined from Eq. (3.1.195). The surface above the triangle $(m+\frac{2}{3}, n+\frac{2}{3})$ is now specified.

Since the edge joining the vertices $(m+1, n)$ and $(m, n+1)$ is shared by the neighboring triangles $(m+\frac{1}{3}, n+\frac{1}{3})$ and $(m+\frac{2}{3}, n+\frac{2}{3})$, the surface is continuous as one passes from one of these triangles to the other across this edge.

Continuing the construction of the surface, the slope of the edge joining the vertices $(m+1, n)$ and $(m+1, n+1)$, namely $(h_{m+1,n+1} - h_{m+1,n})/b$ is $\alpha^{(1)}_{m+\frac{4}{3},n+\frac{1}{3}}$. If this value of $\alpha^{(1)}_{m+\frac{4}{3},n+\frac{1}{3}}$ is used as the value of s_1 in the marginal pdf $\tilde{f}(s_1)$ and in the joint pdf $\tilde{f}(s_1, s_2)$, the use of the conditional pdf $\tilde{f}(s_2|s_1)$ together with the rejection method gives the values of the slope $\alpha^{(2)}_{m+\frac{4}{3}n+\frac{1}{3}}$ in the direction normal to the edge joining $(m+1, n)$ and $(m+1, n+1)$, and hence the height $h_{m+2,n}$. The values of $a^{(1,2)}_{m+\frac{4}{3},n+\frac{1}{3}}$ are then obtained from these values of $\alpha^{(1,2)}_{m+\frac{4}{3},n+\frac{1}{3}}$ by the use of Eqs. (3.1.181). From the heights $h_{m+1,n}$, $h_{m+2,n}$, and $h_{m+1,n+1}$ the coefficient $b^{(0)}_{m+\frac{4}{3},n+\frac{1}{3}}$ is obtained from Eq. (3.1.193). The surface above the triangle $(m+\frac{4}{3}, n+\frac{1}{3})$ is now specified.

The edge joining the vertices $(m+1, n)$ and $(m+1, n+1)$ is shared by the neighboring triangles $(m+\frac{2}{3}, n+\frac{2}{3})$ and $(m+\frac{4}{3}, n+\frac{1}{3})$. Therefore the surface is continuous as one passes from one of these triangles to the other across this edge.

By proceeding in this fashion the contribution to the surface profile function $\zeta(\mathbf{x}_\parallel)$ from the first row of triangles is constructed.

Turning now to the second row of triangles, the slope of the edge joining the vertices $(m, n+1)$ and $(m+1, n+1)$, namely $(h_{m+1,n+1} - h_{m,n+1})/b = a^{(1)}_{m+\frac{1}{3},n+\frac{4}{3}}$, is known because the heights $h_{m,n+1}$ and $h_{m+1,n+1}$ were obtained in constructing the surface above the first row. The conditional pdf $f(s_2|s_1)$ is used with this value of $a^{(1)}_{m+\frac{1}{3},n+\frac{4}{3}}$ as s_1 and the rejection method to obtain the value of the slope $a^{(2)}_{m+\frac{1}{3},n+\frac{4}{3}}$ for the triangle

$(m+\frac{1}{3}, n+\frac{4}{3})$, and hence the height $h_{m,n+2}$. From the heights $h_{m,n+1}$, $h_{m+1,n+1}$, and $h_{m,n+2}$ the coefficient $b^{(0)}_{m+\frac{1}{3},n+\frac{4}{3}}$ is obtained from Eq. (3.1.193). The surface above the triangle $(m+\frac{1}{3}, n+\frac{4}{3})$ is now specified.

Since the edge joining the vertices $(m, n+1)$ and $(m+1, n+1)$ is shared by the neighboring triangles $(m+\frac{1}{3}, n+\frac{1}{3})$ and $(m+\frac{1}{3}, n+\frac{4}{3})$, the surface is continuous as one passes from one of these triangles to the other across this edge.

If the slope of the edge joining the vertices $(m+1, n+1)$ and $(m, n+2)$, namely $(h_{m,n+2} - h_{m+1,n+1})/b = \beta^{(1)}_{m+\frac{1}{3},n+\frac{5}{3}}$, is used as the value of s_1 in the marginal pdf $\tilde{\tilde{f}}(s_1)$ and in the joint pdf $\tilde{\tilde{f}}(s_1, s_2)$, the use of the conditional pdf $\tilde{\tilde{f}}(s_2|s_1)$ in the rejection method gives the value of the slope $\beta^{(2)}_{m+\frac{2}{3},n+\frac{5}{3}}$ in the direction normal to the edge joining $(m+1, n+1)$ and $(m, n+2)$, and therefore the height $h_{m+1,n+2}$. The values of $a^{(1,2)}_{m+\frac{2}{3},n+\frac{4}{3}}$ are then obtained from the values $\beta^{(1,2)}_{m+\frac{2}{3},n+\frac{5}{3}}$ by the use of Eqs. (3.1.183). From the heights $h_{m+1,n+1}$, $h_{m+1,n+2}$, and $h_{m,n+2}$ the coefficient $b^{(0)}_{m+\frac{2}{3},n+\frac{5}{3}}$ is obtained from Eq. (3.1.192). The surface above the triangle $(m+\frac{2}{3}, n+\frac{5}{3})$ is now specified.

The surface is continuous as one passes from the triangle $(m+\frac{1}{3}, n+\frac{4}{3})$ to the triangle $(m+\frac{2}{3}, n+\frac{5}{3})$ across the edge joining the vertices $(m+1, n+1)$ and $(m, n+2)$, since that edge is shared by both triangles.

When we turn to the triangle $(m+\frac{4}{3}, n+\frac{4}{3})$ a new feature arises: the heights of the surface at its three vertices, $h_{m+1,n+1}$, $h_{m+2,n+1}$, and $h_{m+1,n+2}$ have already been determined. The heights $h_{m+1,n+1}$ and $h_{m+2,n+1}$ were determined during the construction of the surface above the first row of triangles, while the height $h_{m+1,n+2}$ was determined in the manner described in the preceding paragraph. From these three heights the amplitude $b^{(0)}_{m+\frac{4}{3},n+\frac{4}{3}}$ and the slopes $a^{(1,2)}_{m+\frac{4}{3},n+\frac{4}{3}}$ are then obtained by means of Eq. (3.1.181). In this way the surface above the triangle $(m+\frac{4}{3}, n+\frac{4}{3})$ is specified.

The surface is continuous as one passes between the triangles $(m+\frac{2}{3}, n+\frac{5}{3})$ and $(m+\frac{4}{3}, n+\frac{4}{3})$ across the edge joining the vertices $(m+1, n+1)$ and $(m+1, n+2)$, since this edge is shared by both triangles. It is also continuous along the edge joining the vertices $(m+1, n+1)$ and $(m+2, n+1)$, since this edge is shared by the neighboring triangles $(m+\frac{4}{3}, n+\frac{4}{3})$ and $(m+\frac{5}{3}, n+\frac{2}{3})$.

The surface above the triangle $(m+\frac{5}{3}, n+\frac{5}{3})$ is determined in the same way as the surface above the triangle $(m+\frac{2}{3}, n+\frac{5}{3})$ was determined.

As one proceeds to determine the surface above the second row of triangles, every successive triangle of the type of $(m+\frac{1}{3}, n+\frac{1}{3})$ will have its heights at its three vertices determined as in the case of the triangle $(m+\frac{4}{3}, n+\frac{4}{3})$. The use of Eq. (3.1.192) then yields the values of the corresponding coefficients $b^{(0)}$, $a^{(1)}$, and $a^{(2)}$.

The surface above the third, fourth, ..., rows of triangles is determined in the same way as the surface above the second row of triangles was determined. In this way, a single realization of a two-dimensional rough surface is constructed. It is a continuous function of $\mathbf{x}_\parallel$ across each edge joining two nearest neighbor vertices because that edge is shared by two neighboring triangles, and the surface therefore is a continuous function of $\mathbf{x}_\parallel$ in its entirety, whose statistical properties are defined by the joint pdf $f(s_1, s_2)$.

It should be noted that the continuity of the surface, which is achieved by construction, determines the value of the amplitude $b^{(0)}$ for each triangle, but does not imply any correlation among the coefficients $a^{(1,2)}$ in one triangle and the coefficients $a^{(1,2)}$ in any of its six neighboring triangles. These coefficients are obtained on the assumption that they are independent, identically distributed random deviates whose joint pdf is $f(s_1, s_2)$.

We have defined the surface above each triangle in terms of the values of $b^{(0)}$, $a^{(1)}$, and $a^{(2)}$ for that triangle because the expression for the mean differential reflection coefficient $\langle \partial R/\partial \Omega_s \rangle$ in the Kirchhoff approximation is conveniently expressed in terms of these amplitudes and slopes. It is to the determination of $\langle \partial R/\partial \Omega_s \rangle$ in this approximation that we now turn.

3.1.4.3. The Kirchhoff Approximation We will illustrate the present approach to the design of random surfaces that scatter an incident field in a prescribed fashion by performing the calculations of the mean differential reflection coefficient in the Kirchhoff approximation, but without passing to the geometrical optical limit of it. The scattering amplitude $R(\mathbf{q}_\parallel|\mathbf{k}_\parallel)$ in the case of non-normal incidence is given by Eq. (3.1.64). We rewrite it as

$$R(\mathbf{q}_\parallel|\mathbf{k}_\parallel) = -\frac{F(\mathbf{q}_\parallel|\mathbf{k}_\parallel)}{\alpha_0(q_\parallel)} r(\mathbf{q}_\parallel|\mathbf{k}_\parallel), \tag{3.1.196}$$

where

$$r(\mathbf{q}_\parallel|\mathbf{k}_\parallel) = \int d^2x_\parallel \exp\left[-i(\mathbf{q}_\parallel - \mathbf{k}_\parallel)\cdot \mathbf{x}_\parallel - ia\zeta(\mathbf{x}_\parallel)\right]. \tag{3.1.197}$$

By the use of the representation of $\zeta(\mathbf{x}_\parallel)$ given by Eqs. (3.1.160) and (3.1.161), we can rewrite this expression in the form

$$\begin{aligned} r(\mathbf{q}_\parallel|\mathbf{k}_\parallel) = \sum_{m=-N}^{N-1}\sum_{n=-N}^{N-1} \Bigg\{ & \exp\left(-iab^{(0)}_{m+\frac{1}{3},n+\frac{1}{3}}\right) \\ & \times \int_{(m+\frac{1}{3},n+\frac{1}{3})} d^2x_\parallel \exp\Big[-i\left(q_1 - k_1 + aa^{(1)}_{m+\frac{1}{3},n+\frac{1}{3}}\right)x_1 \\ & - i\left(q_2 - k_2 + aa^{(2)}_{m+\frac{1}{3},n+\frac{1}{3}}\right)x_2\Big] \\ & + \exp\left(-iab^{(0)}_{m+\frac{2}{3},n+\frac{2}{3}}\right) \int_{(m+\frac{2}{3},n+\frac{2}{3})} d^2x_\parallel \exp\Big[-i\left(q_1 - k_1 + aa^{(1)}_{m+\frac{2}{3},n+\frac{2}{3}}\right)x_1 \\ & - i\left(q_2 - k_2 + aa^{(2)}_{m+\frac{2}{3},n+\frac{2}{3}}\right)x_2\Big]\Bigg\}, \end{aligned} \tag{3.1.198}$$

where $\int_{(m+\frac{1}{3},n+\frac{1}{3})} d^2x_\parallel$, for example, denotes an integral over the area of the triangle $(m+\frac{1}{3}, n+\frac{1}{3})$. The area of the x_1x_2 plane covered by the random surface is therefore $S = 8N^2(\sqrt{3}b^2/4)$. With the aid of the results

$$\int\limits_{(m+\frac{1}{3},n+\frac{1}{3})} d^2x_\parallel = \int\limits_{\frac{\sqrt{3}}{2}nb}^{\frac{\sqrt{3}}{2}(n+1)b} dx_2 \int_{mb+\frac{x_2}{\sqrt{3}}}^{(m+n+1)b-\frac{x_2}{\sqrt{3}}} dx_1, \tag{3.1.199a}$$

$$\int\limits_{(m+\frac{2}{3},n+\frac{2}{3})} d^2x_\parallel = \int\limits_{\frac{\sqrt{3}}{2}nb}^{\frac{\sqrt{3}}{2}(n+1)b} dx_2 \int\limits_{(m+n+1)b-\frac{x_2}{\sqrt{3}}}^{(m+1)b+\frac{x_2}{\sqrt{3}}} dx_1, \tag{3.1.199b}$$

the integrals in Eq. (3.1.198) can be evaluated straightforwardly with the result that

$$r(\mathbf{q}_\parallel|\mathbf{k}_\parallel) = r_1(\mathbf{q}_\parallel|\mathbf{k}_\parallel) + r_2(\mathbf{q}_\parallel|\mathbf{k}_\parallel), \tag{3.1.200}$$

where

$$\begin{aligned}
r_1(\mathbf{q}_\parallel|\mathbf{k}_\parallel) = {} & i\frac{\sqrt{3}}{2}b \sum_{m=-N}^{N-1}\sum_{n=-N}^{N-1} \frac{\exp(-iab^{(0)}_{m+\frac{1}{3},n+\frac{1}{3}})}{q_1 - k_1 + aa^{(1)}_{m+\frac{1}{3},n+\frac{1}{3}}} \\
& \times \left\{\exp\left[-i(q_1 - k_1 + aa^{(1)}_{m+\frac{1}{3},n+\frac{1}{3}})\left(m + \frac{1}{2}n + \frac{1}{2}\right)b \right.\right. \\
& \left.\left. - i\frac{\sqrt{3}}{2}(q_2 - k_2 + aa^{(2)}_{m+\frac{1}{3},n+\frac{1}{3}})\left(n + \frac{1}{2}\right)b\right]\right\} \\
& \left\{\exp\left[-i(q_1 - k_1 + aa^{(1)}_{m+\frac{1}{3},n+\frac{1}{3}})\frac{b}{4}\right] \operatorname{sinc}\left[\frac{\sqrt{3}}{4}b(q_2 - k_2 + aa^{(2)}_{m+\frac{1}{3},n+\frac{1}{3}}) \right.\right. \\
& \left. - \frac{b}{4}(q_1 - k_1 + aa^{(1)}_{m+\frac{1}{3},n+\frac{1}{3}})\right] - \exp\left[i(q_1 - k_1 + aa^{(1)}_{m+\frac{1}{3},n+\frac{1}{3}})\frac{b}{4}\right] \\
& \times \operatorname{sinc}\left[\frac{\sqrt{3}}{4}b(q_2 - k_2 + aa^{(2)}_{m+\frac{1}{3},n+\frac{1}{3}}) \right. \\
& \left.\left. + \frac{b}{4}(q_1 - k_1 + aa^{(1)}_{m+\frac{1}{3},n+\frac{1}{3}})\right]\right\},
\end{aligned} \tag{3.1.201a}$$

$$r_2(\mathbf{q}_\parallel|\mathbf{k}_\parallel) = i\frac{\sqrt{3}}{2}b \sum_{m=-N}^{N-1}\sum_{n=-N}^{N-1} \frac{\exp(-iab^{(0)}_{m+\frac{2}{3},n+\frac{2}{3}})}{q_1 - k_1 + aa^{(1)}_{m+\frac{2}{3},n+\frac{2}{3}}}$$

$$\times \left\{ \exp\left[-i\left(q_1 - k_1 + a a^{(1)}_{m+\frac{2}{3},n+\frac{2}{3}}\right)\left(m + \frac{1}{2}n + 1\right)b \right.\right.$$

$$\left.\left. - i\frac{\sqrt{3}}{2}\left(q_2 - k_2 + a a^{(2)}_{m+\frac{2}{3},n+\frac{2}{3}}\right)\left(n + \frac{1}{2}\right)b\right]\right\}$$

$$\times \left\{ \exp\left[-i\left(q_1 - k_1 + a a^{(1)}_{m+\frac{2}{3},n+\frac{2}{3}}\right)\frac{b}{4}\right] \operatorname{sinc}\left[\frac{\sqrt{3}b}{4}\left(q_2 - k_2 + a a^{(2)}_{m+\frac{2}{3},n+\frac{2}{3}}\right) \right.\right.$$

$$\left. + \frac{b}{4}\left(q_1 - k_1 + a a^{(1)}_{m+\frac{2}{3},n+\frac{2}{3}}\right)\right] - \exp\left[i\left(q_1 - k_1 + a a^{(1)}_{m+\frac{2}{3},n+\frac{2}{3}}\right)\frac{b}{4}\right]$$

$$\times \operatorname{sinc}\left[\frac{\sqrt{3}b}{4}\left(q_2 - k_2 + a a^{(2)}_{m+\frac{2}{3},n+\frac{2}{3}}\right)\right.$$

$$\left.\left. - \frac{b}{4}\left(q_1 - k_1 + a a^{(1)}_{m+\frac{2}{3},n+\frac{2}{3}}\right)\right]\right\}, \tag{3.1.201b}$$

where $\operatorname{sinc} x = \sin x / x$. The mean differential reflection coefficient is now obtained by combining Eqs. (3.1.48) and (3.1.196):

$$\left\langle \frac{\partial R}{\partial \Omega_s} \right\rangle = \frac{1}{4\pi^2 S} \left| F(\mathbf{q}_\| | \mathbf{k}_\|) \right|^2 \frac{\langle |r_1(\mathbf{q}_\| | \mathbf{k}_\|) + r_2(\mathbf{q}_\| | \mathbf{k}_\|)|^2 \rangle}{\cos\theta_0}. \tag{3.1.202}$$

To determine how well the angular dependence of the mean intensity of the field scattered from the random surface generated by the method developed in this section agrees with the mean differential reflection coefficient $\langle \partial R / \partial \Omega_s \rangle$ used as the input to this method, a large number N_p of realizations of the random surface is generated, and for each realization the corresponding value of $|r_1(\mathbf{q}_\| | \mathbf{k}_\|) + r_2(\mathbf{q}_\| | \mathbf{k}_\|)|^2$ is calculated, recalling that $q_1 = (\omega/c)\sin\theta_s \cos\phi_s$, $q_2 = (\omega/c)\sin\theta_s \sin\phi_s$, and $a = (\omega/c)(\cos\theta_0 + \cos\theta_s)$. An arithmetic average of the N_p values of $|r_1(\mathbf{q}_\| | \mathbf{k}_\|) + r_2(\mathbf{q}_\| | \mathbf{k}_\|)|^2$ obtained in this way yields the average indicated in Eq. (3.1.202). It should be noted that $\langle |r_1(\mathbf{q}_\| | \mathbf{k}_\|) + r_2(\mathbf{q}_\| | \mathbf{k}_\|)|^2 \rangle$ should increase quadratically with N if the result for $\langle \partial R / \partial \Omega_s \rangle$ is to be independent of N. (Care should be taken when calculating $r_1(\mathbf{q}_\|)$, Eq. (3.1.201a) and $r_2(\mathbf{q}_\|)$, Eq. (3.1.201b), since the denominators $q_1 - k_1 + a a^{(1)}_{m+\frac{1}{3},n+\frac{1}{3}}$ and $q_1 - k_1 + a a^{(1)}_{m+\frac{2}{3},n+\frac{2}{3}}$ can vanish. However, it is easy to check that the expressions for $r_1(\mathbf{q}_\|)$ and $r_2(\mathbf{q}_\|)$ can be represented through sinc functions of these arguments.)

3.1.4.4. Results: Non-Normal Incidence As it stands, the result for $f(s_1, s_2)$ given by Eq. (3.1.176) is sufficiently complicated that in general the marginal pdf $f(s_1)$, Eq. (3.1.178) and the conditional pdf $f(s_2|s_1)$, Eq. (3.1.179), can be obtained from it only numerically. This is also the case in obtaining $\tilde{f}(s_1, s_2)$, $\tilde{f}(s_1)$, $\tilde{f}(s_2|s_1)$, and $\tilde{\tilde{f}}(s_1, s_2)$, $\tilde{\tilde{f}}(s_1)$, $\tilde{\tilde{f}}(s_2|s_1)$. However, a reasonable approximation yields an expression for $f(s_1, s_2)$

that allows these functions to be obtained analytically in many cases. Thus, we will assume that the polar angles of incidence and scattering are small enough that we can make the replacement $\cos\theta_0 + \cos\theta_s \approx 2$. In this case we find that

$$f(s_1, s_2) \cong 4\left\langle \frac{\partial R}{\partial \Omega_s} \right\rangle \left(k_1 - \frac{2\omega}{c} s_1, k_2 - 2\frac{\omega}{c} s_2 \right). \tag{3.1.203}$$

This is the approximation to $f(s_1, s_2)$ that we will use in the remainder of this section.

As an example of the application of the method for generating a two-dimensional randomly rough Dirichlet surface that scatters a scalar plane wave in a prescribed fashion that has been formulated in this section, we will use it to design a surface that acts as a band-limited uniform diffuser within a square domain of scattering angles when illuminated by a scalar plane wave at non-normal incidence. Thus, the mean differential reflection coefficient that we wish the surface to produce is

$$\left\langle \frac{\partial R}{\partial \Omega_s} \right\rangle (q_1, q_2) = A\theta\big(q_m - |q_1|\big)\theta\big(q_m - |q_2|\big). \tag{3.1.204}$$

We also assume that $k_1 = k_2 = k = k_\parallel/\sqrt{2}$. Therefore $\langle \partial R/\partial \Omega_s \rangle (k_1 - 2(\omega/c)s_1, k_2 - 2(\omega/c)s_2)$ becomes

$$\begin{aligned} \left\langle \frac{\partial R}{\partial \Omega_s} \right\rangle &= A\theta\big(q_m - \big|k - 2(\omega/c)s_1\big|\big)\theta\big(q_m - \big|k - 2(\omega/c)s_2\big|\big) \\ &= A\theta\big((c/2\omega)(k + q_m) - s_1\big)\theta\big(s_1 - (c/2\omega)(k - q_m)\big) \\ &\quad \times \theta\big((c/2\omega)(k + q_m) - s_2\big)\theta\big(s_2 - (c/2\omega)(k - q_m)\big). \end{aligned} \tag{3.1.205}$$

From Eqs. (3.1.203) and (3.1.205) we find that $f(s_1, s_2)$ is given by

$$f(s_1, s_2) = 4A\theta(Q - s_1)\theta(s_1 - P)\theta(Q - s_2)\theta(s_2 - P), \tag{3.1.206}$$

where

$$Q = (c/2\omega)(k + q_m), \qquad P = (c/2\omega)(k - q_m). \tag{3.1.207}$$

From the normalization condition for $f(s_1, s_2)$,

$$4A \int_P^Q ds_1 \int_P^Q ds_2\, f(s_1, s_2) = 4A(Q - P)^2 = 1, \tag{3.1.208}$$

we find that

$$A = \frac{1}{4(Q - P)^2} = \frac{1}{4(cq_m/\omega)^2}. \tag{3.1.209}$$

Therefore we have

$$\left\langle \frac{\partial R}{\partial \Omega_s} \right\rangle = \frac{\theta(q_m - |q_1|)}{2(cq_m/\omega)} \frac{\theta(q_m - |q_2|)}{2(cq_m/\omega)}, \tag{3.1.210}$$

$$f(s_1, s_2) = \frac{\theta(Q - s_1)\theta(s_1 - P)}{(cq_m/\omega)} \frac{\theta(Q - s_2)\theta(s_2 - P)}{(cq_m/\omega)}. \tag{3.1.211}$$

The marginal pdf $f(s_1)$ is

$$f(s_1) = \frac{\theta(Q - s_1)\theta(s_1 - P)}{(cq_m/\omega)}, \tag{3.1.212}$$

while the conditional pdf $f(s_2|s_1)$ is

$$f(s_2|s_1) = \frac{\theta(Q - s_2)\theta(s_2 - P)}{(cq_m/\omega)}. \tag{3.1.213}$$

In what follows, rather than maintaining complete generality we will assume that $Q = \sqrt{3}P$. It follows that $(cq_m/\omega) = Q - P = (\sqrt{3} - 1)P$.

From Eqs. (3.1.185) and (3.1.211) we find that the joint pdf $\tilde{f}(s_1, s_2)$ is given by

$$\tilde{f}(s_1, s_2) = \frac{\theta(\sqrt{3}P - \frac{1}{2}s_1 - \frac{\sqrt{3}}{2}s_2)\theta(\frac{1}{2}s_1 + \frac{\sqrt{3}}{2}s_2 - P)}{(\sqrt{3} - 1)P}$$

$$\times \frac{\theta(\sqrt{3}P - \frac{\sqrt{3}}{2}s_1 + \frac{1}{2}s_2)\theta(\frac{\sqrt{3}}{2}s_1 - \frac{1}{2}s_2 - P)}{(\sqrt{3} - 1)P}. \tag{3.1.214}$$

This expression can be rewritten as

$$\tilde{f}(s_1, s_2) = \frac{1}{(\sqrt{3} - 1)^2 P^2}\theta\left(s_2 - \left(\frac{2}{\sqrt{3}}P - \frac{1}{\sqrt{3}}s_1\right)\right)\theta(-2P + \sqrt{3}s_1 - s_2)$$

$$\frac{1}{2}(\sqrt{3} + 1)P < s_1 < \sqrt{3}P \tag{3.1.215a}$$

$$= \frac{1}{(\sqrt{3} - 1)^2 P^2}\theta\left(s_2 - \left(\frac{2}{\sqrt{3}}P - \frac{1}{\sqrt{3}}s_1\right)\right)\theta\left(\left(2P - \frac{1}{\sqrt{3}}s_1\right) - s_2\right)$$

$$\sqrt{3}P < s_1 < 2P \tag{3.1.215b}$$

$$= \frac{1}{(\sqrt{3} - 1)^2 P^2}\theta(s_2 - (-2\sqrt{3}P + \sqrt{3}s_1))\theta\left(\left(2P - \frac{1}{\sqrt{3}}s_1\right) - s_2\right)$$

$$2P < s_1 < \frac{\sqrt{3}}{2}(\sqrt{3} + 1)P \tag{3.1.215c}$$

$$= 0 \quad \text{otherwise}. \tag{3.1.215d}$$

The marginal pdf $\tilde{f}(s_1)$ is then obtained in the form

$$\tilde{f}(s_1) = \frac{1}{(\sqrt{3}-1)^2 P^2} \int\limits_{\frac{2}{\sqrt{3}}P - \frac{1}{\sqrt{3}}s_1}^{-2P+\sqrt{3}s_1} ds_2$$

$$= -\frac{1}{(\sqrt{3}-1)^2 P^2}\left[-\frac{2}{\sqrt{3}}(\sqrt{3}+1)P + \frac{4}{\sqrt{3}}s_1\right]$$

$$\frac{1}{2}(\sqrt{3}+1)P < s_1 < \sqrt{3}P \tag{3.1.216a}$$

$$= \frac{1}{(\sqrt{3}-1)^2 P^2} \int\limits_{\frac{2}{\sqrt{3}}P - \frac{1}{\sqrt{3}}s_1}^{2P-\frac{1}{\sqrt{3}}s_1} ds_2 = \frac{1}{(\sqrt{3}-1)P}\frac{2}{\sqrt{3}}$$

$$\sqrt{3}P < s_1 < 2P \tag{3.1.216b}$$

$$= \frac{1}{(\sqrt{3}-1)^2 P^2} \int\limits_{-2\sqrt{3}P+\sqrt{3}s_1}^{2P-\frac{1}{\sqrt{3}}s_1} ds_2$$

$$= \frac{1}{(\sqrt{3}-1)^2 P^2}\left[2(\sqrt{3}+1)P - \frac{4}{\sqrt{3}}s_1\right]$$

$$2P < s_1 < \frac{\sqrt{3}}{2}(\sqrt{3}+1)P \tag{3.1.216c}$$

$$= 0 \quad \text{otherwise.} \tag{3.1.216d}$$

The conditional pdf $\tilde{f}(s_2|s_1)$ is given by

$$\tilde{f}(s_2|s_1) = \tilde{f}\frac{(s_1, s_2)}{\tilde{f}(s_1)}. \tag{3.1.217}$$

The joint pdf $\tilde{\tilde{f}}(s_1, s_2)$ is obtained from Eqs. (3.1.188) and (3.1.211) in the form

$$\tilde{\tilde{f}}(s_1, s_2) = \frac{\theta(\sqrt{3}P + \frac{1}{2}s_1 - \frac{\sqrt{3}}{2}s_2)\theta(-\frac{1}{2}s_1 + \frac{\sqrt{3}}{2}s_2 - P)}{(\sqrt{3}-1)P}$$

$$\times \frac{\theta(\sqrt{3}P - \frac{\sqrt{3}}{2}s_1 - \frac{1}{2}s_2)\theta(\frac{\sqrt{3}}{2}s_1 + \frac{1}{2}s_2 - P)}{(\sqrt{3}-1)P}. \tag{3.1.218}$$

This expression can be rewritten as

$$\tilde{\tilde{f}}(s_1, s_2) = \frac{1}{(\sqrt{3}-1)^2 P^2}\theta\left(s_2 - (2P - \sqrt{3}s_1)\right)\theta\left(\left(2P + \frac{1}{\sqrt{3}}s_1\right) - s_2\right)$$

$$0 < s_1 < \frac{1}{2}(\sqrt{3}-1)P \tag{3.1.219a}$$

$$= \frac{1}{(\sqrt{3}-1)^2 P^2}\theta\left(s_2 - \left(\frac{2}{\sqrt{3}}P + \frac{1}{\sqrt{3}}s_1\right)\right)\theta\left(\left(2P + \frac{1}{\sqrt{3}}s_1\right) - s_2\right)$$

$$\frac{1}{2}(\sqrt{3}-1)P < s_1 < \frac{\sqrt{3}}{2}(\sqrt{3}-1)P \tag{3.1.219b}$$

$$= \frac{1}{(\sqrt{3}-1)^2 P^2}\theta\left(s_2 - \left(\frac{2}{\sqrt{3}}P + \frac{1}{\sqrt{3}}s_1\right)\right)\theta\left((2\sqrt{3}P - \sqrt{3}s_1) - s_2\right)$$

$$\frac{\sqrt{3}}{2}(\sqrt{3}-1)P < s_1 < P \tag{3.1.219c}$$

$$= 0 \quad \text{otherwise.} \tag{3.1.219d}$$

The marginal pdf $\tilde{\tilde{f}}(s_1)$ is then

$$\tilde{\tilde{f}}(s_1) = \frac{1}{(\sqrt{3}-1)^2 P^2}\int_{2P-\sqrt{3}s_1}^{2P+\frac{1}{\sqrt{3}}s_1} ds_2 = \frac{1}{(\sqrt{3}-1)^2 P^2}\frac{4}{\sqrt{3}}s_1$$

$$0 < s_1 < \frac{1}{2}(\sqrt{3}-1)P \tag{3.1.220a}$$

$$= \frac{1}{(\sqrt{3}-1)^2 P^2}\int_{\frac{2}{\sqrt{3}}P+\frac{1}{\sqrt{3}}s_1}^{2P+\frac{1}{\sqrt{3}}s_1} ds_2 = \frac{2}{\sqrt{3}}\frac{1}{(\sqrt{3}-1)P}$$

$$\frac{1}{2}(\sqrt{3}-1)P < s_1 < \frac{\sqrt{3}}{2}(\sqrt{3}-1)P \tag{3.1.220b}$$

$$= \frac{1}{(\sqrt{3}-1)^2 P^2}\int_{\frac{2}{\sqrt{3}}P+\frac{1}{\sqrt{3}}s_1}^{2\sqrt{3}P-\sqrt{3}s_1} ds_2 = \frac{1}{(\sqrt{3}-1)^2 P^2}\frac{4}{\sqrt{3}}(P - s_1)$$

$$\frac{\sqrt{3}}{2}(\sqrt{3}-1)P < s_1 < P \tag{3.1.220c}$$

$$= 0 \quad \text{otherwise.} \tag{3.1.220d}$$

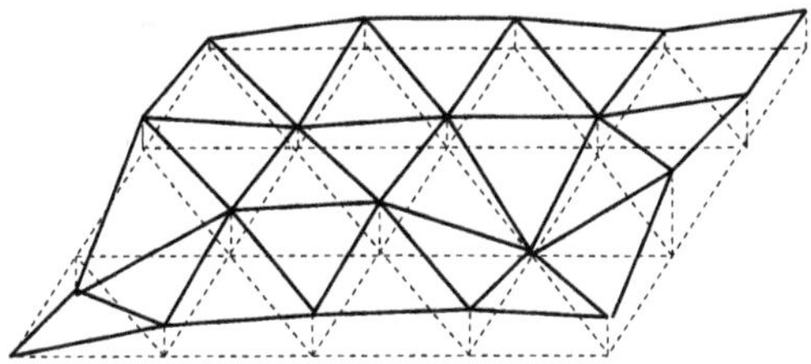

Figure 3.17. A segment of a single realization of the surface profile function for the band-limited uniform diffuser for which $f(s_1, s_2)$ is given by Eq. (3.1.211). The values of the parameters assumed in generating this surface were $q_m = 0.07(\omega/c)$, $k = 0.26(\omega/c)$, and $b = 20$ μm.

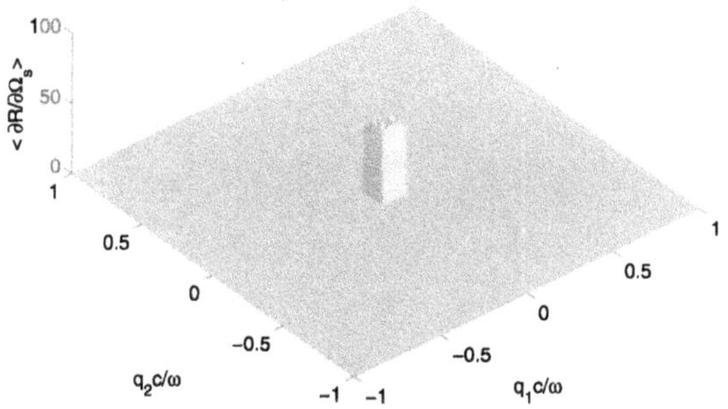

Figure 3.18. The mean differential reflection coefficient, estimated from $N_p = 10\,000$ realizations of the surface profile function for the scattering of a scalar plane wave incident obliquely on a two-dimensional randomly rough Dirichlet surface that has been designed to act as a band-limited uniform diffuser within a square region of scattering angles. The values of the parameters used in this calculation were $\lambda = 632.8$ nm, $q_m = \sin 4^\circ(\omega/c) \approx 0.07(\omega/c)$, $k = \sin 15.1^\circ(\omega/c) \approx 0.26(\omega/c)$, and $b = 20$ μm.

The conditional pdf $\tilde{\tilde{f}}(s_2|s_1)$ is then obtained as

$$\tilde{\tilde{f}}(s_1|s_2) = \frac{\tilde{\tilde{\tilde{f}}}(s_1, s_2)}{\tilde{\tilde{f}}(s_1)}. \tag{3.1.221}$$

In Fig. 3.17 we present a plot of a segment of a single realization of the surface profile function $\zeta(\mathbf{x}_\parallel)$ determined in the manner described in this section, for the case of a surface that acts as a band-limited uniform diffuser within a square region of scattering angles, under non-normal incidence, for which $f(s_1, s_2)$ is given by Eq. (3.1.211).

In Fig. 3.18 we present a plot of $\langle \partial R/\partial \Omega_s \rangle$ calculated on the basis of the Kirchhoff approximation for the surface defined by Eq. (3.1.211). The wavelength of the incident field was $\lambda = 632.8$ nm. The characteristic length b employed was $b = 20$ μm, $q_m = (\omega/c)\sin 4^\circ \approx 0.07(\omega/c)$, and $k = (\omega/c)(\sqrt{3}+1)/(\sqrt{3}-1)\sin 15.1^\circ \approx 0.26(\omega/c)$.

The results for $N_p = 10\,000$ realizations of the surface profile function were averaged to obtain the average indicated in Eq. (3.1.202). It is seen that the dependence of the mean differential reflection coefficient on the angles of scattering reproduces very well the desired result given by Eq. (3.1.210). The intensity of the scattered field vanishes for scattering angles outside the square domain defined by Eq. (3.1.210), and the cutoff is very sharp. For scattering angles within this square domain the intensity of the scattered field is very nearly constant.

3.1.4.5. Results: Normal Incidence The result given by Eq. (3.1.176) simplifies significantly in the case of normal incidence, $\mathbf{k}_\| = 0$:

$$f(s_1, s_2) = \left(\frac{c}{\omega}\right)^2 \left[(\omega/c) + \alpha_0(a_0 s_\|)\right]^2 \left\langle \frac{\partial R}{\partial \Omega_s} \right\rangle (-a_0 s_1, -a_0 s_2), \tag{3.1.222}$$

where $a_0(\theta_s)$ has been defined in Eq. (3.1.75). When $\mathbf{k}_\| = 0$, it follows from Eqs. (3.1.168) and the definitions $q_\| = (\omega/c)\sin\theta_s$, $a_0 = (\omega/c)(1 + \cos\theta_s)$, that

$$\sin\theta_s = \frac{2s_\|}{1 + s_\|^2}, \qquad \cos\theta_s = \frac{1 - s_\|^2}{1 + s_\|^2}, \tag{3.1.223}$$

so that

$$a_0 = (\omega/c)\frac{2}{1 + s_\|^2}. \tag{3.1.224}$$

It follows that $\alpha_0(a_0 s_\|) = (\omega/c)(1 - s_\|^2)/(1 + s_\|^2)$, so that Eq. (3.1.222) finally becomes

$$f(s_1, s_2) = \frac{4}{(1 + s_\|^2)^2} \left\langle \frac{\partial R}{\partial \Omega_s} \right\rangle \left(-\frac{2(\omega/c)s_1}{1 + s_\|^2}, -\frac{2(\omega/c)s_2}{1 + s_\|^2} \right). \tag{3.1.225}$$

The mean differential reflection coefficient, Eq. (3.1.202), now takes the form

$$\left\langle \frac{\partial R}{\partial \Omega_s} \right\rangle = \frac{1}{S}\left(\frac{\omega}{2\pi c}\right)^2 \left\langle \left| r_1(\mathbf{q}_\| | \mathbf{0}) + r_2(\mathbf{q}_\| | \mathbf{0}) \right|^2 \right\rangle, \tag{3.1.226}$$

where $r_1(\mathbf{q}_\| | \mathbf{0})$ and $r_2(\mathbf{q}_\| | \mathbf{0})$ are obtained from Eqs. (3.1.201a) and (3.1.201b), respectively.

To illustrate the method developed in this section for generating a two-dimensional randomly rough Dirichlet surface that scatters a scalar plane wave incident normally on it in such a way that the resulting mean differential reflection coefficient has a prescribed dependence on the scattering angles, we will apply it to the design of a) a surface that acts as a Lambertian diffuser; b) a surface that acts as a band-limited uniform diffuser within a circular region of scattering angles; c) a surface that acts as a band-limited uniform diffuser within an elliptical region of scattering angles; d) a surface that acts as a band-limited uniform diffuser within a rectangular domain of scattering angles; and e) a surface that acts as a band-limited uniform diffuser within a triangular domain of scattering angles. These examples demonstrate the versatility of the present approach to the design problem.

3.1.4.5.1. A Lambertian Diffuser A Lambertian diffuser is defined by the requirement that it produce a mean differential reflection coefficient that is proportional to the cosine of the polar scattering angle θ_s,

$$\left\langle \frac{\partial R}{\partial \Omega_s} \right\rangle (q_1, q_2) = A \frac{a_0^2 - q_\parallel^2}{a_0^2 + q_\parallel^2} \quad 0 \leqslant q_\parallel \leqslant a_0 \tag{3.1.227a}$$

$$= A \cos\theta_s \quad 0 \leqslant \theta_s \leqslant \pi/2, \tag{3.1.227b}$$

where we have used the definitions $a_0 = (\omega/c)(1 + \cos\theta_s)$ and $q_\parallel = (\omega/c)\sin\theta_s$ in obtaining Eq. (3.1.227b). With the use of Eq. (3.1.224) we find that

$$\left\langle \frac{\partial R}{\partial \Omega_s} \right\rangle \left(-\frac{2(\omega/c)s_1}{1 + s_\parallel^2}, -\frac{2(\omega/c)s_2}{1 + s_\parallel^2} \right) = A \frac{1 - s_\parallel^2}{1 + s_\parallel^2} \theta(1 - s_\parallel). \tag{3.1.228}$$

On combining this result with Eq. (3.1.225) we obtain for the joint pdf $f(s_1, s_2)$

$$f(s_1, s_2) = 4A \frac{1 - s_\parallel^2}{(1 + s_\parallel^2)^3} \theta(1 - s_\parallel). \tag{3.1.229}$$

The normalization of $f(s_1, s_2)$,

$$\int_{-\infty}^{\infty} ds_1 \int_{-\infty}^{\infty} ds_2\, f(s_1, s_2) = 1 = 8\pi A \int_0^1 ds_\parallel\, s_\parallel \frac{1 - s_\parallel^2}{(1 + s_\parallel^2)^3} = \pi A, \tag{3.1.230}$$

yields the result that $A = 1/\pi$, so that

$$\left\langle \frac{\partial R}{\partial \Omega_s} \right\rangle = \frac{1}{\pi} \cos\theta_s, \tag{3.1.231}$$

$$f(s_1, s_2) = \frac{4}{\pi} \frac{1 - s_\parallel^2}{(1 + s_\parallel^2)^3} \theta(1 - s_\parallel). \tag{3.1.232}$$

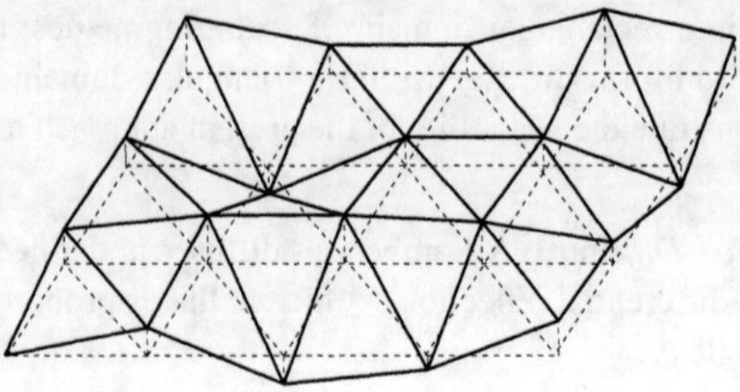

Figure 3.19. A segment of a single realization of the surface profile function for the Lambertian diffuser for which $f(s_1, s_2)$ is given by Eq. (3.1.232). The value of b assumed in generating this surface was $b = 20$ μm.

The marginal pdf $f(s_1)$ is then

$$f(s_1) = \frac{4}{\pi} \int_{-\sqrt{1-s_1^2}}^{\sqrt{1-s_1^2}} ds_2 \frac{1 - s_1^2 - s_2^2}{(1 + s_1^2 + s_2^2)^3} \theta(1 - |s_1|)$$

$$= \frac{1}{\pi} \frac{1}{(1+s_1^2)^{5/2}} \left\{ 2(1 - 2s_1^2) \tan^{-1} \sqrt{\frac{1 - s_1^2}{1 + s_1^2}} + (2 - s_1^2)\sqrt{1 - s_1^4} \right\} \theta(1 - |s_1|). \quad (3.1.233)$$

The conditional pdf $f(s_2|s_1)$ is then obtained from Eqs. (3.1.179) and the expressions given by Eqs. (3.1.232) and (3.1.233).

Due to the result that $f(s_1, s_2)$ is a function of s_1 and s_2 only through the combination $s_\parallel = (s_1^2 + s_2^2)^{\frac{1}{2}}$, the joint pdfs $\tilde{f}(s_1, s_2)$ and $\tilde{\tilde{f}}(s_1, s_2)$ in the present case are equal to $f(s_1, s_2)$. For the same reason the marginal pdfs $\tilde{f}(s_1)$ and $\tilde{\tilde{f}}(s_1)$ are equal to $f(s_1)$. The conditional pdfs $\tilde{f}(s_2|s_1)$ and $\tilde{\tilde{f}}(s_2|s_1)$ are therefore given by $f(s_2|s_1)$, Eq. (3.1.179).

In Fig. 3.19 is plotted a segment of a single realization of the surface profile function $\zeta(\mathbf{x}_\parallel)$ constructed in the manner described in Section 3.1.3.2 for the case of the Lambertian diffuser for which $f(s_1, s_2)$ is given by Eq. (3.1.232).

In Fig. 3.20 we present a plot of $\langle \partial R/\partial \Omega_s \rangle$ calculated on the basis of the Kirchhoff approximation for the Lambertian diffuser defined by Eq. (3.1.231). The wavelength of the incident field was $\lambda = 632.8$ nm. The characteristic length b employed was $b = 20$ μm. The results for $N_p = 10\,000$ realizations of the surface profile function were used in obtaining the average indicated in Eq. (3.1.226). The angular dependence of the mean differential reflection coefficient reproduces well the desired result: for $0 \leqslant \theta_s \leqslant \pi/2$ $\langle \partial R/\partial \Omega_s \rangle$ follows very closely the cosine law (3.1.231).

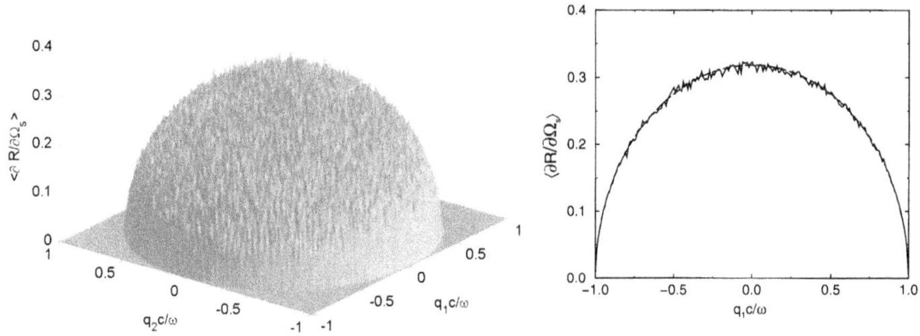

Figure 3.20. The mean differential reflection coefficient estimated from $N_p = 10\,000$ realizations of the surface profile function for the scattering of a scalar plane wave incident normally on a two-dimensional randomly rough Dirichlet surface designed to act as a Lambertian diffuser. The values of the parameters employed in this calculation were $\lambda = 632.8$ nm, and $b = 20$ μm. A typical cross section is also presented.

We note that there is no peak in $\langle \partial R/\partial \Omega_s \rangle$ in the retroreflection direction $\theta_s = 0°$, as was the case when $\langle \partial R/\partial \Omega_s \rangle$ for this diffuser was calculated by the method described in Section 3.1.3, where a circularly symmetric random surface was assumed. Although this peak could be suppressed by modifying slightly the method used to calculate $\langle \partial R/\partial \Omega_s \rangle$, its absence from the result obtained by the present approach is an attractive feature of this approach.

3.1.4.5.2. A Band-Limited Uniform Diffuser Within a Circular Domain of Scattering Angles If we seek to design a two-dimensional randomly rough surface that produces a scattered field that is characterized by a mean differential reflection coefficient that is constant within a circular domain of scattering angles, and vanishes outside this domain, the mean differential reflection coefficient we seek to reproduce can be written

$$\left\langle \frac{\partial R}{\partial \Omega_s} \right\rangle (q_1, q_2) = A\bigl(q_m - \bigl(q_1^2 + q_2^2\bigr)^{\frac{1}{2}}\bigr), \quad 0 < q_m < \omega/c. \tag{3.1.234}$$

By the use of Eq. (3.1.225) we find that the corresponding probability density function $f(s_1, s_2)$ is given by

$$f(s_1, s_2) = \frac{4A}{(1 + s_\parallel^2)^2} \theta\left(\frac{cq_m}{2\omega} - \frac{s_\parallel}{1 + s_\parallel^2}\right). \tag{3.1.235}$$

This result can be rewritten as

$$f(s_1, s_2) = \frac{4A}{(1 + s_\parallel^2)^2} \theta(s_m - s_\parallel), \tag{3.1.236}$$

where

$$s_m = \frac{\omega}{cq_m}\left\{1 - \left[1 - (cq_m/\omega)^2\right]^{\frac{1}{2}}\right\}. \tag{3.1.237}$$

From the normalization of $f(s_1, s_2)$,

$$\int d^2 s_\| \, f(s_1, s_2) = 8\pi A \int_0^{s_m} ds_\| \frac{s_\|}{(1+s_\|^2)^2} = 4\pi A \frac{s_m^2}{1+s_m^2} = 1, \tag{3.1.238}$$

we find that

$$A = \frac{1}{4\pi}\frac{1+s_m^2}{s_m^2}. \tag{3.1.239}$$

It follows, therefore, that

$$\left\langle \frac{\partial R}{\partial \Omega_s} \right\rangle (q_1, q_2) = \frac{1}{4\pi}\frac{1+s_m^2}{s_m^2}\theta(q_m - q_\|), \tag{3.1.240}$$

$$f(s_1, s_2) = \frac{1}{\pi}\frac{1+s_m^2}{s_m^2}\frac{1}{(1+s_\|^2)^2}\theta(s_m - s_\|). \tag{3.1.241}$$

The marginal pdf $f(s_1)$ is then found to be

$$f(s_1) = \frac{1}{\pi}\frac{1+s_m^2}{s_m^2}\left\{\frac{1}{1+s_m^2}\frac{(s_m^2 - s_1^2)^{\frac{1}{2}}}{1+s_1^2} + \frac{1}{(1+s_1^2)^{3/2}}\tan^{-1}\left(\frac{s_m^2 - s_1^2}{1+s_1^2}\right)^{\frac{1}{2}}\right\}\theta\left(s_m - |s_1|\right). \tag{3.1.242}$$

The conditional pdf $f(s_2|s_1)$ is obtained from Eq. (3.1.179) and the results given by Eqs. (3.1.241) and (3.1.242).

Because the joint pdf $f(s_1, s_2)$ is a function of s_1 and s_2 only through the combination $s_\| = (s_1^2 + s_2^2)^{\frac{1}{2}}$, the joint pdfs $\tilde{f}(s_1, s_2)$ and $\tilde{\tilde{f}}(s_1, s_2)$ are equal to $f(s_1, s_2)$, while the marginal pdfs $\tilde{f}(s_1)$ and $\tilde{\tilde{f}}(s_1)$ are equal to $f(s_1)$. The conditional pdfs $\tilde{f}(s_2|s_1)$ and $\tilde{\tilde{f}}(s_2|s_1)$, therefore, are the same as $f(s_2|s_1)$.

In Fig. 3.21 we present a plot of $\langle \partial R/\partial \Omega_s \rangle$ calculated by means of the Kirchhoff approximation for the band-limited uniform diffuser within a circular domain of scattering angles defined by Eq. (3.1.240). The wavelength of the incident field was $\lambda = 632.8$ nm. The characteristic length b employed was $b = 20$ μm, and $q_m = \sin 20°(\omega/c)$. The results for $N_p = 10\,000$ realizations of the surface profile function were averaged to obtain the average indicated in Eq. (3.1.226). It is seen that the angular dependence of the calculated

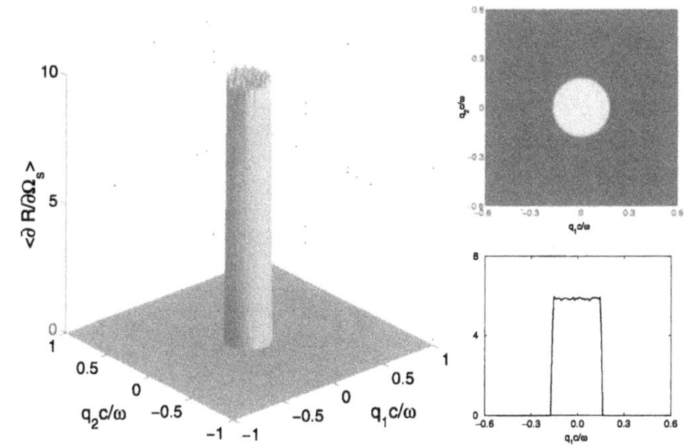

Figure 3.21. The mean differential reflection coefficient estimated from $N_p = 10\,000$ realizations of the surface profile function for the scattering of a scalar plane wave incident normally on a two-dimensional randomly rough Dirichlet surface designed to act as a band-limited uniform diffuser within a circular domain of scattering angles. The values of the parameters assumed in this calculation were $\lambda = 632.8$ nm, $q_m = \sin 20°(\omega/c)$, and $b = 20$ µm. Two cross sections are also presented.

$\langle \partial R/\partial \Omega_s \rangle$ reproduces well the result defined by the expression (3.1.240) that served as the input to the calculation. The mean differential reflection coefficient vanishes for scattering angles outside the circular domain defined by Eq. (3.1.240), and the cut-off is very sharp: the departure of the walls from verticality is by less than one degree. For scattering angles within the circular domain the mean differential reflection coefficient is very nearly constant.

We note that, as in the preceding example, because each realization of the surface profile function is not circularly symmetric there is no peak in $\langle \partial R/\partial \Omega_s \rangle$ in the specular (retroreflection) direction, as was the case when $\langle \partial R/\partial \Omega_s \rangle$ for this diffuser was calculated by the method described in Section 3.1.2, where a circularly symmetric randomly rough surface was assumed.

3.1.4.5.3. A Band-Limited Uniform Diffuser Within an Elliptical Domain We next consider the design of a two-dimensional randomly rough surface that produces a scattered field whose intensity is constant within an elliptical domain of scattering angles and vanishes outside this domain. The mean differential reflection coefficient we seek to produce in this case is

$$\left\langle \frac{\partial R}{\partial \Omega_s} \right\rangle (q_1, q_2) = A\theta\left(1 - \frac{q_1^2}{q_{1m}^2} - \frac{q_2^2}{q_{2m}^2}\right). \tag{3.1.243}$$

From Eqs. (3.1.225) and (3.1.243) we find that the corresponding joint pdf $f(s_1, s_2)$ is given by

$$f(s_1, s_2) = \frac{4A}{(1+s_\parallel^2)^2}\theta\left((1+s_\parallel^2)^2 - \frac{s_1^2}{s_{1m}^2} - \frac{s_2^2}{s_{2m}^2}\right), \tag{3.1.244}$$

where we have introduced the definitions

$$s_{1m} = \frac{c}{2\omega}q_{1m}, \qquad s_{2m} = \frac{c}{2\omega}q_{2m}. \tag{3.1.245}$$

This result simplifies greatly when the range of variation of s_1 and s_2 is sufficiently limited that $s_\parallel^2$ can be neglected compared to unity. This will be the case if $|s_{1m}| \lesssim 0.15$ and $|s_{2m}| \lesssim 0.15$. When this is the case, Eq. (3.1.244) becomes

$$f(s_1, s_2) = 4A\theta\left(1 - \frac{s_1^2}{s_{1m}^2} - \frac{s_2^2}{s_{2m}^2}\right). \tag{3.1.246}$$

It is this form of $f(s_1, s_2)$ that we will work with in what follows, because it allows us to obtain analytic expressions for the marginal pdf $f(s_1)$ and for the joint pdfs $\tilde{f}(s_1, s_2)$ and $\tilde{\tilde{f}}(s_1, s_2)$ and for their associated marginal pdfs $\tilde{f}(s_1)$ and $\tilde{\tilde{f}}(s_1)$. When $s_\parallel^2$ cannot be neglected in comparison with unity, these functions have to be obtained numerically.

The marginal pdf $f(s_1)$ is given by

$$\begin{aligned} f(s_1) &= \int_{-\infty}^{\infty} ds_2\, f(s_1, s_2) \\ &= 4A \int_{-\infty}^{\infty} ds_2\, \theta\left(1 - \frac{s_1^2}{s_{1m}^2} - \frac{s_2^2}{s_{2m}^2}\right) \\ &= 8As_{2m}\left(1 - \frac{s_1^2}{s_{1m}^2}\right)^{\frac{1}{2}} \theta\left(s_{1m} - |s_1|\right). \end{aligned} \tag{3.1.247}$$

The normalization condition for $f(s_1)$ is

$$\begin{aligned} 1 &= 8A\, s_{2m} \int_{-s_{1m}}^{s_{1m}} ds_1 \left(1 - \frac{s_1^2}{s_{1m}^2}\right)^{\frac{1}{2}} \\ &= 4\pi A\, s_{1m} s_{2m}, \end{aligned} \tag{3.1.248}$$

whence it follows that

$$A = \frac{1}{4\pi s_{1m} s_{2m}}. \tag{3.1.249}$$

Therefore, in the present case we have that

$$f(s_1, s_2) = \frac{1}{\pi s_{1m} s_{2m}} \theta\left(1 - \frac{s_1^2}{s_{1m}^2} - \frac{s_2^2}{s_{2m}^2}\right), \tag{3.1.250}$$

$$f(s_1) = \frac{2}{\pi s_{1m}} \left(1 - \frac{s_1^2}{s_{1m}^2}\right)^{\frac{1}{2}} \theta\left(s_{1m} - |s_1|\right), \tag{3.1.251}$$

while

$$\left\langle \frac{\partial R}{\partial \Omega_s} \right\rangle = \frac{1}{4\pi s_{1m} s_{2m}} \theta\left(1 - \frac{q_1^2}{q_{1m}^2} - \frac{q_2^2}{q_{2m}^2}\right). \tag{3.1.252}$$

The joint pdf $\tilde{f}(s_1, s_2)$ is obtained from Eqs. (3.1.185) and (3.1.250) as

$$\begin{aligned} \tilde{f}(s_1, s_2) &= f\left(\frac{1}{2}s_1 + \frac{\sqrt{3}}{2}s_2, \frac{\sqrt{3}}{2}s_1 - \frac{1}{2}s_2\right) \\ &= \frac{1}{\pi s_{1m} s_{2m}} \theta\left(1 - cs_1^2 - bs_1 s_2 - as_2^2\right), \end{aligned} \tag{3.1.253}$$

where

$$a = \frac{1}{4}\left(\frac{3}{s_{1m}^2} + \frac{1}{s_{2m}^2}\right), \tag{3.1.254a}$$

$$b = \frac{\sqrt{3}}{2}\left(\frac{1}{s_{1m}^2} - \frac{1}{s_{2m}^2}\right), \tag{3.1.254b}$$

$$c = \frac{1}{4}\left(\frac{1}{s_{1m}^2} + \frac{3}{s_{2m}^2}\right). \tag{3.1.254c}$$

The marginal pdf $\tilde{f}(s_1)$ is then given by

$$\begin{aligned} \tilde{f}(s_1) &= \frac{1}{\pi s_{1m} s_{2m}} \int_{-\infty}^{\infty} ds_2\, \theta\left(1 - cs_1^2 - bs_1 s_2 - as_2^2\right) \\ &= \frac{1}{\pi s_{1m} s_{2m}} \int_{-\infty}^{\infty} ds_2\, \theta\left(1 - \left(c - \frac{b^2}{4a}\right)s_1^2 - a\left(s_2 + \frac{b}{2a}s_1\right)^2\right) \\ &= \frac{1}{\pi s_{1m} s_{2m}} \int_{-\infty}^{\infty} dx\, \theta\left(1 - \left(c - \frac{b^2}{4a}\right)s_1^2 - ax^2\right) \end{aligned}$$

$$= \frac{2}{\pi}\frac{1}{s_{1m}s_{2m}\sqrt{a}}\left[1-\left(c-\frac{b^2}{4a}\right)s_1^2\right]^{\frac{1}{2}}\theta\left(\frac{2\sqrt{a}}{\sqrt{4ac-b^2}}-|s_1|\right)$$

$$= \frac{8}{\pi}\frac{1}{s_{1m}^2+3s_{2m}^2}\left(\frac{1}{4}\left(s_{1m}^2+3s_{2m}^2\right)-s_1^2\right)^{\frac{1}{2}}\theta\left(\frac{1}{2}\left(s_{1m}^2+3s_{2m}^2\right)^{\frac{1}{2}}-|s_1|\right). \tag{3.1.255}$$

The joint pdf $\tilde{\tilde{f}}(s_1, s_2)$ is obtained from Eqs. (3.1.188) and (3.1.250) as

$$\tilde{\tilde{f}}(s_1, s_2) = f\left(-\frac{1}{2}s_1+\frac{\sqrt{3}}{2}s_2, \frac{\sqrt{3}}{2}s_1+\frac{1}{2}s_2\right)$$

$$= \frac{1}{\pi s_{1m}s_{2m}}\theta\left(1-cs_1^2+bs_1s_2-as_2^2\right). \tag{3.1.256}$$

The marginal pdf $\tilde{\tilde{f}}(s_1)$ is then obtained from Eq. (3.1.255) by noting that the result is independent of the sign of b. The result is

$$\tilde{\tilde{f}}(s_1) = \frac{1}{\pi s_{1m}s_{2m}}\int_{-\infty}^{\infty} ds_2\,\theta\left(1-cs_1^2+bs_1s_2-as_2^2\right)$$

$$= \frac{8}{\pi}\frac{1}{s_{1m}^2+3s_{2m}^2}\left(\frac{1}{4}\left(s_{1m}^2+3s_{2m}^2\right)-s_1^2\right)^{\frac{1}{2}}\theta\left(\frac{1}{2}\left(s_{1m}^2+3s_{2m}^2\right)^{\frac{1}{2}}-|s_1|\right). \tag{3.1.257}$$

In Fig. 3.22, we present results for $\langle \partial R/\partial \Omega_s \rangle$ obtained on the basis of the Kirchhoff approximation for a band-limited uniform diffuser within an elliptical domain of scattering angles defined by Eq. (3.1.252). The wavelength of the incident light was $\lambda = 632.8$ nm. The roughness parameters assumed were $b = 20$ μm, $q_{1m} = 0.1(\omega/c)$, $q_{2m} = 0.05(\omega/c)$. The results for $N_p = 3000$ realizations of the surface profile function were averaged to obtain the average indicated in Eq. (3.1.226). The results presented show that the angular dependence of the intensity of the scattered field is very close to that defined by Eq. (3.1.252). The intensity of the scattered field vanishes for scattering angles outside the elliptical domain defined by Eq. (3.1.252) and the cut-off is very sharp. For scattering angles inside this elliptical domain the scattered intensity is very nearly constant.

3.1.4.5.4. A Band-Limited Uniform Diffuser Within a Rectangular Domain If we wish to design a surface that acts as a band-limited uniform diffuser within a rectangular region of scattering angles, the mean differential reflection coefficient we seek to produce is

$$\left\langle \frac{\partial R}{\partial \Omega_s} \right\rangle (q_1, q_2) = A\theta\left(q_{1m}-|q_1|\right)\theta\left(q_{2m}-|q_2|\right). \tag{3.1.258}$$

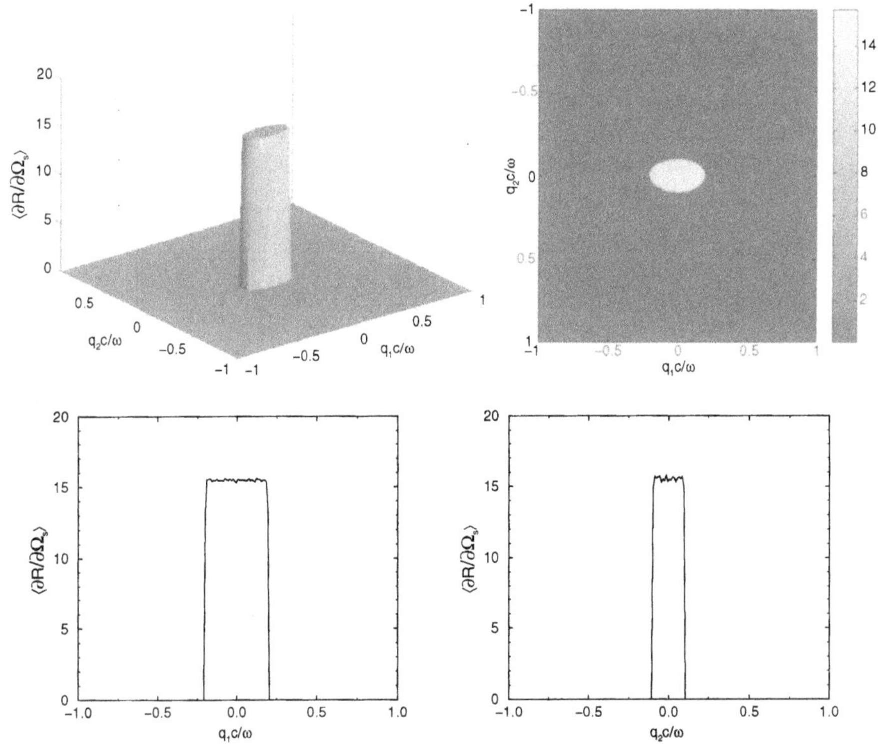

Figure 3.22. The mean differential reflection coefficient estimated from $N_p = 3000$ realizations of the surface profile function for the scattering of a scalar plane wave incident normally on a two-dimensional randomly rough Dirichlet surface designed to act as a band-limited uniform diffuser within an elliptical domain of scattering angles. The values of the parameters used in this calculation were $\lambda = 632.8$ nm, $q_{1m} = 0.1(\omega/c)$, $q_{2m} = 0.05(\omega/c)$, and $b = 20$ μm. The cross sections corresponding to $q_2 = 0$ and $q_1 = 0$ are also presented.

On combining Eqs. (3.1.225) and (3.1.258) we find that the corresponding joint probability density function $f(s_1, s_2)$ is given by

$$f(s_1, s_2) = \frac{4A}{(1+s_\parallel^2)^2}\theta\big(s_{1m}\big(1+s_\parallel^2\big) - |s_1|\big)\theta\big(s_{2m}\big(1+s_\parallel^2\big) - |s_2|\big), \qquad (3.1.259)$$

where s_{1m} and s_{2m} have been defined in Eq. (3.1.245). This result simplifies greatly when the range of variation of s_1 and s_2 is small enough that $s_\parallel^2$ can be neglected compared to unity with little error, e.g., if $|s_{1m}| \lesssim 0.15$, $|s_{2m}| \lesssim 0.15$. When this is the case Eq. (3.1.259) becomes

$$f(s_1, s_2) = 4A\theta\big(s_{1m} - |s_1|\big)\theta\big(s_{2m} - |s_2|\big). \qquad (3.1.260)$$

It is this form of $f(s_1, s_2)$ that we will use in what follows, because it allows us to obtain analytic expressions for the marginal pdf $f(s_1)$ and for the joint pdfs $\tilde{f}(s_1, s_2)$ and $\tilde{\tilde{f}}(s_1, s_2)$ and their associated marginal pdfs $\tilde{f}(s_1)$ and $\tilde{\tilde{f}}(s_2)$. When $s_\parallel^2$ cannot be neglected in comparison with unity, these functions have to be obtained numerically.

The normalization of $f(s_1, s_2)$ leads to the result that

$$A = \frac{1}{16 s_{1m} s_{2m}}. \tag{3.1.261}$$

Consequently, we find that

$$\left\langle \frac{\partial R}{\partial \Omega_s} \right\rangle (q_1, q_2) = \frac{\theta(q_{1m} - |q_1|)}{(c q_{1m}/\omega)} \frac{\theta(q_{2m} - |q_2|)}{(c q_{2m}/\omega)}, \tag{3.1.262}$$

and

$$f(s_1, s_2) = \frac{\theta(s_{1m} - |s_1|)}{2 s_{1m}} \frac{\theta(s_{2m} - |s_2|)}{2 s_{2m}}. \tag{3.1.263}$$

The marginal pdf $f(s_1)$ is

$$f(s_1) = \frac{\theta(s_{1m} - |s_1|)}{2 s_{1m}}, \tag{3.1.264}$$

and the conditional pdf $f(s_2|s_1)$ is

$$f(s_2|s_1) = \frac{\theta(s_{2m} - |s_2|)}{2 s_{2m}} \tag{3.1.265}$$

independent of s_1.

From Eqs. (3.1.185) and (3.1.263) we find that the joint pdf $\tilde{f}(s_1, s_2)$ is given by

$$\tilde{f}(s_1, s_2) = \frac{\theta(s_{1m} - |\frac{1}{2}s_1 + \frac{\sqrt{3}}{2}s_2|)}{2 s_{1m}} \frac{\theta(s_{2m} - |\frac{\sqrt{3}}{2}s_1 - \frac{1}{2}s_2|)}{2 s_{2m}}. \tag{3.1.266}$$

The marginal pdf $\tilde{f}(s_1)$ is then

$$\begin{aligned} \tilde{f}(s_1) &= \frac{1}{4 s_{1m} s_{2m}} \int_{-\infty}^{\infty} ds_2\, \theta\left(s_{1m} - \left|\frac{1}{2}s_1 + \frac{\sqrt{3}}{2}s_2\right|\right) \theta\left(s_{2m} - \left|\frac{\sqrt{3}}{2}s_1 - \frac{1}{2}s_2\right|\right) \\ &= \frac{1}{2 s_{1m} s_{2m}} \int_{-\infty}^{\infty} du\, \theta(s_{2m} - |u|)\theta(s_{1m} - |\sqrt{3}u + 2s_1|). \end{aligned} \tag{3.1.267}$$

In what follows, rather than maintaining complete generality, we will assume that

$$\sqrt{3}s_{2m} > s_{1m}, \tag{3.1.268}$$

an assumption that simplifies the evaluation of $\tilde{f}(s_1)$ without significantly decreasing the generality of the result. The product of two Heaviside unit step functions in the integrand on the right-hand side of Eq. (3.1.267) defines five intervals of the variable s_1 that have to be considered, namely $(-\infty, -b)$, $(-b, -a)$, $(-a, a)$, (a, b), and (b, ∞), where

$$a = \frac{\sqrt{3}}{2}s_{2m} - \frac{1}{2}s_{1m}, \qquad b = \frac{\sqrt{3}}{2}s_{2m} + \frac{1}{2}s_{1m}. \tag{3.1.269}$$

The integral over u vanishes when s_1 is in the first and last of these intervals, because the integrand is zero. The limits within which u varies for s_1 in the second, third, and fourth intervals are $(-\frac{2}{\sqrt{3}}s_1 - \frac{1}{\sqrt{3}}s_{1m}, s_{2m})$, $(-\frac{2}{\sqrt{3}}s_1 - \frac{1}{\sqrt{3}}s_{1m}, -\frac{2}{\sqrt{3}}s_1 + \frac{1}{\sqrt{3}}s_{1m})$, and $(-s_{2m}, -\frac{2}{\sqrt{3}}s_1 + \frac{1}{\sqrt{3}}s_{1m})$, respectively. From these results we obtain

$$\tilde{f}(s_1) = \frac{1}{\sqrt{3}s_{1m}s_{2m}} \begin{cases} 0 & s_1 < -b, \\ (b + s_1) & -b < s_1 < -a, \\ (b - a) & -a < s_1 < a, \\ (b - s_1) & a < s_1 < b, \\ 0 & b < s_1. \end{cases} \tag{3.1.270}$$

The joint pdf $\tilde{\tilde{f}}(s_1, s_2)$ is obtained from Eqs. (3.1.188) and (3.1.263) in the form

$$\tilde{\tilde{f}}(s_1, s_2) = \frac{\theta(s_{1m} - |-\frac{1}{2}s_1 + \frac{\sqrt{3}}{2}s_2|)}{2s_{1m}} \frac{\theta(s_{2m} - |\frac{\sqrt{3}}{2}s_1 + \frac{1}{2}s_2|)}{2s_{2m}}. \tag{3.1.271}$$

The marginal pdf $\tilde{\tilde{f}}(s_1)$ is therefore given by

$$\begin{aligned} \tilde{\tilde{f}}(s_1) &= \frac{1}{4s_{1m}s_{2m}} \int_{-\infty}^{\infty} ds_2\, \theta\left(s_{1m} - \left|-\frac{1}{2}s_1 + \frac{\sqrt{3}}{2}s_2\right|\right)\theta\left(s_{2m} - \left|\frac{\sqrt{3}}{2}s_1 + \frac{1}{2}s_2\right|\right) \\ &= \frac{1}{2s_{1m}s_{2m}} \int_{-\infty}^{\infty} du\, \theta(s_{2m} - |u|)\theta(s_{1m} - |\sqrt{3}u - 2s_1|). \end{aligned} \tag{3.1.272}$$

If we compare this expression for $\tilde{\tilde{f}}(s_1)$ with the expression for $\tilde{f}(s_1)$ given by Eq. (3.1.267), we observe the relation $\tilde{\tilde{f}}(s_1) = \tilde{f}(-s_1)$. Since $\tilde{f}(s_1)$ is an even function

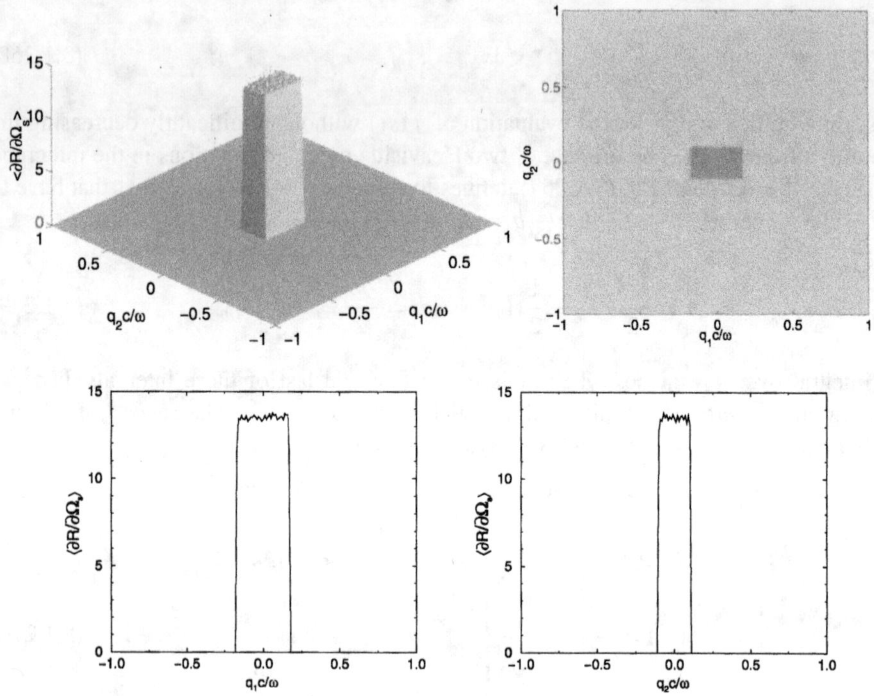

Figure 3.23. The mean differential reflection coefficient estimated from $N_p = 3000$ realizations of the surface profile function for the scattering of a scalar plane wave incident normally on a two-dimensional randomly rough Dirichlet surface designed to act as a band-limited uniform diffuser within a rectangular domain of scattering angles. The values of the parameters used in this calculation were $\lambda = 632.8$ nm, $q_{1m} = 0.1736(\omega/c)$, $q_{2m} = 0.1045(\omega/c)$, and $b = 20$ μm. The cross sections corresponding to $q_2 = 0$ and $q_1 = 0$ are also presented.

of s_1, we obtain finally the result that

$$\tilde{\tilde{f}}(s_1) = \tilde{f}(s_1). \tag{3.1.273}$$

In Fig. 3.23 we present results for $\langle \partial R/\partial \Omega_s \rangle$, calculated by means of the Kirchhoff approximation, for the band-limited uniform diffuser within a rectangular domain of scattering angles defined by Eq. (3.1.262). The wavelength of the incident light was $\lambda = 632.8$ nm. The roughness parameters assumed were $b = 20$ μm, $q_{1m} = (\omega/c)\sin 10° = 0.1736(\omega/c)$, $q_{2m} = (\omega/c)\sin 6° = 0.1045(\omega/c)$. The results for $N_p = 3000$ realizations of the surface profile function were used to obtain the average indicated in Eq. (3.1.226). The results presented show that the angular dependence of the intensity of the scattered field is in very good agreement with that defined by Eq. (3.1.262). The intensity of the scattered field vanishes for scattering angles outside the rectangular domain defined by

Eq. (3.1.262), and the cutoff is very sharp. For scattering angles within this rectangular domain the intensity of the scattered field is very closely constant.

3.1.4.5.5. A Band-Limited Uniform Diffuser Within a Triangular Domain As a final example, we apply the approach developed in the preceding sections of this chapter to the design of a random surface that acts as a band-limited uniform diffuser within a triangular region of scattering angles. The mean differential reflection coefficient in this case is chosen to have the form

$$\left\langle \frac{\partial R}{\partial \Omega_s} \right\rangle (q_1, q_2) = A\theta\left(\frac{1}{\sqrt{3}}(m - q_2) - |q_1| \right)\theta(m - q_2)\theta(q_2). \tag{3.1.274}$$

On combining Eqs. (3.1.225) and (3.1.274) we find that the corresponding joint probability density function $f(s_1, s_2)$ is given by

$$f(s_1, s_2) = \frac{4A}{(1 + s_\parallel^2)^2}\theta\left(\frac{1}{\sqrt{3}}\left(\frac{M}{2}(1 + s_\parallel^2) + s_2 \right) - |s_1| \right)\theta\left(\frac{M}{2}(1 + s_\parallel^2) + s_2 \right)\theta(-s_2), \tag{3.1.275}$$

where we have introduced the definition

$$M = (c/\omega)m. \tag{3.1.276}$$

The expression given by Eq. (3.1.275) simplifies greatly if the range of variation of s_1 and s_2 is sufficiently small that $s_\parallel^2$ can be neglected in comparison with unity. This will be the case, for example, if $M \lesssim 0.35$. In this limit Eq. (3.1.275) takes the form

$$f(s_1, s_2) = 4A\theta\left(\frac{M}{2\sqrt{3}} + \frac{s_2}{\sqrt{3}} - |s_1| \right)\theta\left(\frac{M}{2} + s_2 \right)\theta(-s_2). \tag{3.1.277}$$

This is the expression for $f(s_1, s_2)$ we will use in what follows, because it allows obtaining the marginal pdf $f(s_1)$ as well as the joint pdfs $\tilde{f}(s_1, s_2)$ and $\tilde{\tilde{f}}(s_1, s_2)$ and the corresponding marginal pdfs $\tilde{f}(s_1)$ and $\tilde{\tilde{f}}(s_1)$, analytically, in closed form, rather than numerically. If $s_\parallel^2$ is not small compared to unity, these functions have to be obtained numerically.

The marginal pdf $f(s_1)$ is given by

$$\begin{aligned} f(s_1) &= 4A \int_{-\infty}^{\infty} ds_2\, \theta\left(\frac{M}{2\sqrt{3}} + \frac{s_2}{\sqrt{3}} - |s_1| \right)\theta\left(\frac{M}{2} + s_2 \right)\theta(-s_2) \\ &= 4A\theta\left(\frac{M}{2\sqrt{3}} - |s_1| \right) \int_{\sqrt{3}|s_1| - \frac{M}{2}}^{0} ds_2 \\ &= 4\sqrt{3}A\left(\frac{M}{2\sqrt{3}} - |s_1| \right)\theta\left(\frac{M}{2\sqrt{3}} - |s_1| \right). \end{aligned} \tag{3.1.278}$$

The normalization of $f(s_1)$ yields the result that $A = \sqrt{3}/M^2$. Thus, we have the results that

$$\left\langle \frac{\partial R}{\partial \Omega_s} \right\rangle (q_1, q_2) = \frac{\sqrt{3}}{M^2}\theta\left(\frac{M}{\sqrt{3}} - \frac{c}{\omega}\frac{q_2}{\sqrt{3}} - \frac{c}{\omega}|q_1|\right)\theta\left(M - \frac{c}{\omega}q_2\right)\theta\left(\frac{c}{\omega}q_2\right), \tag{3.1.279}$$

$$f(s_1, s_2) = \frac{4\sqrt{3}}{M^2}\theta\left(\frac{M}{2\sqrt{3}} + \frac{s_2}{\sqrt{3}} - |s_1|\right)\theta\left(\frac{M}{2} + s_2\right)\theta(-s_2), \tag{3.1.280}$$

and

$$f(s_1) = \frac{12}{M^2}\left(\frac{M}{2\sqrt{3}} - |s_1|\right)\theta\left(\frac{M}{2\sqrt{3}} - |s_1|\right). \tag{3.1.281}$$

The joint pdf $\tilde{f}(s_1, s_2)$ in the present case is obtained from Eqs. (3.1.185) and (3.1.280) as

$$\tilde{f}(s_1, s_2) = \frac{4\sqrt{3}}{M^2}\theta\left(\frac{M}{2\sqrt{3}} + \frac{1}{\sqrt{3}}\left(\frac{\sqrt{3}}{2}s_1 - \frac{1}{2}s_2\right) - \left|\frac{1}{2}s_1 + \frac{\sqrt{3}}{2}s_2\right|\right) \\ \times \theta\left(\frac{M}{2} + \frac{\sqrt{3}}{2}s_1 - \frac{1}{2}s_2\right)\theta\left(\frac{1}{2}s_2 - \frac{\sqrt{3}}{2}s_1\right). \tag{3.1.282}$$

The marginal pdf $\tilde{f}(s_1)$ is then given by

$$\tilde{f}(s_1) = \frac{4\sqrt{3}}{M^2}\int_{-\infty}^{\infty} ds_2\, \theta\left(\frac{M}{2\sqrt{3}} + \frac{1}{\sqrt{3}}\left(\frac{\sqrt{3}}{2}s_1 - \frac{1}{2}s_2\right) - \left|\frac{1}{2}s_1 + \frac{\sqrt{3}}{2}s_2\right|\right) \\ \times \theta\left(\frac{M}{2} + \frac{\sqrt{3}}{2}s_1 - \frac{1}{2}s_2\right)\theta\left(\frac{1}{2}s_2 - \frac{\sqrt{3}}{2}s_1\right). \tag{3.1.283}$$

The constraints imposed by the step functions in the integrand of this expression yield the result that

$$\tilde{f}(s_1) = \frac{4\sqrt{3}}{M^2}\int_{-\frac{M}{2}-\sqrt{3}s_1}^{\frac{M}{4}} ds_2 \qquad -\frac{\sqrt{3}}{4}M < s_1 < -\frac{M}{4\sqrt{3}} \tag{3.1.284a}$$

$$= \frac{4\sqrt{3}}{M^2}\int_{\sqrt{3}s_1}^{\frac{M}{4}} ds_2 \qquad -\frac{M}{4\sqrt{3}} < s_1 < \frac{M}{4\sqrt{3}}. \tag{3.1.284b}$$

Thus we obtain

$$\tilde{f}(s_1) = \begin{cases} \frac{4\sqrt{3}}{M^2}\left(\frac{3}{4}M + \sqrt{3}s_1\right) & -\frac{\sqrt{3}}{4}M < s_1 < -\frac{M}{4\sqrt{3}}, \\ \frac{4\sqrt{3}}{M^2}\left(\frac{1}{4}M - \sqrt{3}s_1\right) & -\frac{M}{4\sqrt{3}} < s_1 < \frac{M}{4\sqrt{3}}. \end{cases} \tag{3.1.285}$$

From Eqs. (3.1.188) and (3.1.280) we find that the joint pdf $\tilde{\tilde{f}}(s_1, s_2)$ is

$$\tilde{\tilde{f}}(s_1, s_2) = \frac{4\sqrt{3}}{M^2}\theta\left(\frac{M}{2\sqrt{3}} + \frac{1}{\sqrt{3}}\left(\frac{\sqrt{3}}{2}s_1 + \frac{1}{2}s_2\right) - \left|-\frac{1}{2}s_1 + \frac{\sqrt{3}}{2}s_2\right|\right)$$
$$\times \theta\left(\frac{M}{2} + \frac{\sqrt{3}}{2}s_1 + \frac{1}{2}s_2\right)\theta\left(-\frac{\sqrt{3}}{2}s_1 - \frac{1}{2}s_2\right). \tag{3.1.286}$$

The marginal pdf $\tilde{\tilde{f}}(s_1)$ is then given by

$$\tilde{\tilde{f}}(s_1) = \frac{4\sqrt{3}}{M^2}\int_{-\infty}^{\infty} ds_2\,\theta\left(\frac{M}{2\sqrt{3}} + \frac{1}{\sqrt{3}}\left(\frac{\sqrt{3}}{2}s_1 + \frac{1}{2}s_2\right) - \left|-\frac{1}{2}s_1 + \frac{\sqrt{3}}{2}s_2\right|\right)$$
$$\times \theta\left(\frac{M}{2} + \frac{\sqrt{3}}{2}s_1 + \frac{1}{2}s_2\right)\theta\left(-\frac{\sqrt{3}}{2}s_1 - \frac{1}{2}s_2\right). \tag{3.1.287}$$

If we replace s_2 by $-s_2$ as the integration variable in Eq. (3.1.287), and compare the resulting integral with the one in Eq. (3.1.283), we obtain the result that

$$\tilde{\tilde{f}}(s_1) = \tilde{f}(s_1). \tag{3.1.288}$$

In Fig. 3.24 we present a plot of $\langle \partial R/\partial \Omega_s \rangle$, calculated by means of the Kirchhoff approximation, for the band-limited uniform diffuser within a triangular region of scattering angles defined by Eq. (3.1.279). The wavelength of the incident field was $\lambda = 632.8$ nm. The parameters characterizing the surface were $b = 20$ μm and $M = \sin 20° = 0.342$. The results for $N_p = 10\,000$ realizations of the surface profile function were averaged to obtain $\langle \partial R/\partial \Omega_s \rangle$. The angular dependence of $\langle \partial R/\partial \Omega_s \rangle$ is very close to that defined by Eq. (3.1.279). There is no scattered intensity for scattering outside the triangular domain defined by Eq. (3.1.279), and the cutoff is very sharp. For scattering angles inside this triangular domain $\langle \partial R/\partial \Omega_s \rangle$ is very closely constant.

3.2. Random Diffusers that Extend the Depth of Focus

So far in this chapter we have considered the design of a two-dimensional randomly rough surface that scatters a scalar plane wave incident on it in such a way that the scattered

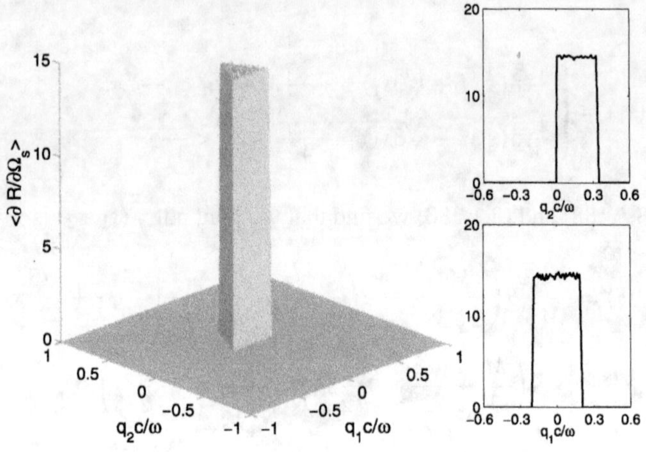

Figure 3.24. The mean differential reflection coefficient estimated from $N_p = 10\,000$ realizations of the surface profile function for the scattering of a scalar plane wave incident normally on a two-dimensional randomly rough Dirichlet surface designed to act as a band-limited uniform diffuser within a triangular region of scattering angles. The values of the parameters assumed in this calculation were $\lambda = 632.8$ nm, $M = \sin 20° = 0.342$, and $b = 20$ μm. The cross sections corresponding to $q_1 = 0$ and $q_2 = 0$ are also presented.

field has a prescribed *angular* dependence of the mean differential reflection coefficient. However, a two-dimensional randomly rough surface that produces a scattered field with a mean intensity that has a specified *spatial* dependence is also of interest, and such a surface can also be designed. To make this point, in the present section we consider a particular example of this type of design problem. Specifically, we consider the problem of designing diffusers that, when illuminated by a converging beam, produce a constant distribution of intensity along a segment of the optical axis. The interest in such diffusers is due to the fact that a common criterion used in evaluating masks for extending the depth of focus of optical systems is the rate of decay of the intensity along the optical axis as one moves away from the focus: the slower this rate of decay the better [3.12,3.13]. This suggests that diffusers that produce a constant intensity along a specified segment of the optical axis that includes the focus can be used as focal depth extenders. In this section we present an approach to the production of such diffusers that is based on designing a two-dimensional circularly symmetric randomly rough surface of the diffuser that produces a field of this nature.

3.2.1. Diffusers that Produce a Prescribed Distribution of the Mean Intensity Along the Optical Axis

In designing a random surface that produces a specified distribution of intensity along the optical axis, we assume that we are dealing with a circularly symmetric aberration-free imaging system. For a point source object the system (Fig. 3.25) produces a converging

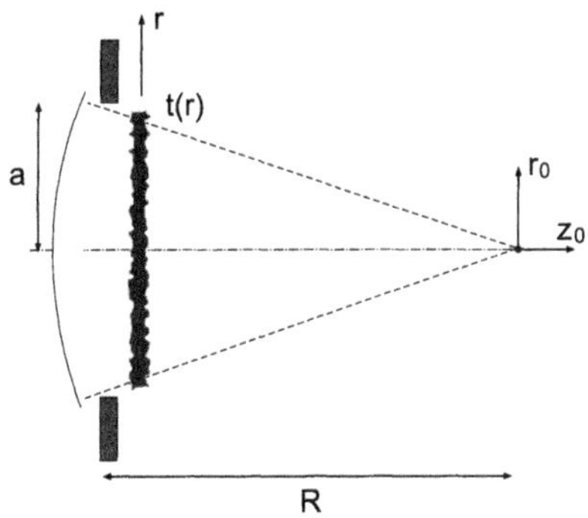

Figure 3.25. Diffraction of a converging spherical wave transmitted through a circularly symmetric diffuser.

spherical wave. After passing through a diffuser with a circularly symmetric amplitude transmittance function $t(r)$, the complex amplitude in the neighborhood of focus can be expressed as [3.14]

$$\psi(z_0, r_0) = -ik_0\frac{A}{f^2}\exp(ik_0z_0)\int_0^a t(r)J_0\left(\frac{k_0r_0}{R}r\right)\exp\left\{-ik_0\frac{z_0r^2}{2R^2}\right\}r\,dr. \tag{3.2.1}$$

In this expression A is a constant amplitude, and $k_0 = (\omega/c)$ is the wave number in vacuum, where ω is the frequency of the wave and c is the speed of light in vacuum. The length a is the radius of the pupil, R is the distance from the pupil plane to the best focus, z_0 is the defocus distance, r_0 is the radial distance from the optical axis, $J_0(z)$ is a Bessel function of the first kind and zero order, and $t(r)$ is a complex function of the form

$$t(r) = \exp\bigl[-iv_3H(r)\bigr], \tag{3.2.2}$$

where $H(r)$ is the surface profile function of the diffuser. We assume that $H(r)$ is a single-valued function of r that is differentiable and constitutes a random process, but not necessarily a stationary one. In the case of small angles of incidence and scattering we have that $v_3 = \Delta n(\omega/c)$ for the transmission geometry depicted in Fig. 3.25 [3.15], where Δn is the difference between the refractive index of the material from which the diffuser is fabricated and the refractive index of the surrounding medium.

We consider first the complex amplitude along the optical axis ($r_0 = 0$),

$$\psi(z_0, 0) = C_0\int_0^a \exp\bigl[-iv_3H(r)\bigr]\exp\bigl(-i\kappa z_0r^2\bigr)r\,dr, \tag{3.2.3}$$

where

$$C_0 = -ik_0\left(A/f^2\right)\exp(ik_0z_0), \tag{3.2.4a}$$

$$\kappa = k_0/2R^2. \tag{3.2.4b}$$

With the change of variable $t = r^2$ Eq. (3.2.3) becomes

$$\psi(z_0, 0) = \frac{1}{2}C_0 \int_0^{a^2} dt\, \exp\left\{-i\left[\nu_3 h(t) + \kappa z_0 t\right]\right\}, \tag{3.2.5}$$

where

$$h(t) = H(\sqrt{t}). \tag{3.2.6}$$

Equation (3.2.5) is a special case of a more general result obtained by McCutchen [3.16]. The mean intensity along the optical axis is therefore

$$\begin{aligned}\left\langle I(z_0, 0)\right\rangle &= \left\langle\left|\psi(z_0, 0)\right|^2\right\rangle \\ &= \frac{1}{4}|C_0|^2 \int_0^{a^2} dt \int_0^{a^2} dt'\, \exp\left[-i\kappa z_0(t-t')\right]\left\langle\exp\left\{-i\nu_3\left[h(t) - h(t')\right]\right\}\right\rangle,\end{aligned} \tag{3.2.7}$$

where the angle brackets denote an average over the ensemble of realizations of the surface profile function $H(r)$. Our goal, now, is to find the function $h(t)$, and hence the function $H(r)$, that produces a specified form for $\langle I(z_0, 0)\rangle$.

As it stands, the expression for $\langle I(z_0, 0)\rangle$ given by Eq. (3.2.7) is too difficult to invert to obtain $h(t)$ in terms of $\langle I(z_0, 0)\rangle$. To obtain an equation that can be inverted we use an approximation that is equivalent to passing to the geometrical optics limit of the right-hand side of Eq. (3.2.7). Namely, we expand $h(t)$ about $t = t'$, $h(t) = h(t') + (t - t')h'(t) + \cdots$, and retain only terms through the leading nonzero order in $t - t'$. In this way we obtain

$$\left\langle I(z_0, 0)\right\rangle = \frac{1}{4}|C_0|^2 \int_0^{a^2} dt \int_0^{a^2} dt'\, \exp\left[-i\kappa z_0(t-t')\right]\left\langle\exp\left[-i\nu_3(t-t')h'(t')\right]\right\rangle. \tag{3.2.8}$$

To evaluate the double integral in Eq. (3.2.8) we first introduce a characteristic length b through the definition $a^2 = Nb^2$, where N is a large integer. We then represent the function $h(t)$ in the form

$$h(t) = \frac{a_n}{b}t + b_n, \quad nb^2 \leqslant t \leqslant (n+1)b^2,\ n = 0, 1, 2, \ldots, N-1. \tag{3.2.9}$$

The $\{a_n\}$ are assumed to be independent identically distributed random deviates. The probability density function of a_n, $f(\gamma) = \langle\delta(\gamma - a_n)\rangle$, is therefore independent of n. The continuity of the surface at, say, $t = (n+1)b^2$ requires the satisfaction of the relation

$$b_{n+1} = b_n - (n+1)(a_{n+1} - a_n)b. \qquad (3.2.10)$$

The $\{b_n\}$ can be determined from a knowledge of the $\{a_n\}$ by means of this recurrence relation, provided that an initial value, e.g., b_0 is specified. With no loss of generality we can assume that $b_0 = 0$, and we do so. The solution of Eq. (3.2.10) is then

$$b_n = [a_0 + a_1 + \cdots + a_{n-1} - na_n]b, \quad n \geqslant 1. \qquad (3.2.11)$$

With the representation of $h(t)$ given by Eq. (3.2.9) the double integral in Eq. (3.2.8) becomes

$$\begin{aligned}
&\int_0^{a^2} dt \int_0^{a^2} dt' \exp\bigl[-i\kappa z_0(t-t')\bigr]\bigl\langle\exp\bigl[-i\nu_3(t-t')h'(t')\bigr]\bigr\rangle \\
&= \int_0^{a^2} dt \sum_{n=0}^{N-1} \int_{nb^2}^{(n+1)b^2} dt' \exp\bigl[-i\kappa z_0(t-t')\bigr]\bigl\langle\exp\bigl[-i\nu_3(t-t')(a_n/b)\bigr]\bigr\rangle \\
&= \int_0^{a^2} dt \sum_{n=0}^{N-1} \int_{nb^2}^{(n+1)b^2} dt' \exp\bigl[-i\kappa z_0(t-t')\bigr] \int_{-\infty}^{\infty} d\gamma\, f(\gamma) \exp\bigl[-i\nu_3(t-t')(\gamma/b)\bigr].
\end{aligned}$$

We used the independence of the $\{a_n\}$ and the fact that they are identically distributed in the last step. Since the integrands in this expression now are independent of n, the sum on n can be carried out immediately, with the result that the double integral becomes

$$\begin{aligned}
&\int_{-\infty}^{\infty} d\gamma\, f(\gamma) \left| \int_0^{a^2} dt \exp\bigl\{-i\bigl[\kappa z_0 + (\nu_3\gamma/b)\bigr]t\bigr\} \right|^2 \\
&= 4 \int_{-\infty}^{\infty} d\gamma\, f(\gamma) \frac{\sin^2[\kappa z_0 + (\nu_3\gamma/b)](a^2/2)}{[\kappa z_0 + (\nu_3\gamma/b)]^2}.
\end{aligned} \qquad (3.2.12)$$

Equations (3.2.8) and (3.2.12) together give us

$$\bigl\langle I(z_0, 0)\bigr\rangle = |C_0|^2 \frac{a^4}{4} \int_{-\infty}^{\infty} d\gamma\, f(\gamma) \operatorname{sinc}^2 \frac{\nu_3 a^2}{2b}\left(\gamma + \frac{\kappa b}{\nu_3} z_0\right). \qquad (3.2.13)$$

The function $\text{sinc}^2((\nu_3 a^2/2b)(\gamma + (\kappa b z_0/\nu_3)))$ is sharply peaked about $\gamma = -(\kappa b z_0/\nu_3)$, and the larger $(\nu_3 a^2/2b) = \pi \Delta n N b/\lambda$ is the more sharply peaked it is. We assume that $Nb \gg \lambda$, and use the result that in the limit as $A \to \infty$

$$\text{sinc}^2 Ax \to \frac{\pi}{A}\delta(x), \tag{3.2.14}$$

where $\delta(x)$ is the Dirac delta function, to rewrite Eq. (3.2.13) as

$$\langle I(z_0, 0)\rangle = \pi a^2 |C_0|^2 \frac{b}{2\nu_3} f\left(-\frac{\kappa b}{\nu_3} z_0\right). \tag{3.2.15}$$

On inverting Eq. (3.2.15) we obtain the pdf of a_n in terms of $\langle I(z_0, 0)\rangle$ in the form

$$f(\gamma) = \frac{2}{\pi} \frac{1}{|C_0|^2} \frac{\nu_3}{a^2 b} \left\langle I\left(-\frac{\nu_3}{\kappa b}\gamma, 0\right)\right\rangle. \tag{3.2.16}$$

From the result given by Eq. (3.2.16) a long sequence of $\{a_n\}$ is generated by the rejection method [3.8], and a single realization of the surface profile function $H(r)$ is constructed on the basis of Eqs. (3.2.9), (3.2.11), and the fact that $t = r^2$. Thus, the surface profile function can be written as

$$H(r) = \frac{a_n}{b} r^2 + b_n, \quad \sqrt{n}b \leqslant r \leqslant \sqrt{n+1}b, \; n = 0, 1, 2, \ldots, N-1. \tag{3.2.17}$$

In what follows we consider the design of a diffuser that produces a constant mean intensity along the optical axis in the interval $-z_m < z_0 < z_m$, and zero mean intensity along the rest of the optical axis, and examine the possibility of using it to extend the depth of focus of imaging systems. The mean axial intensity we wish the diffuser to produce is therefore

$$\langle I(z_0, 0)\rangle = I_0 \theta(z_m - |z_0|), \tag{3.2.18}$$

where I_0 is a constant. We obtain from Eq. (3.2.16) the result that

$$f(\gamma) = \frac{2}{\pi} \frac{\nu_3 I_0}{a^2 b |C_0|^2} \theta\left(\frac{\kappa b}{\nu_3} z_m - |\gamma|\right). \tag{3.2.19}$$

The constant I_0 is obtained from the normalization of $f(\gamma)$, and is found to be

$$I_0 = \frac{\pi}{4} \frac{a^2 |C_0|^2}{\kappa z_m}. \tag{3.2.20}$$

It follows that

$$\langle I(z_0, 0)\rangle = \frac{\pi}{4} \frac{a^2 |C_0|^2}{\kappa z_m} \theta(z_m - |z_0|) \tag{3.2.21}$$

and

$$f(\gamma) = \frac{\nu_3}{2b} \frac{1}{\kappa z_m} \theta\left(\frac{\kappa b}{\nu_3} z_m - |\gamma|\right). \tag{3.2.22}$$

3.2.2. Three-Dimensional Distribution of the Mean Intensity in the Neighborhood of the Focus

If the diffuser designed in Section 3.2.1 is to be used in an imaging system, not only the axial intensity distribution is of interest: it is also important to know the intensity distribution in the radial direction about the optical axis. Therefore, in this section we calculate the mean intensity in the focal region. We return to Eq. (3.2.1), which gives the field in this region as

$$\psi(z_0, r_0) = C_0 \int_0^a J_0\left(\frac{k_0 r_0}{R} r\right) \exp\left[-i\nu_3 H(r) - i\kappa z_0 r^2\right] r \, dr. \tag{3.2.23}$$

When the form of $H(r)$ given by Eq. (3.2.17) is substituted into Eq. (3.2.23) and we make the change of variable $t = r^2$, $\psi(z_0, r_0)$ becomes

$$\psi(z_0, r_0) = C_0 \sum_{n=0}^{N-1} \exp(-i\nu_3 b_n) \psi_n(z_0, r_0; a_n), \tag{3.2.24}$$

where

$$\psi_n(z_0, r_0; a_n) = \frac{1}{2} \int_{nb^2}^{(n+1)b^2} dt \, J_0(\beta\sqrt{t}) \exp(-i\alpha_n t), \tag{3.2.25}$$

with

$$\alpha_n = \nu_3 \frac{a_n}{b} + \kappa z_0, \qquad \beta = \frac{k_0 r_0}{R}. \tag{3.2.26}$$

The integral in Eq. (3.2.25) can be evaluated numerically, or it can be evaluated analytically by expanding $J_0(\beta\sqrt{t})$ in powers of $\beta^2 t$ and integrating the resulting series term-by-term [3.17]. The results presented below were obtained by evaluating the integral numerically.

The mean intensity obtained from Eq. (3.2.24) is

$$\begin{aligned} \langle I(z_0, r_0)\rangle &= \langle|\psi(z_0, r_0)|^2\rangle \\ &= |C_0|^2 \sum_{m=0}^{N-1} \sum_{n=0}^{N-1} \langle \exp\left[-i\nu_3(b_m - b_n)\right] \psi_m(z_0, r_0; a_m) \psi_n^*(z_0, r_0; a_n)\rangle. \end{aligned} \tag{3.2.27}$$

We now recall that the random numbers $\{a_n\}$ are statistically independent. If we now assume that the coherent component of this mean intensity is negligible, i.e., that $\langle\psi_n(z_0, r_0; a_n)\rangle = 0$, Eq. (3.2.27) simplifies to

$$\langle I(z_0, r_0)\rangle = |C_0|^2 \sum_{n=0}^{N-1} \langle|\psi_n(z_0, r_0; a_n)|^2\rangle$$

$$= \frac{\nu_3}{2b}\frac{|C_0|^2}{\kappa z_m} \sum_{n=0}^{N-1} \int_{-\frac{\kappa b}{\nu_3} z_m}^{\frac{\kappa b}{\nu_3} z_m} d\gamma\, |\psi_n(z_0, r_0; \gamma)|^2, \tag{3.2.28}$$

where we have used the result for $f(\gamma)$ given by Eq. (3.2.22) in the last step. Calculations, whose results we do not show here, indicate that the relative error made in using Eq. (3.2.28) instead of Eq. (3.2.27) is of the order of one part in 10^4 for the values of all the parameters, and the values of z_0 and r_0, assumed here. Consequently, in what follows we will use the simpler expression (3.2.28) in calculating the mean intensity $\langle I(z_0, r_0)\rangle$.

In concluding this section we note that under normal experimental conditions the estimation of the ensemble average is not a practical consideration, and one has to work with a single realization of the diffuser. For any realization different diffuser rings focus light on different points along the optical axis with equal probability within the design region. For monochromatic illumination the interference between all these randomly phased contributions produces speckle, which manifests itself as random variations in the axial intensity. At the same time, due to the circular symmetry of the system about the optical axis, the transverse intensity pattern contains rings that change rapidly with changing z_0. These fluctuations can be smoothed through the use of broadband illumination instead of monochromatic illumination (see Section 2.9).

3.2.3. *Example*

We consider, as an example, a diffuser that produces a uniform distribution of intensity in the range -2 cm $< z_0 < 2$ cm of the optical axis, and produces zero intensity along the rest of this axis. The pdf of a_n for this case is given by Eq. (3.2.22). We assume the values $a = 2$ cm, $R = 15$ cm, $\lambda = 632.8$ nm, and $\Delta n = 0.6$. The diffuser is assumed to have 99 zones ($N = 100$), so that the characteristic length $b = 0.2$ cm.

The result for $\langle I(z_0, r_0)\rangle$ is shown in Fig. 3.26. In Fig. 3.26(a) we show the axial intensity distribution. It is seen that the intensity is fairly constant in the design region -2 cm $< z_0 < 2$ cm of the optical axis, and decreases rapidly along the optical axis away from this region. In Fig. 3.26(b) we show the intensity distribution in the best focus plane $z_0 = 0$ as a function of the radial coordinate r_0. This distribution is fairly invariant within

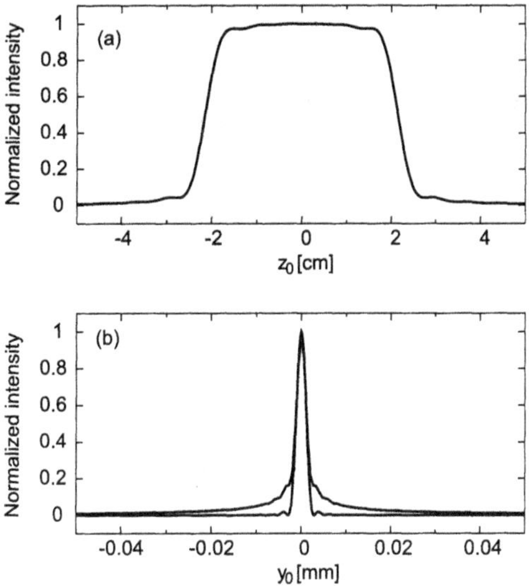

Figure 3.26. (a) The mean intensity distribution along the optical axis; (b) the mean intensity in the transverse direction in the best focus plane $z_0 = 0$. The axial response of the system with a clear aperture is too narrow to be shown in (a). The lower curve in (b) represents the response of the system with a clear aperture.

the design region. It consists of a bright central core surrounded by a halo. A consequence of this halo is that the high frequency components of the object will be transferred to the image with low contrast. Despite this, the sharpness of the central core suggests that such diffusers can be useful in imaging applications.

Circularly symmetric randomly rough surfaces of the type considered here can be fabricated on photoresist by the approach described in Section 3.4. We now present some experimental results obtained by using these diffusers in the formation of extended images.

In Fig. 3.27 we show images of a wheel-like object taken under white light illumination. The photographs on the top row correspond to the best-focus condition. The image (a) was formed by an aberration-free optical system. Image (b) was formed by the same system, but with the inclusion of a diffuser in the pupil plane. It is seen that, due to the halo in the mean transverse intensity distribution, the edges of the image show a reduction in contrast. Images corresponding to an out-of-focus situation are shown in the bottom row of Fig. 3.27. The image (c) is noticeably blurred, and its quality has been degraded with respect to that of the image (a). In contrast, the image (d), obtained with the system that includes the diffuser in the pupil plane, has changed only slightly. The resolution of the images can be assessed by inspection of the fine details in the circled regions of images (a)–(d). These results confirm the expectation that the optical system with the diffuser has a greater depth of focus than the system without it.

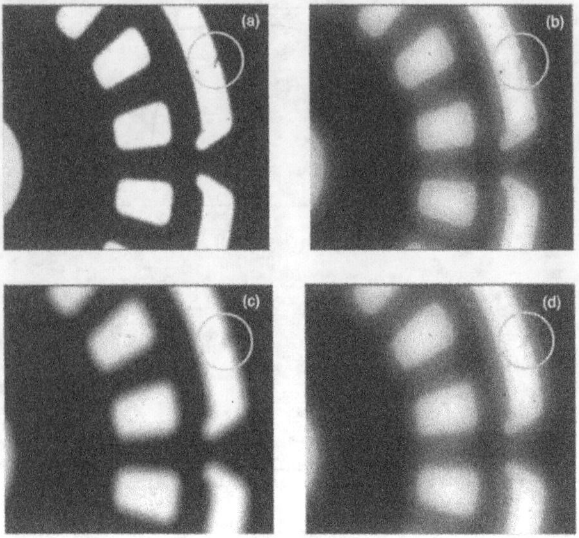

Figure 3.27. Images of a wheel-like binary object under white light illumination in a well-corrected 4f imaging system with $a = 0.5$ cm and $f = 30$ cm. (a) clear-aperture system, image in the best-focus plane ($z_0 = 0$); (b) system with diffuser, image in the best-focus plane; (c) clear-aperture system, image in an out-of-focus plane ($z_0 = 8$ mm); (d) system with diffuser, image in the same out-of-focus plane. The images show a region of about 4 mm × 4 mm of the object.

3.3. A Two-Dimensional Randomly Rough Surface that Acts as a Gaussian Schell-Model Source

Up to this point in the present chapter we have considered only the design of two-dimensional randomly rough surfaces that produce scattered fields with specified forms of their mean intensity. A mean scattered intensity is a two-point amplitude correlation function of the scattered field in the limit as the two points at which the field and its complex conjugate are determined merge. However, the method developed in Section 3.1.4 can be used to design surfaces that produce scattered fields with prescribed genuine two-point amplitude correlation functions, at least of a particular form. We illustrate this in the present section by considering the design of such a surface whose scattering properties are of considerable interest.

Thus, we consider a secondary-fluctuating planar source situated in the plane $x_3 = 0$ that produces an electromagnetic field of frequency ω in the region $x_3 > 0$. We denote a component of this field by $U(\mathbf{x}|\omega)$. Its value in the plane of the source will be denoted by $U(\mathbf{x}_\parallel|\omega)$. The cross-spectral density of the radiated field in the source plane, $W^{(0)}(\mathbf{x}_\parallel, \mathbf{x}'_\parallel|\omega)$, is defined by

$$W^{(0)}(\mathbf{x}_\parallel, \mathbf{x}'_\parallel|\omega) = \langle U(\mathbf{x}_\parallel\omega)U^*(\mathbf{x}'_\parallel|\omega)\rangle, \tag{3.3.1}$$

where the angle brackets denote an average over the ensemble of realizations of the functions $\{U(\mathbf{x}_\parallel|\omega)\}$.

A particular interest attaches to secondary fluctuating planar sources that are characterized by a cross-spectral density of the form [3.18]

$$W^{(0)}(\mathbf{x}_\parallel, \mathbf{x}'_\parallel|\omega) = \left[S^{(0)}(\mathbf{x}_\parallel|\omega)\right]^{\frac{1}{2}} g^{(0)}(\mathbf{x}_\parallel - \mathbf{x}'_\parallel|\omega)\left[S^{(0)}(\mathbf{x}'_\parallel|\omega)\right]^{\frac{1}{2}}. \tag{3.3.2}$$

In this expression $S^{(0)}(\mathbf{x}_\parallel|\omega)$ is the spectral density (intensity) of the light at a typical point in the source plane, and is given by

$$S^{(0)}(\mathbf{x}_\parallel|\omega) = W^{(0)}(\mathbf{x}_\parallel, \mathbf{x}_\parallel|\omega) = \left\langle\left|U(\mathbf{x}_\parallel|\omega)\right|^2\right\rangle, \tag{3.3.3}$$

while $g^{(0)}(\mathbf{x}_\parallel - \mathbf{x}'_\parallel|\omega)$ is the spectral degree of coherence of the source in the plane of the source. The spectral degree of coherence possesses the following important properties [3.19],

$$g^{(0)}(\mathbf{x}_\parallel - \mathbf{x}'_\parallel|\omega) = g^{(0)}(\mathbf{x}'_\parallel - \mathbf{x}_\parallel|\omega)^*, \tag{3.3.4a}$$

$$0 \leqslant \left|g^{(0)}(\mathbf{x}_\parallel - \mathbf{x}'_\parallel|\omega)\right| \leqslant 1, \tag{3.3.4b}$$

$$g^{(0)}(\mathbf{0}|\omega) = 1. \tag{3.3.4c}$$

Sources defined by these properties are called Schell-model sources, and are partially coherent sources.

When the spatial distributions of the spectral density and of the spectral degree of coherence are both Gaussian, namely if

$$S^{(0)}(\mathbf{x}_\parallel|\omega) = A^2(\omega)\exp\left[-x_\parallel^2/2\sigma_s^2(\omega)\right], \tag{3.3.5}$$

$$g^{(0)}(\mathbf{x}_\parallel|\omega) = \exp\left[-x_\parallel^2/2\sigma_g^2(\omega)\right], \tag{3.3.6}$$

where $A(\omega)$, $\sigma_s(\omega)$ and $\sigma_g(\omega)$ are all positive, the corresponding source is called a Gaussian Schell-model source. It is also a partially coherent source [3.20]. The interest in Gaussian Schell-model sources stems from an equivalence theorem that states that two different Gaussian Schell-model sources for which the quantities $(1/4\sigma_s^2) + (1/\sigma_g^2)$ have the same value, and for which the products $A\sigma_s$ are the same, generate the same distribution of the radiant intensity [3.20].

A particularly interesting limiting case of this equivalence theorem is one where $\sigma_g \to \infty$. From Eq. (3.3.6) we see that this limit characterizes a Gaussian Schell-model source that is spatially coherent at the frequency ω, i.e., one for which $g^{(0)}(\mathbf{x}_\parallel|\omega) \equiv 1$ for all $\mathbf{x}_\parallel$. An example of such a source is a laser with a flat output mirror, which operates in its lowest-order mode, provided that effects arising from diffraction at the edge of the mirror are neglected. The form that Eq. (3.3.5) takes for such a source can be written [3.21]

$$S^{(0)}(\mathbf{x}_\parallel|\omega) = A_L^2 \exp\left[-2x_\parallel^2/\delta_L^2\right]. \tag{3.3.7}$$

It follows from the aforementioned equivalence theorem that Gaussian Schell-model sources for which $\delta_L = 2\sigma_s\sigma_g/(\sigma_g^2 + 4\sigma_s^2)^{\frac{1}{2}}$ and $A_L\delta_L = 2A\sigma_s$ will generate the same angular distribution of radiant intensity as the laser source whose intensity is given by Eq. (3.3.7). If the source is characterized by a coherence length σ_g that is much smaller than the linear dimensions of the source, i.e., if $\sigma_g \ll \sigma_s$, such a source is practically homogeneous, and the angular divergence of the beam produced can be made quite small (the equivalent laser beam width in this limit is $\delta_L = \sigma_g$). Such sources will be called Collett–Wolf sources [3.22].

In addition to mimicking the radiation from a laser, the radiation from a Collett–Wolf source produces no speckle when it is incident on a surface [3.23], and propagates farther through a turbulent atmosphere, and spreads less, than does a laser beam [3.24].

A Gaussian Schell-model source can be produced by focusing a beam of laser light on a rotating ground glass plate that is located in the focal plane of a lens that is followed by a transmission filter that has a Gaussian transmission function. The light in a plane that follows the filter is a Gaussian Schell-model source [3.22]. By choosing suitable values for the rms width of the surface roughness correlation function of the rotating ground glass plate, the rms width of the transmission function of the amplitude filter, and the focal length of the lens behind the ground glass plate, one can produce Gaussian Schell-model sources with desired parameters, including those corresponding to a Collett–Wolf source.

Several other methods have been devised for producing sources of partially coherent light, such as, for example, the application of a dc electric field to a liquid crystal that scatters a laser beam [3.25], and the use of various transmission filters or holographic filters to change the coherence properties of the light that passes through them [3.26].

These attractive properties of the radiation from a partially coherent Collett–Wolf source prompt the question of whether it is possible for the partially coherent light scattered from a static suitably designed randomly rough surface to mimic the radiation produced by a laser. This question was answered affirmatively in the case that the randomly rough surface is a one-dimensional randomly rough surface [3.27]. In this section we examine this possibility in the case that the randomly rough surface is a two-dimensional random surface on which the Dirichlet boundary condition is satisfied.

3.3.1. The Cross-Spectral Density in Fourier Space for a Gaussian Schell-Model Source

The physical system we consider here is the one described in Section 3.1. We begin by assuming that the surface $x_3 = \zeta(\mathbf{x}_\parallel)$ is illuminated from the vacuum by a scalar plane wave that satisfies Eq. (3.1.3) and can be written in the form

$$\Psi(\mathbf{x}|\omega)_{\mathrm{inc}} = \exp\left[i\mathbf{k}_\parallel \cdot \mathbf{x}_\parallel - i\alpha_0(k_\parallel)x_3\right]. \tag{3.3.8}$$

The scattered field in the region $x_3 > \zeta(\mathbf{x}_\parallel)_{\max}$ is a superposition of outgoing plane waves, which also satisfies Eq. (3.1.3) and can be written as

$$\Psi(\mathbf{x}|\omega)_{\mathrm{sc}} = \int \frac{d^2q_\parallel}{(2\pi)^2} R(\mathbf{q}_\parallel|\mathbf{k}_\parallel) \exp\big[i\mathbf{q}_\parallel \cdot \mathbf{x}_\parallel + i\alpha_0(q_\parallel)x_3\big]. \tag{3.3.9}$$

We now invoke the Rayleigh hypothesis [3.28], which is the assumption that the sum $\Psi(\mathbf{x}|\omega)_{\mathrm{inc}} + \Psi(\mathbf{x}_\parallel|\omega)_{\mathrm{sc}}$ which, strictly speaking is a valid solution of the scattering problem only for $x_3 > \zeta(\mathbf{x}_\parallel)_{\max}$, can be used in satisfying the Dirichlet boundary condition (3.1.4) on the surface itself. It then follows that the scattering amplitude $R(\mathbf{q}_\parallel|\mathbf{k}_\parallel)$ is the solution of the integral equation

$$\int \frac{d^2q_\parallel}{(2\pi)^2} R(\mathbf{q}_\parallel|\mathbf{k}_\parallel) \exp\big[i\mathbf{q}_\parallel \cdot \mathbf{x}_\parallel + i\alpha_0(q_\parallel)\zeta(\mathbf{x}_\parallel)\big] = -\exp\big[i\mathbf{k}_\parallel \cdot \mathbf{x}_\parallel - i\alpha_0(k_\parallel)\zeta(\mathbf{x}_\parallel)\big]. \tag{3.3.10}$$

We will return to this equation presently.

Let us now consider the case in which the surface $x_3 = \zeta(\mathbf{x}_\parallel)$ is illuminated from the vacuum side at normal incidence by a scalar beam. Such a beam is given by

$$\psi(\mathbf{x}|\omega)_{\mathrm{inc}} = \int\limits_{k_\parallel < \frac{\omega}{c}} \frac{d^2k_\parallel}{(2\pi)^2} F(k_\parallel) \exp\big[i\mathbf{k}_\parallel \cdot \mathbf{x}_\parallel - i\alpha_0(k_\parallel)x_3\big], \tag{3.3.11}$$

where the function $F(k_\parallel)$ is

$$F(k_\parallel) = 2\pi w^2 \exp\big(-k_\parallel^2 w^2/2\big). \tag{3.3.12}$$

In the limit $w \gg (c/\omega)$ Eq. (3.3.11) takes the form

$$\psi(\mathbf{x}|\omega)_{\mathrm{inc}} \cong \exp\big[-x_\parallel^2/2w^2\big] \exp\big[-i(\omega/c)x_3\big]. \tag{3.3.13}$$

Due to the linearity of the scattering problem the scattered field becomes

$$\psi(\mathbf{x}|\omega)_{\mathrm{sc}} = \int \frac{d^2q_\parallel}{(2\pi)^2} R(\mathbf{q}_\parallel) \exp\big[i\mathbf{q}_\parallel \cdot \mathbf{x}_\parallel + i\alpha_0(q_\parallel)x_3\big], \tag{3.3.14}$$

where

$$R(\mathbf{q}_\parallel) = \int\limits_{k_\parallel < \frac{\omega}{c}} \frac{d^2k_\parallel}{(2\pi)^2} R(\mathbf{q}_\parallel|\mathbf{k}_\parallel) F(k_\parallel). \tag{3.3.15}$$

Let us now introduce the function

$$\psi(\mathbf{x}_\parallel|\omega)_{\mathrm{sc}} \equiv \psi(\mathbf{x}|\omega)_{\mathrm{sc}}\big|_{x_3=0}. \tag{3.3.16}$$

We then seek to determine the surface profile function $\zeta(\mathbf{x}_\parallel)$ for which the correlation function of the scattered field at two arbitrary points in the plane $x_3 = 0$,

$$W^{(0)}(\mathbf{x}_\parallel, \mathbf{x}'_\parallel|\omega) = \langle \psi(\mathbf{x}_\parallel|\omega)_{\mathrm{sc}} \psi^*(\mathbf{x}'_\parallel|\omega)_{\mathrm{sc}} \rangle, \tag{3.3.17}$$

has the form (Eqs. (3.3.2), (3.3.5), and (3.3.6))

$$W^{(0)}(\mathbf{x}_\parallel, \mathbf{x}'_\parallel|\omega) = A^2 \exp(-x_\parallel^2/4\sigma_s^2) \exp[-(\mathbf{x}_\parallel - \mathbf{x}'_\parallel)^2/2\sigma_g^2] \exp(-x_\parallel'^2/4\sigma_s^2). \tag{3.3.18}$$

The angle brackets in Eq. (3.3.17) denote an average over the ensemble of realizations of $\zeta(\mathbf{x}_\parallel)$.

The use of Eq. (3.3.14) together with Eqs. (3.3.16)–(3.3.18) yields

$$\begin{aligned} &\int \frac{d^2q_\parallel}{(2\pi)^2} \int \frac{d^2q'_\parallel}{(2\pi)^2} \langle R(\mathbf{q}_\parallel) R^*(\mathbf{q}'_\parallel) \rangle \exp[i\mathbf{q}_\parallel \cdot \mathbf{x}_\parallel - i\mathbf{q}'_\parallel \cdot \mathbf{x}'_\parallel] \\ &= A^2 \exp(-x_\parallel^2/4\sigma_s^2) \exp[-(\mathbf{x}_\parallel - \mathbf{x}'_\parallel)^2/2\sigma_g^2] \exp(-x_\parallel'^2/4\sigma_s^2). \end{aligned} \tag{3.3.19}$$

By inverting this equation we find that the correlation function $\langle R(\mathbf{q}_\parallel) R^*(\mathbf{q}'_\parallel)\rangle$ must have the form

$$\langle R(\mathbf{q}_\parallel) R^*(\mathbf{q}'_\parallel) \rangle = \frac{4\pi^2 A^2}{4\alpha^2 - \beta^2} \exp\left[-\frac{1}{2} \frac{(\mathbf{q}_\parallel - \mathbf{q}'_\parallel)^2}{2\alpha - \beta}\right] \exp\left[-\frac{1}{4} \frac{(\mathbf{q}_\parallel + \mathbf{q}'_\parallel)^2}{2\alpha + \beta}\right], \tag{3.3.20}$$

where

$$\alpha = \frac{1}{4\sigma_s^2} + \frac{1}{2\sigma_g^2}, \qquad \beta = \frac{1}{\sigma_g^2}. \tag{3.3.21}$$

Returning now to Eq. (3.3.10), we solve it for $R(\mathbf{q}_\parallel|\mathbf{k}_\parallel)$ as an expansion in powers of the surface profile function. To first order in $\zeta(\mathbf{x}_\parallel)$ the solution is

$$R(\mathbf{q}_\parallel|\mathbf{k}_\parallel) = -(2\pi)^2 \delta(\mathbf{q}_\parallel - \mathbf{k}_\parallel) + 2i\hat{\zeta}(\mathbf{q}_\parallel - \mathbf{k}_\parallel)\alpha_0(k_\parallel), \tag{3.3.22}$$

where

$$\hat{\zeta}(\mathbf{Q}_\parallel) = \int d^2x_\parallel\, \zeta(\mathbf{x}_\parallel) \exp(-i\mathbf{Q}_\parallel \cdot \mathbf{x}_\parallel). \tag{3.3.23}$$

With the aid of Eq. (3.3.23) we rewrite Eq. (3.3.22) as

$$\begin{aligned} R(\mathbf{q}_\parallel|\mathbf{k}_\parallel) &= -\int d^2x_\parallel\, [1 - 2i\alpha_0(k_\parallel)\zeta(\mathbf{x}_\parallel)] \exp[-i(\mathbf{q}_\parallel - \mathbf{k}_\parallel) \cdot \mathbf{x}_\parallel] \\ &\cong -\int d^2x_\parallel\, \exp[-i(\mathbf{q}_\parallel - \mathbf{k}_\parallel) \cdot \mathbf{x}_\parallel] \exp[-2i\alpha_0(k_\parallel)\zeta(\mathbf{x}_\parallel)]. \end{aligned} \tag{3.3.24}$$

The result given by Eq. (3.3.24) is the phase perturbation theory expression for $R(\mathbf{q}_\parallel|\mathbf{k}_\parallel)$. We will use it here due to its simplicity.

From the result given by Eqs. (3.3.24) and (3.3.12) we find that in the limit $w \gg (c/\omega)$ the scattering amplitude $R(\mathbf{q}_\parallel)$ becomes

$$R(\mathbf{q}_\parallel) \cong -\int d^2x_\parallel \exp\bigl[-(x_\parallel^2/2w^2) - i\mathbf{q}_\parallel \cdot \mathbf{x}_\parallel - i2(\omega/c)\zeta(\mathbf{x}_\parallel)\bigr]. \tag{3.3.25}$$

It follows that the correlation function $\langle R(\mathbf{q}_\parallel)R^*(\mathbf{q}'_\parallel)\rangle$ is

$$\begin{aligned}\bigl\langle R(\mathbf{q}_\parallel)R^*(\mathbf{q}'_\parallel)\bigr\rangle &= \int d^2x_\parallel \int d^2x'_\parallel \exp\bigl[-(x_\parallel^2 + x_\parallel'^2)/2w^2\bigr] \\ &\quad \times \exp\bigl[-i(\mathbf{q}_\parallel \cdot \mathbf{x}_\parallel - \mathbf{q}'_\parallel \cdot \mathbf{x}'_\parallel)\bigr]\bigl\langle \exp\bigl\{-i2(\omega/c)\bigl[\zeta(\mathbf{x}_\parallel) - \zeta(\mathbf{x}'_\parallel)\bigr]\bigr\}\bigr\rangle.\end{aligned} \tag{3.3.26}$$

As it stands, Eq. (3.3.26) is too complicated to invert to obtain $\zeta(\mathbf{x}_\parallel)$ in terms of $\langle R(\mathbf{q}_\parallel)R^*(\mathbf{q}'_\parallel)\rangle$. However, if we pass to the geometrical optics limit of this equation, we obtain an equation we can use to determine $\zeta(\mathbf{x}_\parallel)$. To accomplish this we make the change of variable $\mathbf{x}'_\parallel = \mathbf{x}_\parallel + \mathbf{u}_\parallel$ and expand the difference $\zeta(\mathbf{x}_\parallel) - \zeta(\mathbf{x}_\parallel + \mathbf{u}_\parallel)$ in powers of $\mathbf{u}_\parallel$ through terms linear in $\mathbf{u}_\parallel$. In this way we obtain

$$\begin{aligned}\bigl\langle R(\mathbf{q}_\parallel)R^*(\mathbf{q}'_\parallel)\bigr\rangle &= \int d^2x_\parallel \exp\bigl[-i(\mathbf{q}_\parallel - \mathbf{q}'_\parallel) \cdot \mathbf{x}_\parallel - x_\parallel^2/w^2\bigr] \\ &\quad \times \int d^2u_\parallel \exp\bigl[-u_\parallel^2/(2w^2)\bigr]\exp\bigl[-(\mathbf{x}_\parallel \cdot \mathbf{u}_\parallel)/w^2 + i\mathbf{q}'_\parallel \cdot \mathbf{u}_\parallel\bigr] \\ &\quad \times \bigl\langle \exp\bigl[i2(\omega/c)\mathbf{u}_\parallel \cdot \nabla\zeta(\mathbf{x}_\parallel)\bigr]\bigr\rangle.\end{aligned} \tag{3.3.27}$$

In the remainder of this section we present two quite different approaches to the determination of a surface profile function $\zeta(\mathbf{x}_\parallel)$ which when substituted into the right-hand side of Eq. (3.3.27) yields the expression for $\langle R(\mathbf{q}_\parallel)R^*(\mathbf{q}'_\parallel)\rangle$ given by Eq. (3.3.20).

3.3.2. *A Surface Formed from Triangular Facets*

To evaluate the double integral on the right-hand side of Eq. (3.3.27), as in Section 3.1.4 we begin by covering the x_1x_2 plane by equilateral triangles of edge b, whose vertices are given by Eqs. (3.1.158)–(3.1.159). Each triangle is again labeled by the coordinates of its center of gravity, and the surface profile function $\zeta(\mathbf{x}_\parallel)$ for $\mathbf{x}_\parallel$ within a given triangle is given by Eqs. (3.1.160) and (3.1.161). Equation (3.3.27) then takes the form

$$
\begin{aligned}
\langle R(\mathbf{q}_\parallel)R^*(\mathbf{q}'_\parallel)\rangle &= \int d^2u_\parallel \exp(-u_\parallel^2/2w^2 + i\mathbf{q}'_\parallel \cdot \mathbf{u}_\parallel) \\
&\quad \times \sum_{m=-N}^{N-1} \sum_{n=-N}^{N-1} \Biggl\{ \int_{(m+\frac{1}{3},n+\frac{1}{3})} d^2x_\parallel \exp\{-x_\parallel^2/w^2 \\
&\quad - [\mathbf{u}_\parallel/w^2 + i(\mathbf{q}_\parallel - \mathbf{q}'_\parallel)] \cdot \mathbf{x}_\parallel\} \\
&\quad \times \langle \exp\{i2(\omega/c)[u_1 a^{(1)}_{m+\frac{1}{3},n+\frac{1}{3}} + a^{(2)}_{m+\frac{1}{3},n+\frac{1}{3}}]\}\rangle \\
&\quad + \int_{(m+\frac{2}{3},n+\frac{2}{3})} d^2x_\parallel \exp\{-x_\parallel^2/w^2 - [\mathbf{u}_\parallel/w^2 + i(\mathbf{q}_\parallel - \mathbf{q}'_\parallel)] \cdot \mathbf{x}_\parallel\} \\
&\quad \times \langle \exp\{i2(\omega/c)[u_1 a^{(1)}_{m+\frac{2}{3},n+\frac{2}{3}} + u_2 a^{(2)}_{m+\frac{2}{3},n+\frac{2}{3}}]\}\rangle \Biggr\} \\
&= \int d^2u_\parallel \exp(-u_\parallel^2/2w^2 + i\mathbf{q}'_\parallel \cdot \mathbf{u}_\parallel) \\
&\quad \times \sum_{m=-N}^{N-1} \sum_{n=-N}^{N-1} \Biggl\{ \int_{(m+\frac{1}{3},n+\frac{1}{3})} d^2x_\parallel + \int_{(m+\frac{2}{3},n+\frac{2}{3})} d^2x_\parallel \Biggr\} \\
&\quad \times \exp\{-x_\parallel^2/w^2 - [\mathbf{u}_\parallel/w^2 + i(q_\parallel - q'_\parallel)] \cdot \mathbf{x}_\parallel\} \\
&\quad \times \int d^2\gamma_\parallel f(\boldsymbol{\gamma}_\parallel) \exp[i2(\omega/c)\boldsymbol{\gamma}_\parallel \cdot \mathbf{u}_\parallel] \\
&= \int d^2\gamma_\parallel f(\boldsymbol{\gamma}_\parallel) \int d^2u_\parallel \exp[-u_\parallel^2/2w^2 + i\mathbf{q}'_\parallel \cdot \mathbf{u}_\parallel + i2(\omega/c)\gamma_\parallel \cdot \mathbf{u}_\parallel] \\
&\quad \times \int d^2x_\parallel \exp\{-x_\parallel^2/w^2 - [\mathbf{u}_\parallel/w^2 + i(\mathbf{q}_\parallel - \mathbf{q}'_\parallel)] \cdot \mathbf{x}_\parallel\} \\
&= \pi w^2 \exp[-(w^2/4)(\mathbf{q}_\parallel - \mathbf{q}'_\parallel)^2] \int d^2\gamma_\parallel f(\boldsymbol{\gamma}_\parallel) \\
&\quad \times \int d^2u_\parallel \exp\{-u_\parallel^2/4w^2 + i\mathbf{u}_\parallel \cdot [2(\omega/c)\boldsymbol{\gamma}_\parallel + (1/2)(\mathbf{q}_\parallel + \mathbf{q}'_\parallel)]\} \\
&= 4\pi^2 w^4 \exp[-(w^2/4)(\mathbf{q}_\parallel - \mathbf{q}'_\parallel)^2] \int d^2\gamma_\parallel f(\boldsymbol{\gamma}_\parallel) \\
&\quad \times \exp\{-w^2[2(\omega/c)\gamma_\parallel + (1/2)(\mathbf{q}_\parallel + \mathbf{q}'_\parallel)]^2\}, \qquad (3.3.28)
\end{aligned}
$$

where we have passed to the limit $N \to \infty$. In writing Eq. (3.3.28) we have introduced the vector $\boldsymbol{\gamma}_\parallel = (\gamma_1, \gamma_2, 0)$, and the function $f(\boldsymbol{\gamma}_\parallel)$ is the joint pdf of the slopes of the triangular facets that form the surface and is defined by Eq. (3.1.162).

The integral equation satisfied by $f(\boldsymbol{\gamma}_{\|})$ is obtained from Eqs. (3.3.20) and (3.3.28) as

$$\int d^2\gamma_{\|}\, f(\boldsymbol{\gamma}_{\|})\exp\{-w^2[2(\omega/c)\boldsymbol{\gamma}_{\|}+(1/2)(\mathbf{q}_{\|}+\mathbf{q}'_{\|})]^2\}$$
$$=\frac{2A^2}{(4\alpha^2-\beta^2)w^4}\exp\left\{\frac{1}{4}\left[w^2-\frac{1}{2\alpha-\beta}\right](\mathbf{q}_{\|}-\mathbf{q}'_{\|})^2\right\}\exp\left[-\frac{1}{4}\frac{(\mathbf{q}_{\|}+\mathbf{q}'_{\|})^2}{2\alpha+\beta}\right]. \tag{3.3.29}$$

Since the left-hand side of this equation is a function of $\mathbf{q}_{\|}$ and $\mathbf{q}'_{\|}$ only through the combination $\mathbf{q}_{\|}+\mathbf{q}'_{\|}$ whatever the function $f(\boldsymbol{\gamma}_{\|})$ is, this equation has a solution only if the right-hand side is a function of $\mathbf{q}_{\|}$ and $\mathbf{q}'_{\|}$ only through the combination $\mathbf{q}_{\|}+\mathbf{q}'_{\|}$. This requires that the condition

$$w^2=\frac{1}{2\alpha-\beta} \tag{3.3.30}$$

be satisfied which, with the aid of Eq. (3.3.21), yields the relation

$$\sigma_s=\frac{w}{\sqrt{2}}. \tag{3.3.31}$$

The solution of the resulting integral equation for $f(\boldsymbol{\gamma}_{\|})$,

$$\int d^2\gamma_{\|}\, e^{-w^2[2(\omega/c)\boldsymbol{\gamma}_{\|}+\frac{1}{2}(\mathbf{q}_{\|}+\mathbf{q}'_{\|})]^2} f(\boldsymbol{\gamma}_{\|})=\frac{2A^2}{(4\alpha^2-\beta^2)w^4}e^{-\frac{1}{4}\frac{(\mathbf{q}_{\|}+\mathbf{q}'_{\|})^2}{2\alpha+\beta}}, \tag{3.3.32}$$

is

$$f(\boldsymbol{\gamma}_{\|})=\frac{2}{\pi}\left(\frac{\omega}{c}\right)^2\sigma_g^2\exp[-2(\omega/c)^2\sigma_g^2\gamma_{\|}^2], \tag{3.3.33}$$

provided that

$$A=\frac{1}{\sqrt{2}}. \tag{3.3.34}$$

The marginal pdf of $a^{(1)}_{m+\frac{1}{3},n+\frac{1}{3}}$ (and of $a^{(1)}_{m+\frac{2}{3},n+\frac{2}{3}}$) $f(\gamma_1)$ is

$$f(\gamma_1)=\int_{-\infty}^{\infty} d\gamma_2\, f(\boldsymbol{\gamma}_{\|})=\left(\frac{2}{\pi}\right)^{\frac{1}{2}}\left(\frac{\omega}{c}\right)\sigma_g\exp[-2(\omega/c)^2\sigma_g^2\gamma_1^2], \tag{3.3.35}$$

and the conditional pdf of $a^{(2)}_{m+\frac{1}{3},n+\frac{1}{3}}$ $(a^{(2)}_{m+\frac{2}{3},n+\frac{2}{3}})$ given $a^{(1)}_{m+\frac{1}{3},n+\frac{1}{3}}$ $(a^{(1)}_{m+\frac{2}{3},n+\frac{2}{3}})$ is

$$f(\gamma_2|\gamma_1) = \frac{f(\gamma_1,\gamma_2)}{f(\gamma_1)} = \left(\frac{2}{\pi}\right)^{\frac{1}{2}}\left(\frac{\omega}{c}\right)\sigma_g \exp[-2(\omega/c)^2\sigma_g^2\gamma_2^2]. \tag{3.3.36}$$

Because $f(\boldsymbol{\gamma}_\parallel)$ depends on γ_1 and γ_2 only in the form $\gamma_\parallel = (\gamma_1^2+\gamma_2^2)^{\frac{1}{2}}$, the additional joint pdfs $\tilde{f}(\boldsymbol{\gamma}_\parallel)$, $\tilde{\tilde{f}}(\boldsymbol{\gamma}_\parallel)$, and the associated marginal pdfs $\tilde{f}(\gamma_1)$, $\tilde{\tilde{f}}(\gamma_1)$, needed for constructing the surface profile function are equal to $f(\boldsymbol{\gamma}_\parallel)$ and $f(\gamma_1)$, respectively.

The construction of the surface profile function on the basis of the preceding results is carried out in the manner described in Section 3.1.4.2. Does this surface produce the same far field as a laser? To answer this question we note that the intensity of the scattered field in the far zone is related to the correlation function of the scattered field at $x_3 = x_3' = 0$. To see this, we use the expression for the scattered field given by Eq. (3.3.14) to write

$$\begin{aligned}\langle|\psi(\mathbf{x}|\omega)_{\rm sc}|^2\rangle = \int \frac{d^2q_\parallel}{(2\pi)^2}\int \frac{d^2q_\parallel'}{(2\pi)^2}\langle R(\mathbf{q}_\parallel)R^*(\mathbf{q}_\parallel')\rangle \exp[i(\mathbf{q}_\parallel - \mathbf{q}_\parallel')\cdot\mathbf{x}_\parallel]\\ \times \exp[i(\alpha_0(q_\parallel) - \alpha_0^*(q_\parallel'))x_3].\end{aligned} \tag{3.3.37}$$

We also obtain from Eq. (3.3.14) the result that

$$R(\mathbf{q}_\parallel) = \int d^2x_\parallel\, \psi(\mathbf{x}_\parallel|\omega)_{\rm sc}\exp(-i\mathbf{q}_\parallel\cdot\mathbf{x}_\parallel). \tag{3.3.38}$$

On substituting Eq. (3.3.38) into Eq. (3.3.37) we obtain

$$\begin{aligned}\langle|\psi(\mathbf{x}|\omega)_{\rm sc}|^2\rangle &= \int d^2x_\parallel'\int d^2x_\parallel''\langle\psi(\mathbf{x}_\parallel'|\omega)_{\rm sc}\psi^*(\mathbf{x}_\parallel''|\omega)_{\rm sc}\rangle\\ &\quad\times\int\frac{d^2q_\parallel}{(2\pi)^2}\exp[-i\mathbf{q}_\parallel\cdot(\mathbf{x}_\parallel'-\mathbf{x}_\parallel)+i\alpha_0(q_\parallel)x_3]\\ &\quad\times\int\frac{d^2q_\parallel'}{(2\pi)^2}\exp[i\mathbf{q}_\parallel'\cdot(\mathbf{x}_\parallel''-\mathbf{x}_\parallel)-i\alpha_0^*(q_\parallel')x_3].\end{aligned} \tag{3.3.39}$$

To evaluate the integrals over $\mathbf{q}_\parallel$ and $\mathbf{q}_\parallel'$ we use the parabolic approximation, namely

$$\alpha_0(q_\parallel) = [(\omega/c)^2 - q_\parallel^2]^{\frac{1}{2}} \cong (\omega/c) - (c/2\omega)q_\parallel^2. \tag{3.3.40}$$

In making this approximation we are assuming that x_3 is sufficiently large that only small values of $q_\parallel$ contribute significantly to the integral. With this approximation we obtain

$$\int \frac{d^2q_\parallel}{(2\pi)^2} \exp\left[i\mathbf{q}_\parallel \cdot (\mathbf{x}_\parallel - \mathbf{x}'_\parallel) + i\alpha_0(q_\parallel)x_3\right]$$

$$\cong \frac{1}{4\pi^2} \exp\left[i(\omega/c)x_3\right] \int_{-\infty}^{\infty} dq_1 \exp\left[iq_1(x_1 - x'_1) - i(cx_3/2\omega)q_1^2\right]$$

$$\times \int_{-\infty}^{\infty} dq_2 \exp\left[iq_2(x_2 - x'_2) - i(cx_3/2\omega)q_2^2\right]$$

$$= \left(\frac{\omega}{2\pi c x_3}\right) \exp(-i\pi/2) \exp\left[i(\omega/c)x_3\right] \exp\left[i(\omega/2cx_3)(\mathbf{x}_\parallel - \mathbf{x}'_\parallel)^2\right]. \tag{3.3.41}$$

The use of this result in Eq. (3.3.37) gives us finally

$$\left\langle \left|\psi(\mathbf{x}|\omega)_{\mathrm{sc}}\right|^2 \right\rangle = \left(\frac{\omega}{2\pi c x_3}\right)^2 \int d^2x'_\parallel \int d^2x''_\parallel \exp\left[i(\omega/2cx_3)(\mathbf{x}_\parallel - \mathbf{x}'_\parallel)^2\right]$$
$$\times \exp\left[-i(\omega/2cx_3)(\mathbf{x}_\parallel - \mathbf{x}''_\parallel)^2\right] \left\langle \psi(\mathbf{x}'_\parallel|\omega)_{\mathrm{sc}} \psi^*(\mathbf{x}''_\parallel|\omega)_{\mathrm{sc}} \right\rangle. \tag{3.3.42}$$

With the use of Eqs. (3.3.17)–(3.3.18) we find that

$$\left\langle \left|\psi(\mathbf{x}|\omega)_{\mathrm{sc}}\right| \right\rangle^2 = 2A^2 \frac{\sigma_s^2}{\sigma_{\mathrm{eff}}^2(x_3)} \exp\left[-\frac{x_\parallel^2}{\sigma_{\mathrm{eff}}^2(x_3)}\right], \tag{3.3.43a}$$

where

$$\sigma_{\mathrm{eff}}^2(x_3) = 2\sigma_s^2 + \left(\frac{cx_3}{\omega}\right)^2 \left(\frac{1}{2\sigma_s^2} + \frac{2}{\sigma_g^2}\right). \tag{3.3.43b}$$

Thus $\sqrt{2}\sigma_s$ is the initial rms halfwidth of the beam in the plane of the mean surface, and the quantity

$$\Delta = \frac{c}{\omega}\left(\frac{1}{2\sigma_s^2} + \frac{2}{\sigma_g^2}\right)^{\frac{1}{2}} \tag{3.3.44}$$

is the angle of the beam divergence.

For a laser with a Gaussian distribution of intensity in the source plane $S^{(0)}(\mathbf{x}_\parallel|\omega) = A_L^2 \exp(-2x_\parallel^2/\delta_L^2)$, the far field intensity is given by

$$\left|\psi(\mathbf{x}|\omega)_L\right|^2 = \frac{A_L^2 \delta_L^2}{2\sigma_L^2(x_3)} \exp\left(-\frac{x_\parallel^2}{\sigma_L^2(x_3)}\right), \tag{3.3.45a}$$

where

$$\sigma_L^2(x_3) = \frac{1}{2}\delta_L^2 + \left(\frac{cx_3}{\omega}\right)^2 \frac{2}{\delta_L^2}. \tag{3.3.45b}$$

This result is readily obtained from Eq. (3.3.42) if the cross-spectral density function $\langle\psi(\mathbf{x}'_\parallel|\omega)_{\text{sc}}\psi^*(\mathbf{x}''_\parallel|\omega)_{\text{sc}}\rangle = W^{(0)}(\mathbf{x}'_\parallel, \mathbf{x}''_\parallel|\omega)$ is replaced by (Eq. (3.3.2)) $[S^{(0)}(\mathbf{x}'_\parallel|\omega)]^{\frac{1}{2}} \times [S^{(0)}(\mathbf{x}''_\parallel|\omega)]^{\frac{1}{2}}$, since for a laser the spectral degree of coherence $g^{(0)}(\mathbf{x}'_\parallel - \mathbf{x}''_\parallel|\omega)$ equals unity for all $\mathbf{x}'_\parallel - \mathbf{x}''_\parallel$. Therefore, the far-field intensity of the laser beam scattered from the random surface, Eq. (3.3.43a) is the same as that of the laser beam, Eq. (3.3.45a), if $4/\delta_L^2 = 1/\sigma_s^2 + 4/\sigma_g^2$ and $A_L^2\delta_L^2 = 4A^2\sigma_s^2$. In the limit $\sigma_s \gg \sigma_g$, i.e., when the scattered field is practically incoherent, the far-field intensity of the scattered beam is the same as the far-field intensity of a laser beam with $\delta_L = \sigma_g$ and $A_L\delta_L = 2A\sigma_s$, as should be the case for a two-dimensional Collett–Wolf source.

An alternative expression for $\langle|\psi(\mathbf{x}|\omega)_{\text{sc}}|^2\rangle$ that is useful for numerical simulation calculations of it is obtained by using the parabolic approximation (3.3.40) in Eq. (3.3.37). The result is

$$\begin{aligned}\langle|\psi(\mathbf{x}|\omega)_{\text{sc}}|^2\rangle = &\int \frac{d^2q_\parallel}{(2\pi)^2} \exp\left[i\mathbf{q}_\parallel \cdot \mathbf{x}_\parallel - i(cx_3/2\omega)q_\parallel^2\right] \\ &\times \int \frac{d^2q'_\parallel}{(2\pi)^2} \exp\left[-i\mathbf{q}'_\parallel \cdot \mathbf{x}_\parallel + i(cx_3/2\omega)q'^2_\parallel\right]\langle R(\mathbf{q}_\parallel)R^*(\mathbf{q}'_\parallel)\rangle. \end{aligned} \tag{3.3.46}$$

3.3.2.1. Solution of the Scattering Problem Because rigorous numerical simulation calculations of the scattering of a scalar beam from a two-dimensional randomly rough Dirichlet surface are computationally intensive, to determine the correlation function $\langle R(\mathbf{q}_\parallel)R^*(\mathbf{q}'_\parallel)\rangle$ the scattering amplitude $R(\mathbf{q}_\parallel)$, Eq. (3.3.15), is calculated on the basis of the Kirchhoff approximation for $R(\mathbf{q}_\parallel|\mathbf{k}_\parallel)$, Eqs. (3.1.196), (3.1.197), (3.1.200), and (3.1.201), for a large number N_p of realizations of the surface profile function. The expression for $R(\mathbf{q}_\parallel)$ in this approximation becomes

$$R(\mathbf{q}_\parallel) = R_1(\mathbf{q}_\parallel) + R_2(\mathbf{q}_\parallel), \tag{3.3.47}$$

where

$$\begin{aligned} R_1(\mathbf{q}_\parallel) = -i\frac{\sqrt{3}}{2}b \int_{k_\parallel < \frac{\omega}{c}} \frac{d^2k_\parallel}{(2\pi)^2} G(\mathbf{q}_\parallel|\mathbf{k}_\parallel)F(k_\parallel) \sum_{m=-N}^{N-1}\sum_{n=-N}^{N-1} \frac{\exp(-iab^{(0)}_{m+\frac{1}{3},n+\frac{1}{3}})}{q_1 - k_1 + aa^{(1)}_{m+\frac{1}{3},n+\frac{1}{3}}} \\ \times \exp\Bigg[-i\left(q_1 - k_1 + aa^{(1)}_{m+\frac{1}{3},n+\frac{1}{3}}\right)\left(m + \frac{n}{2} + \frac{1}{2}\right)b \\ -i\frac{\sqrt{3}}{2}\left(q_2 - k_2 + aa^{(2)}_{m+\frac{1}{3},n+\frac{1}{3}}\right)\left(n + \frac{1}{2}\right)b\Bigg]\Bigg\{\exp\left[-i\left(q_1 - k_1 + aa^{(1)}_{m+\frac{1}{3},n+\frac{1}{3}}\right)\frac{b}{4}\right] \end{aligned}$$

$$\times \operatorname{sinc}\left[\frac{\sqrt{3}}{4}b\left(q_2-k_2+aa^{(2)}_{mfa,n+\frac{1}{3}}\right)-\frac{b}{4}\left(q_1-k_1+aa^{(1)}_{m+\frac{1}{3},n+\frac{1}{3}}\right)\right]$$

$$-\exp\left[i\left(q_1-k_1+aa^{(1)}_{m+\frac{1}{3}n+\frac{1}{3}}\right)\frac{b}{4}\right]$$

$$\times \operatorname{sinc}\left[\frac{\sqrt{3}}{4}b\left(q_2-k_2+aa^{(2)}_{m+\frac{1}{3},n+\frac{1}{3}}\right)+\frac{b}{4}\left(q_1-k_1+aa^{(1)}_{m+\frac{1}{3},n+\frac{1}{3}}\right)\right]\Bigg\}, \quad (3.3.48)$$

$$R_2(\mathbf{q}_\parallel)=-i\frac{\sqrt{3}}{2}b\int\limits_{k_\parallel<\frac{w}{c}}\frac{d^2k_\parallel}{2(\pi)^2}G(\mathbf{q}_\parallel|\mathbf{k}_\parallel)F(k_\parallel)\sum_{m=-N}^{N-1}\sum_{n=-N}^{N-1}\frac{\exp(-iab^{(0)}_{m+\frac{2}{3},n+\frac{2}{3}})}{q_1-k_1+aa^{(1)}_{m+\frac{2}{3},n+\frac{2}{3}}}$$

$$\times \exp\left[-i\left(q_1-k_1+aa^{(1)}_{m+\frac{2}{3},n+\frac{2}{3}}\right)\left(m+\frac{n}{2}+1\right)b\right.$$

$$\left.-i\frac{\sqrt{3}}{2}\left(q_2-k_2+aa^{(2)}_{m+\frac{2}{3},n+\frac{2}{3}}\right)\left(n+\frac{1}{2}\right)b\right]$$

$$\times\Bigg\{\exp\left[-i\left(q_1-k_1+a^{(1)}_{m+\frac{2}{3},n+\frac{2}{3}}\right)\frac{b}{4}\right]$$

$$\times \operatorname{sinc}\left[\frac{\sqrt{3}}{4}b\left(q_2-k_2+aa^{(2)}_{m+\frac{2}{3},n+\frac{2}{3}}\right)+\frac{b}{4}\left(q_1-k_1+aa^{(1)}_{m+\frac{2}{3},n+\frac{2}{3}}\right)\right]$$

$$-\exp\left[i\left(q_1-k_1+aa^{(1)}_{m+\frac{2}{3},n+\frac{2}{3}}\right)\frac{b}{4}\right]$$

$$\times \operatorname{sinc}\left[\frac{\sqrt{3}}{4}b\left(q_2-k_2+aa^{(2)}_{m+\frac{2}{3},n+\frac{2}{3}}\right)-\frac{b}{4}\left(q_1-k_1+aa^{(1)}_{m+\frac{2}{3},n+\frac{2}{3}}\right)\right]\Bigg\}, \quad (3.3.49)$$

with

$$G(\mathbf{q}_\parallel|\mathbf{k}_\parallel)=\frac{(\omega/c)^2+\alpha_0(q_\parallel)\alpha_0(k_\parallel)-\mathbf{q}_\parallel\cdot\mathbf{k}_\parallel}{\alpha_0(q_\parallel)(\alpha_0(q_\parallel)+\alpha_0(k_\parallel))} \quad (3.3.50)$$

and $a=\alpha_0(q_\parallel)+\alpha_0(k_\parallel)$. The integrals in this expression have to be evaluated numerically. (Care should be taken when calculating $R_1(\mathbf{q}_\parallel)$, Eq. (3.3.48), and $R_2(\mathbf{q}_\parallel)$, Eq. (3.3.49), since the denominators $q_1-k_1+aa^{(1)}_{m+\frac{1}{3},n+\frac{1}{3}}$ and $q_1-k_1+aa^{(1)}_{m+\frac{2}{3},n+\frac{2}{3}}$ can vanish. However, it is easy to check that the expressions for $R_1(\mathbf{q}_\parallel)$ and $R_2(\mathbf{q}_\parallel)$ can be represented through sinc functions of these arguments.) For each realization of the surface profile function the product $R(\mathbf{q}_\parallel)R^*(\mathbf{q}'_\parallel)$ is calculated from this expression for a large number of values of $\mathbf{q}_\parallel$ and $\mathbf{q}'_\parallel$. An arithmetic average of the values of these products obtained from the N_p realizations of the surface profile function yields the correlation function $\langle R(\mathbf{q}_\parallel)R^*(\mathbf{q}'_\parallel)\rangle$. This result is then substituted into the integrand on the right-hand side of

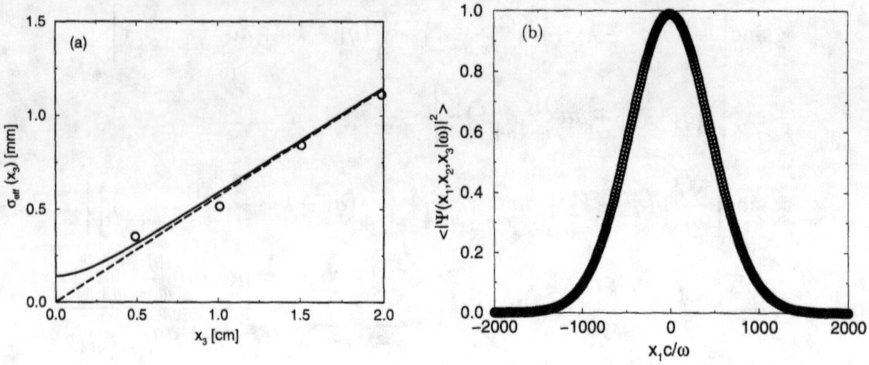

Figure 3.28. (a) A plot of the halfwidth of the scattered beam, as a function of x_3. The solid curve is a plot of $\sigma_{\text{eff}}(x_3)$, obtained from Eq. (3.3.43b). The dashed curve is a plot of $\sigma_L(x_3)$, obtained from Eq. (3.3.45b). The open circles are obtained by means of calculations of $\langle R(\mathbf{q}_\parallel)R^*(\mathbf{q}'_\parallel)\rangle$ on the basis of the Kirchhoff approximation, Eqs. (3.3.47)–(3.3.50). (b) The beam profile obtained in the calculations based on the Kirchhoff approximation (open circles), and the result given by Eqs. (3.3.43a) (solid curve), for a fixed distance from the mean surface $x_3 = 8000(c/\omega) \approx 1273.89\lambda$. The parameters employed are $\lambda = 632.8$ nm, $\sigma_s = 106$ μm, $\sigma_g = 25$ μm, $w = 150$ μm, $b = 20$ μm, $N = 40$, and $N_p = 10000$.

Eq. (3.3.46) and the result of the fourfold integration is then compared with the results given by Eqs. (3.3.43a) and (3.3.45a).

3.3.2.2. Results In Fig. 3.28(a) we present a plot of $\sigma_{\text{eff}}(x_3)$ as a function of the distance x_3 from the mean surface. The solid curve is obtained from Eq. (3.3.43b). The dashed curve is a plot of $\sigma_L(x_3)$, obtained from Eq. (3.3.45b). For large values of x_3 $\sigma_{\text{eff}}(x_3)$ approaches $\sigma_L(x_3)$ asymptotically. However, we see that for all practical purposes the two functions coincide for x_3 as small as 1.5 cm. The open circles show the results obtained from calculations of the intensity of the field scattered from a Dirichlet surface by means of the Kirchhoff approximation, Eqs. (3.3.47)–(3.3.50). They are seen to lie on the solid curve. In Fig. 3.28(b), as an example we show a beam profile (open circles) obtained in the numerical simulations, together with the result given by Eqs. (3.3.43a) (solid curve) for a fixed distance from the mean surface $x_3 = 8000(c/\omega) \simeq 1273.89\lambda$. The agreement between the two beam profiles is seen to be excellent. Consequently, we conclude that the surface generated in the manner described here indeed produces the same far field as does a laser.

3.3.3. A Gaussian Random Surface

The construction of the two-dimensional randomly rough surface described in Section 3.3.2 is based on a joint probability density function of two orthogonal slopes of each of the triangular facets from which the surface is constructed, which has a zero-mean

Gaussian form. The resulting surface is a zero-mean random surface, but it is not a stationary surface, and the surface profile function is not a Gaussian random process. It can also be shown that $\langle\zeta^2(\mathbf{x}_\parallel)\rangle$ increases monotonically with increasing $|\mathbf{x}_\parallel|$. Nevertheless, the result that the joint pdf of two orthogonal slopes of each facet is a zero-mean Gaussian function, and that derivatives of Gaussian random processes are also Gaussian random processes [3.29], prompts the conjecture that there exists a two-dimensional random surface defined by a surface profile function $\zeta(\mathbf{x}_\parallel)$ that is a stationary, zero-mean, isotropic, Gaussian random process, that also acts as a Gaussian Schell-model source. If this is the case, such a surface would be easier to fabricate by well-known methods [3.30,3.31] than the surface generated by the method of Section 3.3.2. In this section we show that such a surface indeed exists.

Our starting point is Eq. (3.3.27). But now we assume that the surface profile function $\zeta(\mathbf{x}_\parallel)$ is a single-valued function of $\zeta(\mathbf{x}_\parallel)$ that is differentiable with respect to x_1 and x_2 as many times as is necessary, and constitutes a stationary, zero-mean, isotropic Gaussian random process, defined by

$$\big\langle\zeta(\mathbf{x}_\parallel)\zeta(\mathbf{x}'_\parallel)\big\rangle=\delta^2 W\big(|\mathbf{x}_\parallel-\mathbf{x}'_\parallel|\big), \tag{3.3.51}$$

where $\delta=\langle\zeta^2(\mathbf{x}_\parallel)\rangle^{\frac{1}{2}}$ is the rms height of the surface.

Since $\zeta(\mathbf{x}_\parallel)$ is a Gaussian random process, so is $\mathbf{u}_\parallel\cdot\nabla\zeta(\mathbf{x}_\parallel)$ [3.29]. It follows, therefore that

$$\big\langle\exp\big[i2(\omega/c)\mathbf{u}_\parallel\cdot\nabla\zeta(\mathbf{x}_\parallel)\big]\big\rangle=\exp\big[-2(\omega/c)^2\big\langle\big(\mathbf{u}_\parallel\cdot\nabla\zeta(\mathbf{x}_\parallel)\big)^2\big\rangle\big]. \tag{3.3.52}$$

For a stationary, zero-mean, isotropic, Gaussian random process we have

$$\left\langle\left(\frac{\partial\zeta(\mathbf{x}_\parallel)}{\partial x_1}\right)^2\right\rangle=-\delta^2\frac{\partial^2}{\partial x_1^2}W\big(|\mathbf{x}_\parallel|\big)\Big|_{\mathbf{x}_\parallel=0} \tag{3.3.53a}$$

$$\left\langle\frac{\partial\zeta(\mathbf{x}_\parallel)}{\partial x_1}\frac{\partial\zeta(\mathbf{x}_\parallel)}{\partial x_2}\right\rangle=0, \tag{3.3.53b}$$

$$\left\langle\left(\frac{\partial\zeta(\mathbf{x}_\parallel)}{\partial x_2}\right)^2\right\rangle=-\delta^2\frac{\partial^2}{\partial x_2^2}W\big(|\mathbf{x}_\parallel|\big)\Big|_{\mathbf{x}_\parallel=0}. \tag{3.3.53c}$$

It follows, therefore, that

$$\big\langle\big(\mathbf{u}_\parallel\cdot\nabla\zeta(\mathbf{x}_\parallel)\big)^2\big\rangle=-\delta^2\left[u_1^2\frac{\partial^2 W(|\mathbf{x}_\parallel|)}{\partial x_1^2}+u_2^2\frac{\partial^2 W(|\mathbf{x}_\parallel|)}{\partial x_2^2}\right]_{\mathbf{x}_\parallel=0}. \tag{3.3.54}$$

We assume that the surface height autocorrelation function $W(|\mathbf{x}_\parallel|)$ has the form

$$W\big(|\mathbf{x}_\parallel|\big)=1-\frac{x_\parallel^2}{a^2}+o\big(x_\parallel^2\big) \tag{3.3.55}$$

as $|\mathbf{x}_\parallel| \to 0$. The characteristic length a in this expression is the transverse correlation length of the surface roughness, which is approximately the mean distance from a maximum on the surface to its nearest minimum [3.32]. The absence of a term linear in $|\mathbf{x}_\parallel|$ in this expansion means that we are excluding surfaces with fractal dimensions from consideration. With the form for $W(|\mathbf{x}_\parallel|)$ given by Eq. (3.3.55), Eq. (3.3.54) becomes

$$\langle (\mathbf{u}_\parallel \cdot \nabla\zeta(\mathbf{x}_\parallel))^2 \rangle = 2\delta^2 u_\parallel^2 / a^2. \tag{3.3.56}$$

We thus obtain finally

$$\langle \exp[i2(\omega/c)\mathbf{u}_\parallel \cdot \nabla\zeta(\mathbf{x}_\parallel)] \rangle = \exp[-(2\omega\delta/ca)^2 u_\parallel^2]. \tag{3.3.57}$$

When this result is substituted into Eq. (3.3.27) the resulting Gaussian integrals are readily evaluated, with the result

$$\langle R(\mathbf{q}_\parallel) R^*(\mathbf{q}'_\parallel) \rangle = \frac{4\pi^2 w^4}{1+16D} \exp\{-(w^2/4)[(\mathbf{q}_\parallel - \mathbf{q}'_\parallel)^2 + (\mathbf{q}_\parallel + \mathbf{q}'_\parallel)^2/(1+16D)]\}, \tag{3.3.58}$$

where $D = (w\omega\delta/ca)^2$. On equating the right-hand sides of Eqs. (3.3.20) and (3.3.58), and making use of Eqs. (3.3.21), we obtain the relations

$$\sigma_s = \frac{w}{2}, \tag{3.3.59a}$$

$$\sigma_g = \frac{1}{2\sqrt{2}} \frac{c}{\omega} \frac{a}{\delta}, \tag{3.3.59b}$$

$$A^2 = 1. \tag{3.3.59c}$$

In order that the two-dimensional Gaussian random surface with a Gaussian surface height autocorrelation function defined by Eqs. (3.3.51) and (3.3.55) act as a Collett–Wolf source, the $1/e$ half-width $\sqrt{2}w$ of the incident beam, its wavelength $\lambda = 2\pi c/\omega$, and the surface roughness parameters δ and a, have to be chosen to satisfy two conditions. The first is that $\sigma_s \gg \sigma_g$ which translates into

$$w \gg \frac{1}{\sqrt{2}} \frac{c}{\omega} \frac{a}{\delta} = \frac{\lambda}{2\sqrt{2}\pi} \frac{a}{\delta}. \tag{3.3.60}$$

The second is that the transverse correlation length of the surface roughness be significantly larger than the wavelength λ of the incident beam, $a \gg \lambda$. When this condition is satisfied the angular dependence of the intensity of the scattered field does not display the enhanced backscattering effect [3.31], namely a well-defined peak in the retroreflection direction, which would distort the profile of the scattered beam from that of the laser source we wish to reproduce. In practice a value of $a \gtrsim 15\lambda$ suffices for this purpose.

3.3.3.1. The Mean Intensity of the Scattered Field To examine whether a surface of the type obtained here produces a scattered field that in the far zone is the same as the field produced by a laser, we return to the expression for $\langle|\psi(\mathbf{x}|\omega)_{sc}|^2\rangle$ given by Eq. (3.3.46). Because numerical simulation calculations of the scattering of a scalar beam from a two-dimensional randomly rough Dirichlet surface are computationally intensive, to determine the correlation function $\langle R(\mathbf{q}_\parallel)R^*(\mathbf{q}'_\parallel)\rangle$ we calculate the scattering amplitude $R(\mathbf{q}_\parallel)$ on the basis of Eqs. (3.3.12) and (3.3.15), with the plane wave scattering amplitude $R(\mathbf{q}_\parallel|\mathbf{k}_\parallel)$ given by the Kirchhoff approximation (3.1.62). The result can be rewritten as

$$\begin{aligned}\langle R(\mathbf{q}_\parallel)R^*(\mathbf{q}'_\parallel)\rangle = &\int_{k_\parallel<\frac{\omega}{c}} \frac{d^2k_\parallel}{(2\pi)^2} G(\mathbf{q}_\parallel|\mathbf{k}_\parallel)F(k_\parallel) \int_{k'_\parallel<\frac{\omega}{c}} \frac{d^2k'_\parallel}{(2\pi)^2} G^*(\mathbf{q}'_\parallel|\mathbf{k}'_\parallel)F(k'_\parallel)\\ &\times \int d^2x_\parallel \exp\left[-i(\mathbf{q}_\parallel-\mathbf{k}_\parallel)\cdot\mathbf{x}_\parallel\right] \int d^2x'_\parallel \exp\left[i(\mathbf{q}'_\parallel-\mathbf{k}'_\parallel)\cdot\mathbf{x}'_\parallel\right]\\ &\times \left\langle \exp\left\{-i\left[\alpha_0(q_\parallel)+\alpha_0(k_\parallel)\right]\zeta(\mathbf{x}_\parallel) + i\left[\alpha_0^*(q'_\parallel)+\alpha_0(k'_\parallel)\right]\zeta(\mathbf{x}'_\parallel)\right\}\right\rangle,\end{aligned} \tag{3.3.61}$$

where $G(\mathbf{q}_\parallel|\mathbf{k}_\parallel)$ has been defined in Eq. (3.3.50). The average over the ensemble of realizations of the surface profile function $\zeta(\mathbf{x}_\parallel)$ can now be carried out analytically, since $\zeta(\mathbf{x}_\parallel)$ has been assumed to be a Gaussian random process, so that

$$\begin{aligned}&\int d^2x_\parallel \exp\left[-i(\mathbf{q}_\parallel-\mathbf{k}_\parallel)\cdot\mathbf{x}_\parallel\right] \int d^2x'_\parallel \exp\left[i(\mathbf{q}'_\parallel-\mathbf{k}'_\parallel)\cdot\mathbf{x}'_\parallel\right]\left\langle \exp\left\{-i\left[m\zeta(\mathbf{x}_\parallel)-n\zeta(\mathbf{x}'_\parallel)\right]\right\}\right\rangle\\ &\quad= \exp\left[-(1/2)(m^2+n^2)\delta^2\right]\int d^2x_\parallel \exp\left[-i(\mathbf{q}_\parallel-\mathbf{k}_\parallel)\cdot\mathbf{x}_\parallel\right]\\ &\qquad\times \int d^2x'_\parallel \exp\left[i(\mathbf{q}'_\parallel-\mathbf{k}'_\parallel)\cdot\mathbf{x}'_\parallel\right]\exp\left[mn\delta^2 W(|\mathbf{x}_\parallel-\mathbf{x}'_\parallel|)\right]\\ &\quad= \exp\left[-(1/2)(m^2+n^2)\delta^2\right]\Bigg\{(2\pi)^2\delta(\mathbf{q}_\parallel-\mathbf{k}_\parallel)(2\pi)^2\delta(\mathbf{q}'_\parallel-\mathbf{k}'_\parallel)\\ &\qquad+(2\pi)^2\delta(\mathbf{q}_\parallel-\mathbf{k}_\parallel-\mathbf{q}'_\parallel+\mathbf{k}'_\parallel)\sum_{\ell=1}^{\infty}\frac{(mn\delta^2)^\ell}{\ell!}\int d^2u_\parallel\, W^\ell(|\mathbf{u}_\parallel|)\\ &\qquad\times \exp\left[-i(\mathbf{q}_\parallel-\mathbf{k}_\parallel)\cdot\mathbf{u}_\parallel\right]\Bigg\},\end{aligned} \tag{3.3.62}$$

where, to simplify the notation, we have defined

$$m = \alpha_0(q_\parallel)+\alpha_0(k_\parallel), \qquad n = \alpha_0^*(q'_\parallel)+\alpha_0(k'_\parallel). \tag{3.3.63}$$

If we assume a Gaussian form for the surface height autocorrelation function,

$$W(|\mathbf{x}_\parallel|) = \exp(-x_\parallel^2/a^2), \tag{3.3.64}$$

which is consistent with Eq. (3.3.55), Eq. (3.3.62) becomes

$$\begin{aligned}
&\int d^2x_\parallel \exp[-i(\mathbf{q}_\parallel - \mathbf{k}_\parallel)\cdot\mathbf{x}_\parallel] \int d^2x'_\parallel \exp[i(\mathbf{q}'_\parallel - \mathbf{k}'_\parallel)\cdot\mathbf{x}'_\parallel]\langle \exp\{-i[m\zeta(\mathbf{x}_\parallel) - n\zeta(\mathbf{x}'_\parallel)]\}\rangle \\
&\quad = \exp\left[-\frac{1}{2}(m^2+n^2)\delta^2\right]\Big\{(2\pi)^2\delta(\mathbf{q}_\parallel - \mathbf{k}_\parallel)(2\pi)^2\delta(\mathbf{q}'_\parallel - \mathbf{k}'_\parallel) \\
&\qquad + (2\pi)^2\delta(\mathbf{q}_\parallel - \mathbf{k}_\parallel - \mathbf{q}'_\parallel + \mathbf{k}'_\parallel)\pi a^2 \sum_{\ell=1}^{\infty} \frac{(mn\delta^2)^\ell}{\ell\cdot\ell!} \exp[-a^2(\mathbf{q}_\parallel - \mathbf{k}_\parallel)^2/4\ell]\Big\}.
\end{aligned} \tag{3.3.65}$$

When this result is substituted into Eq. (3.3.61), the result that $G(\mathbf{q}_\parallel|\mathbf{q}_\parallel) = 1$ is used, and $(2\pi)^2\ \delta(\mathbf{q}_\parallel - \mathbf{k}_\parallel - \mathbf{q}'_\parallel + \mathbf{k}'_\parallel)$ is replaced by its Fourier integral representation, we obtain a compact expression for $\langle R(\mathbf{q}_\parallel)R^*(\mathbf{q}'_\parallel)\rangle$:

$$\begin{aligned}
\langle R(\mathbf{q}_\parallel)R^*(\mathbf{q}'_\parallel)\rangle &= F(q_\parallel)F(q'_\parallel)\exp\{-2\delta^2[\alpha_0^2(q_\parallel) + \alpha_0^2(q'_\parallel)]\} \\
&\quad + \pi a^2 \sum_{\ell=1}^{\infty} \frac{\delta^{2\ell}}{\ell\cdot\ell!} \int d^2x_\parallel\, H_\ell(\mathbf{q}_\parallel|\mathbf{x}_\parallel) H_\ell^*(\mathbf{q}'_\parallel|\mathbf{x}_\parallel),
\end{aligned} \tag{3.3.66}$$

where

$$\begin{aligned}
H_\ell(\mathbf{q}_\parallel|\mathbf{x}_\parallel) &= \exp(-i\mathbf{q}_\parallel\cdot\mathbf{x}_\parallel) \int\limits_{k_\parallel < \frac{\omega}{c}} \frac{d^2k_\parallel}{(2\pi)^2} G(\mathbf{q}_\parallel|\mathbf{k}_\parallel)F(k_\parallel) \\
&\quad \times \exp\{-(\delta^2/2)[\alpha_0(q_\parallel) + \alpha_0(k_\parallel)]^2\} \\
&\quad \times [\alpha_0(q_\parallel) + \alpha_0(k_\parallel)]^\ell \exp[-(a^2/8\ell)(\mathbf{q}_\parallel - \mathbf{k}_\parallel)^2]\exp(i\mathbf{k}_\parallel\cdot\mathbf{x}_\parallel).
\end{aligned} \tag{3.3.67}$$

In interpreting the first term on the right-hand side of Eq. (3.3.66) it should be kept in mind that $F(q_\parallel)$ is nonzero only for $q_\parallel < \omega/c$. The two-dimensional integrals in Eqs. (3.3.66)–(3.3.67) have to be evaluated numerically. The substitution of Eq. (3.3.66) into Eq. (3.3.46) yields $\langle|\psi(\mathbf{x}|\omega)_{sc}|^2\rangle$ in the form

$$\langle|\psi(\mathbf{x}|\omega)_{sc}|^2\rangle = |M(\mathbf{x}|\omega)|^2 + \pi a^2 \sum_{\ell=1}^{\infty} \frac{\delta^{2\ell}}{\ell\cdot\ell!} \int d^2u_\parallel\, |N_\ell(\mathbf{x}|\mathbf{u}_\parallel)|^2, \tag{3.3.68}$$

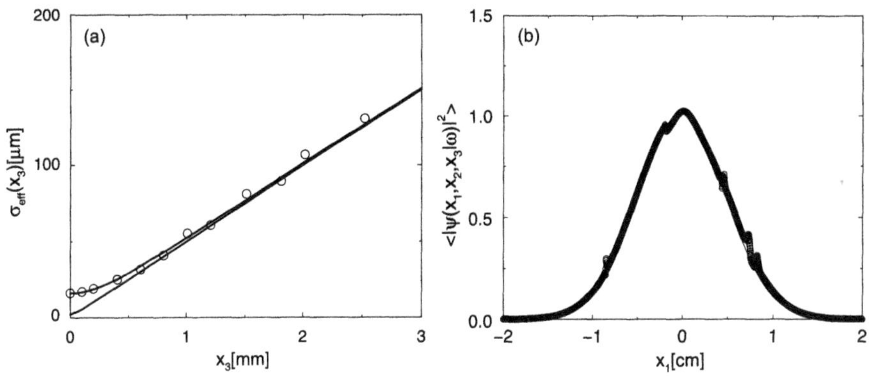

Figure 3.29. (a) A plot of the half width of the scattered beam as a function of x_3. The solid curve is a plot of $\sigma_{\text{eff}}(x_3)$, obtained from Eq. (3.3.43b). The dashed curve is a plot of $\sigma_L(x_3)$ obtained from Eq. (3.3.45b). The open circles show the results obtained on the basis of the Kirchhoff approximation, Eqs. (3.3.68)–(3.3.69). (b) The beam profile obtained in calculations based on the Kirchhoff approximation, Eqs. (3.3.68)–(3.3.69) (open circles), and the result for a laser source given by Eq. (3.3.45a) (solid curve), for a fixed distance from the mean surface $x_3 = 3$ cm. The parameters employed are $\lambda = 632.8$ nm, $\delta = 0.6$ µm, $a = 12$ µm, and $N_p = 10\,000$.

where

$$M(\mathbf{x}|\omega) = \int\limits_{q_\parallel < \frac{\omega}{c}} \frac{d^2 q_\parallel}{(2\pi)^2} F(q_\parallel) \exp\{-2\delta^2 \alpha_0^2(q_\parallel) + i[\mathbf{q}_\parallel \cdot \mathbf{x}_\parallel - (cx_3/2\omega) q_\parallel^2]\}, \tag{3.3.69a}$$

$$N_\ell(\mathbf{x}|\mathbf{u}_\parallel) = \int \frac{d^2 q_\parallel}{(2\pi)^2} H_\ell(\mathbf{q}_\parallel|\mathbf{u}_\parallel) \exp\{i[\mathbf{q}_\parallel \cdot \mathbf{x}_\parallel - (cx_3/2\omega) q_\parallel^2]\}. \tag{3.3.69b}$$

3.3.3.2. Results In Fig. 3.29(a) we present a plot of $\sigma_{\text{eff}}(x_3)$ as a function of the distance x_3 from the source plane. The solid curve is obtained from Eq. (3.3.43b). The dashed curve is a plot of $\sigma_L(x_3)$, obtained from Eq. (3.3.45b). The open circles show the results obtained from the intensity of the field scattered from the two-dimensional randomly rough Dirichlet surface assumed here, calculated in the Kirchhoff approximation, Eqs. (3.3.68)–(3.3.69). For large values of x_3 they are seen to lie on the solid curve. In Fig. 3.29(b) we show a beam profile obtained on the basis of the Kirchhoff approximation Eqs. (3.3.68)–(3.3.69) (open circles), together with the result for a laser source given by Eq. (3.3.45a) (solid curve), for a fixed distance from the source plane $x_3 = 3$ cm. The agreement between the two beam profiles is excellent.

We conclude, therefore, that a two-dimensional randomly rough surface of the type designed here produces a partially coherent scattered field that in the far zone is the same as the one produced by a fully coherent laser, i.e., it acts as a Gaussian Schell-model source.

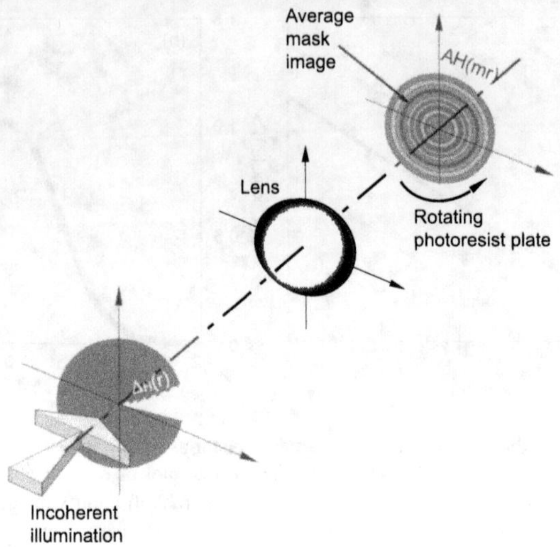

Figure 3.30. Schematic diagram of the setup employed for the fabrication of the circularly symmetric random diffusers.

An attractive feature of the present approach to the design of such a source is that methods exist for fabricating on photoresist two-dimensional randomly rough surfaces characterized by a surface profile function that is a Gaussian random process defined by a Gaussian surface height autocorrelation function with specified roughness parameters [3.30,3.31]. If a highly reflecting surface is desired, as in the present case, a metallic layer can be deposited on the photoresist surface.

3.4. Fabrication of Circularly-Symmetric Surfaces

Circularly symmetric surfaces of the kind discussed in the previous sections can be fabricated on photoresist using a variation of the method described in Section 2.10. A schematic diagram of the experimental setup employed in the fabrication is shown in Fig. 3.30. An incoherent image of a disk-shaped mask is formed on the rotating photoresist-coated plate by a well-corrected imaging system with magnification $M = 1/m$. The illumination is provided by a He–Cd laser beam ($\lambda = 442$ nm) that is transmitted through a rotating ground glass to reduce its coherence. The plate is exposed during a time T_e, during which it executes a large number of revolutions. As explained below, the arrangement is such that the total exposure of the plate is a scaled version of the profile function employed in the generation of the mask.

To produce a mask, we first generate a function $H(r)$ with the kind of statistical properties required. An example is shown in Fig. 3.31, where we show the profile of a diffuser that, when inserted in the pupil of an imaging system, should lead to a uniform mean in-

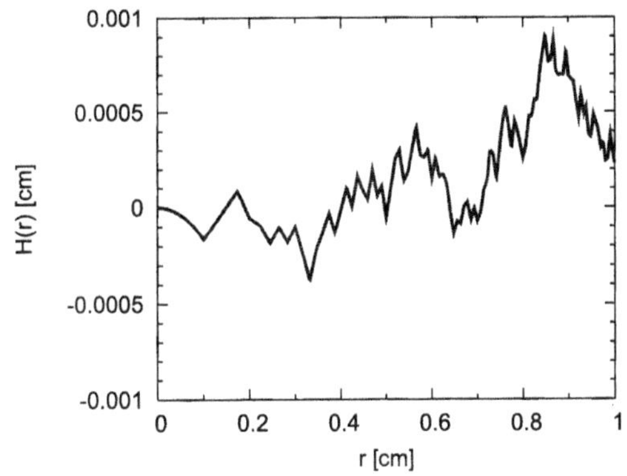

Figure 3.31. Realization of a random surface profile $H(r)$ with 100 zones.

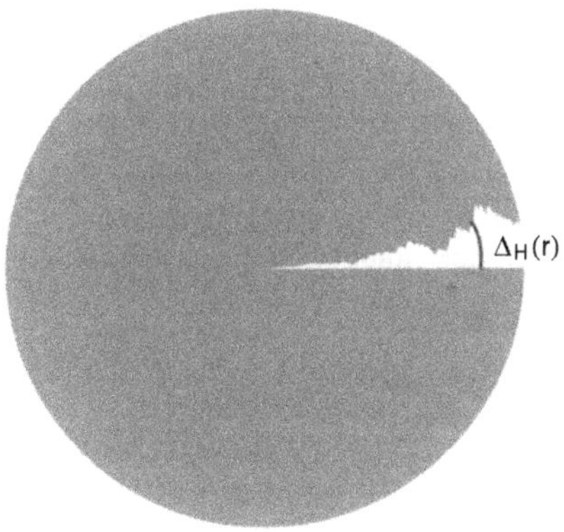

tensity in a region of the optical axis. We now define the function

$$\Delta_H(r) = K H(r). \tag{3.4.1}$$

With an appropriate choice of units of the constant K, $\Delta_H(r)$ can be interpreted as an angle. With reference to Fig. 3.32, for a given radius, the angles θ that fall in the transparent section of the mask are defined by the condition $\Delta_H(r) + \Delta_0 > \theta > 0$, where the positive constant Δ_0 is larger than the largest negative value of $\Delta_H(r)$. That is, the intensity transmittance of the binary mask can be written in the form

$$T(r,\theta) = \theta(\theta)\theta\big(\Delta_H(r) + \Delta_0 - \theta\big), \tag{3.4.2}$$

where $\theta(\theta)$ is the Heaviside unit step function.

The incoherent image of the mask formed on the surface of the rotating photoresist plate produces an exposure that is circularly symmetric with a radial dependence of the form

$$E(r) = \frac{I_0 T_e}{2\pi}\left[\Delta_H(mr) + \Delta_0\right], \tag{3.4.3}$$

where I_0 is a constant related to the intensity of the illumination. If the response of the photoresist is linear, the resulting radial profile will have the generic form

$$h(r) = h_0 + \alpha H(mr), \tag{3.4.4}$$

where h_0 and α are constants that one can adjust by varying the intensity of the light reaching the plate, the aperture of the mask, and the exposure time. The parameter m can be changed with the magnification.

As we have seen, through the magnification of the system and the time of exposure, the original profile $H(r)$ can be scaled in both, the vertical and horizontal directions. These transformations are almost inevitable in our fabrication scheme. We now consider the consequences of this. We then ask what is the mean intensity produced by a diffuser with a surface profile function $\alpha H(mr)$, where α and m are dimensionless constants, with reference to the mean intensity obtained with the original function $H(r)$. From Eq. (3.2.17) we see that the transformation is equivalent to choosing new random deviates $a_n' = \alpha m^2 a_n$. In consequence, the parameter that defines the new region of constant intensity is $z_m' = \alpha m^2 z_m$. In other words, by scaling the function $H(r)$ in the vertical or horizontal direction one can change the length of the region of constant intensity.

References

[3.1] A.E. Danese, Advanced Calculus, Allyn and Bacon, Boston, 1965, p. 152.

[3.2] M. Born, E. Wolf, Principles of Optics, seventh (expanded) ed., Cambridge University Press, New York, 2002, p. 18.

[3.3] Ref. [3.2], pp. 918–920.

[3.4] See, for example, A.A. Maradudin, Interaction of surface polaritons and plasmons with surface roughness, in: V.M. Agranovich, D.L. Mills (Eds.), Surface Polaritons, North-Holland, Amsterdam, 1982, pp. 405–510.

[3.5] A.G. Voronovich, Wave Scattering from Rough Surfaces, second ed., Springer-Verlag, New York, 1999 (Sections 5.1, 5.2).

[3.6] E.R. Méndez, E.E. García Guerrero, H.M. Escamilla, A.A. Maradudin, T.A. Leskova, A.V. Shchegrov, Photofabrication of random achromatic optical diffusers for uniform illumination, Appl. Opt. 40 (2001) 1098–1108.

[3.7] T.A. Leskova, A.A. Maradudin, E.R. Méndez, J. Muñoz-Lopez, The design and photofabrication of random achromatic optical diffusers for uniform illumination, in: P. Kuchment (Ed.), Wave Propagation in Periodic and Random Media, American Mathematical Society, Providence, RI, 2003, pp. 117–140.

[3.8] W.H. Press, S.A. Teukolsky, W.T. Vetterling, B.P. Flannery, Numerical Recipes in Fortran, second ed., Cambridge University Press, New York, 1992, pp. 281–282.

[3.9] P. Tran, A.A. Maradudin, Scattering of a scalar beam from a two-dimensional randomly rough hard wall: Enhanced backscattering, Phys. Rev. B 45 (1992) 3936–3939.

[3.10] C. Macaskill, B.J. Kachoyan, Iterative approach for the numerical simulation of scattering from one- and two-dimensional rough surfaces, Appl. Opt. 32 (1993) 2839–2847.

[3.11] P. Tran, A.A. Maradudin, Scattering of a scalar beam from a two-dimensional randomly rough hardwall: Dirichlet and Neumann boundary conditions, Appl. Opt. 32 (1993) 2848–2851.

[3.12] J. Ojeda-Castaneda, P. Andrés, E. Montes, Phase space representation of the Strehl ration: Ambiguity function, J. Opt. Soc. Am. A 4 (1987) 313–317.

[3.13] J. Ojeda-Castaneda, J.E.A. Landgrave, H.M. Escamilla, Annular phase-only mask for high focal depth, Opt. Lett. 30 (2005) 1647–1649.

[3.14] M. Born, E. Wolf, Principles of Optics, seventh (expanded) ed., Cambridge University Press, New York, 2001 (Section 8.8).

[3.15] Z.H. Gu, H.M. Escamilla, E.R. Méndez, A.A. Maradudin, J.Q. Lu, T. Michel, M. Nieto-Vesperinas, Interaction of two optical beams at a symmetric random surface, Appl. Opt. 31 (1992) 5878–5879.

[3.16] C.W. McCutchen, Generalized aperture and the three-dimensional diffraction image, J. Opt. Soc. Am. 54 (1964) 240–244.

[3.17] A.A. Maradudin, T.A. Leskova, E.R. Méndez, Pseudo-nondiffracting beams from rough surface scattering, in: Y.-Q. Jin (Ed.), Wave Propagation, Scattering, and Emission in Complex Media, World Scientific, Singapore, 2004, pp. 100–118.

[3.18] A.C. Schell, The Multiple Plate Antenna, Ph.D. Dissertation, Massachusetts Institute of Technology, 1961 (Section 7.5).

[3.19] L. Mandel, E. Wolf, Optical Coherence and Quantum Optics, Cambridge University Press, New York, 1995 (Section 4.3.2).

[3.20] E. Collett, E. Wolf, Is complete spatial coherence necessary for the generation of highly directional light beams?, Opt. Lett. 2 (1978) 27–29;
E. Wolf, E. Collett, Partially coherent sources which produce the same far-field intensity distribution as a laser, Opt. Commun. 25 (1978) 293–296.

[3.21] Ref. [3.19], p. 256.

[3.22] P. de Santis, F. Gori, G. Guattari, C. Palma, An example of a Collett–Wolf source, Opt. Commun. 29 (1979) 256–260.

[3.23] Ref. [3.19], pp. 259–260.

[3.24] M. Salem, T. Shirai, A. Dogariu, E. Wolf, Long-distance propagation of partially coherent beams through atmospheric turbulence, Opt. Commun. 216 (2003) 261–265.

[3.25] F. Scudieri, M. Bertolotti, R. Bartolino, Light scattered by a liquid crystal: A new quasi-thermal source, Appl. Opt. 13 (1974) 181–185;
W.H. Carter, M. Bertolotti, An analysis of the far-field coherence and radiant intensity of light scattered from liquid crystals, J. Opt. Soc. Am. 68 (1978) 329–333.

[3.26] D. Courjon, J. Bulabois, W.H. Carter, Use of a holographic filter to modify the coherence of a light field, J. Opt. Soc. Am. 71 (1981) 469–473.

[3.27] T.A. Leskova, A.A. Maradudin, J. Muñoz-Lopez, The design of one dimensional randomly rough surfaces that act as Collett–Wolf sources, Opt. Commun. 234 (2004) 123–133.

[3.28] Lord Rayleigh, The Theory of Sound, vol. II, second ed., Macmillan, London, 1896, pp. 89, 297–311.

[3.29] D. Middleton, An Introduction to Statistical Communication Theory, McGraw–Hill, New York, 1960, p. 370.

[3.30] P.F. Gray, A method of forming optical diffusers of simple known statistical properties, Optica Acta 25 (1978) 765–775.

[3.31] E.R. Méndez, K.A. O'Donnell, Observation of depolarization and backscattering enhancement in light scattering from Gaussian random surfaces, Opt. Commun. 61 (1987) 91–95.

[3.32] S. Simeonov, A.R. McGurn, A.A. Maradudin, Transverse correlation length for randomly rough surfaces: Two-dimensional roughness, SPIE 3141 (1997) 152–163.

Chapter 4

CONCLUSIONS AND OUTLOOK

The main conclusion we would like a reader to take away from this introduction to the subject is that it is possible to design, and to fabricate, one- and two-dimensional rough surfaces that scatter or transmit electromagnetic waves to produce fields with specified angular, spatial, coherence, and wavelength properties.

The approach to the design of surfaces with the properties presented in this book is largely based on the use of single-scattering approximations in solving the direct problem of scattering from a rough surface. These include the Kirchhoff approximation and phase perturbation theory, or their geometrical optics limits, and the Fraunhofer approximation; and the use of the Rayleigh hypothesis and small-amplitude perturbation theory, together with the assumption of weak surface roughness. The assumptions and approximations are often specific to the problem being considered.

The one-dimensional surface profiles considered are represented as continuous functions formed from straight line segments, whose slopes are independent, identically distributed random deviates drawn from an initially unknown probability density function. The two-dimensional surface profile functions for the most part are represented as continuous functions formed from triangular facets, whose slopes along two mutually perpendicular directions are drawn from an initially unknown joint probability density function that is the same for each facet. Within the single-scattering approximation assumed in solving the direct problem of scattering from such surfaces, the average of the property of the scattered field of interest—for example its intensity, cross-spectral density, or spectrum—averaged over the ensemble of realizations of the surface profile function is expressed in terms of these probability density functions. This expression can be inverted analytically or numerically, to yield the probability density function in terms of the averaged property of interest. From a knowledge of the probability density function a single realization or large number (an ensemble) of realizations of the surface profile function can be generated by standard methods, such as the rejection method.

However, once a surface has been designed for a specific application, by the use of the assumptions and approximations mentioned above, its quality is tested by computer simulations employing more rigorous methods to solve the scattering problem than those used in solving the design problem, and by experimental results. These have shown that the fields scattered from these surfaces possess the specified properties, even in parameter regimes where the methods used in their design might not be expected to be valid. In this sense the designed surfaces are remarkably robust.

The use of an average over the ensemble of realizations of the surface profile function in the design of rough surfaces with specified scattering properties seems to be inevitable in the stochastic approach to the design problem adopted in this book. However, it has been shown that the use of this ensemble average in the more rigorous calculations and experiments carried out to validate the solutions to this design problem, can be replaced by calculations or measurements in which a single realization of the random surface is illuminated by a broadband beam.

One conclusion that we would not want a reader to reach from this book is that there is little more that needs to be done before the design of surfaces with specified scattering or transmission properties becomes a mature field.

There are several directions in which theoretical studies of designer surfaces can go in the future. One of them is the design of surfaces that produce a scattered field with a more complex angular or spatial distribution of its intensity than do the surfaces considered in this book. Surfaces that generate intensity distributions that reproduce the letters of the alphabet, for example, can be designed, and could be useful in projection systems for display purposes.

The polarization of the incident and scattered light also offers opportunities for designing surfaces with useful properties. As an example, let us consider a one-dimensional surface that is illuminated by monochromatic light whose plane of incidence is perpendicular to the generators of the surface and is a linear combination of p- and s-polarized light. Can such a surface be designed that produces a scattered field, whose plane of scattering coincides with the plane of incidence, that consists of only p-polarized light or only of s-polarized light? If so, such a surface could be used as a polarizer. Polarizers operating in the infrared are expensive and not very effective. A surface of the type described here could serve as an effective and inexpensive polarizer in this region of wavelengths.

Much of the discussion in this book has focused on the design of surfaces that scatter an incident electromagnetic field into a field with specified properties. The problem of designing surfaces that transmit an electromagnetic field to produce a field that possesses desired properties merits greater attention, particularly in the case of two-dimensional randomly rough surfaces. A useful starting point for solving this design problem would seem to be transmission through a thin film whose rough surface is treated on the basis of a random phase screen model [4.1].

The design of random surfaces that produce propagation-invariant beams in scattering or transmission could be useful for a variety of applications including optical interconnection, high precision autocollimation, optical alignment, laser machining, and laser surgery [4.2–4.7]. These are beams for which their cross-spectral density is independent of the coordinate in the direction of their propagation [4.8].

On the experimental side, a challenging problem is the fabrication of two-dimensional randomly rough surfaces formed from triangular facets of the kind described in Section 3.1.4, when the characteristic length scale b is of the order of tens of microns. Such surfaces could be used in the visible region of the optical spectrum. For applications involving millimeter waves, where the characteristic sizes of the surface features are measured in centimeters, such surfaces can be fabricated by the computer-controlled milling method employed by Phu et al. [4.9] for the creation of a two-dimensional randomly rough surface that obeys Gaussian statistics with a Gaussian surface height autocorrelation function. This method has yet to be applied to the fabrication of surfaces formed from triangular facets.

This listing hardly exhausts the possible directions research on designer surfaces could take in the future. If anyone is stimulated to tackle one or another of these problems, or others, by reading this book, our efforts in writing it will have been rewarded.

References

[4.1] W.T. Welford, Optical estimation of statistics of surface roughness from light scattering measurements, Opt. Quant. Electron. 9 (1977) 269–387.

[4.2] J. Durnin, Exact solutions for non-diffracting beams, I. The scalar theory, J. Opt. Soc. Am. A 4 (1987) 651–654.

[4.3] J. Durnin, J.J. Miceli Jr., J.H. Eberly, Diffraction-free beams, Phys. Rev. Lett. 58 (1987) 1499–1501.

[4.4] C. Ozkul, S. Leroux, N. Anthore, M.K. Amara, S. Rasset, Optical amplification of diffraction-free beams by photorefractive two-wave mixing and its application to laser Doppler velocimetry, Appl. Opt. 34 (1995) 5485–5491.

[4.5] J. Turunen, A. Vasara, A.T. Friberg, Holographic generation of diffraction-free beams, Appl. Opt. 27 (1988) 3959–3962.

[4.6] R.P. McDonald, J. Chrostowski, S.A. Boothroyd, B.A. Syrett, Holographic formation of a diode laser non-diffracting beam, Appl. Opt. 32 (1993) 6470–6474.

[4.7] R.P. McDonald, S.A. Boothroyd, T. Okamoto, J. Chrostowski, B.A. Syrett, Interboard optical data distribution by Bessel beam shadowing, Opt. Commun. 122 (1996) 169–177.

[4.8] A. Ostrovsky, Coherent Mode Representations in Optics, SPIE, Bellingham, WA, 2006 (Chapter 3).

[4.9] P. Phu, A. Ishimaru, Y. Kuga, Copolarized and cross-polarized enhanced backscattering from two-dimensional very rough surfaces at millimeter wave frequencies, Radio Sci. 29 (1994) 1275–1291.

APPENDIX A. THE KERNELS IN THE INTEGRAL EQUATIONS (2.8.77)–(2.8.80)

The kernels entering the integral equations (2.8.77)–(2.8.80) are tabulated here

$$
\begin{aligned}
H^{(1)}(x_1|x_1') &= (1/4\pi)\big[(\partial/\partial x_3')G_1(x_3|x_1',x_3')\big]_{\substack{x_3=d+\eta\\ x_3'=d}} \\
&= -\frac{i}{4}\big(n_1(\omega/c)\big)^2 \frac{H_1^{(1)}(n_1(\omega/c)[(x_1-x_1')^2+\eta^2]^{\frac{1}{2}})}{n_1(\omega/c)[(x_1-x_1')^2+\eta^2]^{\frac{1}{2}}}\eta, \\
L^{(1)}(x_1|x_1') &= (1/4\pi)\big[G_1(x_1,x_3|x_1',x_3')\big]_{\substack{x_3=d+\eta\\ x_3'=d}} \\
&= \frac{i}{4}H_0^{(1)}\big(n_1(\omega/c)\big[(x_1-x_1')^2+\eta^2\big]^{\frac{1}{2}}\big), \\
H^{(11)}(x_1|x_1') &= (1/4\pi)\big[(\partial/\partial x_3')G_2(x_3|x_1',x_3')\big]_{\substack{x_3=d+\eta\\ x_3'=d}} \\
&= -\frac{i}{4}\big(n_2(\omega/c)\big)^2 \frac{H_1^{(1)}(n_2(\omega/c)[(x_1-x_1')^2+\eta^2]^{\frac{1}{2}})}{n_2(\omega/c)[(x_1-x_1')^2+\eta^2]^{\frac{1}{2}}}\eta, \\
L^{(11)}(x_1|x_1') &= (1/4\pi)\big[G_2(x_1,x_3|x_1',x_3')\big]_{\substack{x_3=d+\eta\\ x_3'=d}} \\
&= \frac{i}{4}H_0^{(1)}\big(n_2(\omega/c)\big[(x_1-x_1')^2+\eta^2\big]^{\frac{1}{2}}\big), \\
H^{(12)}(x_1|x_1') &= (1/4\pi)\big[(\partial/\partial N')G_2(x_1,x_3|x_1',x_3')\big]_{\substack{x_3=d\\ x_3'=\zeta(x_1')}} \\
&= \frac{i}{4}\big(n_2(\omega/c)\big)^2 \frac{H^{(1)}(n_2(\omega/c)[(x_1-x_1')+(d-\zeta(x_1'))^2]^{\frac{1}{2}})}{n_2(\omega/c)[(x_1-x_1')+(d-\zeta(x_1'))^2]^{\frac{1}{2}}} \\
&\quad\times\big[-(x_1-x_1')\zeta'(x_1')+\big(d-\zeta(x_1')\big)\big],
\end{aligned}
$$

$$
\begin{aligned}
L^{(12)}(x_1|x_1') &= (1/4\pi)\big[G_2(x_1,x_3|x_1',x_3')\big]_{\substack{x_3=d\\ x_3'=\zeta(x_1')}}\\
&= \frac{i}{4}H_0^{(1)}\big(n_2(\omega/c)\big[(x_1-x_1')+\big(d-\zeta(x_1')\big)^2\big]^{\frac{1}{2}}\big),\\
H^{(21)}(x_1|x_1') &= (1/4\pi)\big[(\partial/\partial x_3')G_2(x_1,x_3|x_1',x_3')\big]_{\substack{x_3=\zeta(x_1)\\ x_3'=d}}\\
&= i\frac{i}{4}\big(n_2(\omega/c)\big)^2\frac{H_1^{(1)}(n_2(\omega/c)[(x_1-x_1')^2+(\zeta(x_1)-d)^2]^{\frac{1}{2}})}{n_2(\omega/c)[(x_1-x_1')^2+(\zeta(x_1)-d)^2]^{\frac{1}{2}}}\big(\zeta(x_1)-d\big),\\
L^{(21)}(x_1|x_1') &= (1/4\pi)\big[G_2(x_1,x_3|x_1',x_3')\big]_{\substack{x_3=\zeta(x_1)\\ x_3'=d}}\\
&= \frac{i}{4}H_0^{(1)}\big(n_2(\omega/c)\big[(x_1-x_1')+\big(\zeta(x_1)-d\big)^2\big]^{\frac{1}{2}}\big),\\
H^{(22)}(x_1|x_1') &= (1/4\pi)\big[(\partial/\partial N')G_2(x_1,x_3|x_1',x_3')\big]_{\substack{x_3=\zeta(x_1)+\eta\\ x_3'=\zeta(x_1')}}\\
&= \frac{i}{4}\big(n_2(\omega/c)\big)^2\frac{H_1^{(1)}(n_2(\omega/c)[(x_1-x_1')-(\zeta(x_1')+\eta)^2]^{\frac{1}{2}})}{n_2(\omega/c)[(x_1-x_1')^2+(\zeta(x_1)-\zeta(x_1')+\eta)^2]^{\frac{1}{2}}}\\
&\quad\times\big[-(x_1-x_1')\zeta'(x_1')+\big(\zeta(x_1)-\zeta(x_1')+\eta\big)\big],\\
L^{(22)}(x_1|x_1') &= (1/4\pi)\big[G_2(x_1,x_3|x_1',x_3')\big]_{\substack{x_3=\zeta(x_1)+\eta\\ x_3'=\zeta(x_1')}}\\
&= \frac{i}{4}H_0^{(1)}\big(n_2(\omega/c)\big[(x_1-x_1')^2+\big(\zeta(x_1)-\zeta(x_1')+\eta\big)\big]^{\frac{1}{2}}\big),\\
H^{(3)}(x_1|x_1') &= (1/4\pi)\big[(\partial/\partial N')G_3(x_1,x_3|x_1',x_3')\big]_{\substack{x_3=\zeta(x_1)+\eta\\ x_3'=\zeta(x_1')}}\\
&= \frac{i}{4}\big(n_3(\omega/c)\big)^2\frac{H_1^{(1)}(n_3(\omega/c)[(x_1-x_1')^2+(\zeta(x_1)-\zeta(x_1')+\eta)^2]^{\frac{1}{2}})}{n_3(\omega/c)\big[(x_1-x_1')+(\zeta(x_1)-\zeta(x_1')+\eta)^2\big]^{\frac{1}{2}}}\\
&\quad\times\big[-(x_1-x_1')\zeta'(x_1')+\big(\zeta(x_1)-\zeta(x_1')+\eta\big)\big],\\
L^{(3)}(x_1|x_1') &= (1/4\pi)\big[G_3(x_1,x_3|x_1',x_3')\big]_{\substack{x_3=\zeta(x_1)+\eta\\ x_3'=\zeta(x_1')}}\\
&= \frac{i}{4}H_0^{(1)}\big(n_3(\omega/c)\big[(x_1-x_1')^2+\big(\zeta(x_1)-\zeta(x_1')+\eta\big)^2\big]^{\frac{1}{2}}\big).
\end{aligned}
$$

APPENDIX B. THE MATRIX ELEMENTS ENTERING EQS. (2.8.77)–(2.8.80)

In this appendix we tabulate the matrix elements appearing in Eqs. (2.8.77)–(2.8.80), calculated to first order in Δx:

$$
\begin{aligned}
H_{mn}^{(1)} &= \frac{1}{2}\delta_{mn}, \\
L_{mn}^{(1)} &= \Delta x(i/4)H_0^{(1)}\left(n_1\frac{\omega}{c}|x_m - n_n|\right) \quad m \neq n \\
&= \Delta x(i/4)H_0^{(1)}\left(n_1\frac{\omega}{c}\frac{\Delta x}{2e}\right) \quad m = n, \\
H_{mn}^{(11)} &= \frac{1}{2}\delta_{mn}, \\
L_{mn}^{(11)} &= \Delta x(i/4)H_0^{(1)}\left(n_2\frac{\omega}{c}|x_m - x_n|\right) \quad m \neq n \\
&= \Delta x(i/4)H_0^{(1)}\left(n_2\frac{\omega}{c}\frac{\Delta x}{2e}\right) \quad m = n, \\
H_{mn}^{(12)} &= \Delta x(i/4)\left(n_2\frac{\omega}{c}\right)^2 \frac{H_1^{(1)}(n_2\frac{\omega}{c}[(x_m - x_n)^2 + (d - \zeta(x_n))^2]^{\frac{1}{2}})}{n_2\frac{\omega}{c}[(x_m - x_n)^2 + (d - \zeta(x_n))^2]^{\frac{1}{2}}} \\
&\quad \times \left[-(x_m - x_n)\zeta'(x_n) + d - \zeta(x_n)\right], \\
L_{mn}^{(12)} &= \Delta x(i/4)H_0^{(1)}\left(n_2\frac{\omega}{c}\left[(x_m - x_n)^2 + \left(d - \zeta(x_n)\right)^2\right]^{\frac{1}{2}}\right), \\
H_{mn}^{(21)} &= \Delta x(i/4)\left(n_2\frac{\omega}{c}\right)^2 \frac{H_1^{(1)}(n_2\frac{\omega}{c}[(x_m - x_n)^2 + (\zeta(x_m) - d)^2]^{\frac{1}{2}})}{n_2\frac{\omega}{c}[(x_m - x_n)^2 + (\zeta(x_m) - d)^2]^{\frac{1}{2}}}\left(\zeta(x_m) - d\right), \\
L_{mn}^{(21)} &= \Delta x(i/4)H_0^{(1)}\left(n_2\frac{\omega}{c}\left[(x_m - x_n)^2 + \left(\zeta(x_m) - d\right)^2\right]^{\frac{1}{2}}\right),
\end{aligned}
$$

$$H_{mn}^{(22)} = \Delta x(i/4)\left(n_2\frac{\omega}{c}\right)^2 \frac{H_1^{(1)}(n_2\frac{\omega}{c}[(x_m - x_n)^2 + (\zeta(x_m) - \zeta(x_n))^2]^{\frac{1}{2}})}{n_2\frac{\omega}{c}[(x_m - x_n)^2 + (\zeta(x_m) - \zeta(x_n))^2]^{\frac{1}{2}}}$$

$$\times\big[-(x_m - x_n)\big(\zeta'(x_n) + \zeta(x_n)\big)\big] \quad m \neq n$$

$$= \frac{1}{2} + \frac{\Delta x}{4\pi}\frac{\zeta''(x_n)}{\phi_n^2} \quad m = n,$$

$$L_{mn}^{(22)} = \Delta x(i/4)H_0^{(1)}\left(n_2\frac{\omega}{c}\big[(x_m - x_n)^2 + \big(\zeta(x_m) - \zeta(x_n)\big)^2\big]^{\frac{1}{2}}\right) \quad m \neq n$$

$$= \Delta x(i/4)H_0^{(1)}\left(n_2\frac{\omega}{c}\frac{\phi_n \Delta x}{2e}\right) \quad m = n,$$

$$H_{mn}^{(3)} = \Delta x(i/4)\left(n_3\frac{\omega}{c}\right)^2 \frac{H_1^{(1)}(n_3\frac{\omega}{c}[(x_m - x_n)^2 + (\zeta(x_m) - \zeta(x_n))^2]^{\frac{1}{2}})}{n_3\frac{\omega}{c}[(x_m - x_n)^2 + (\zeta(x_m) - \zeta(x_n))^2]^{\frac{1}{2}}}$$

$$\times\big[-(x_m - x_n)\zeta(x_n) + \big(\zeta(x_m) - \zeta(x_n)\big)\big] \quad m \neq n$$

$$= \frac{1}{2} + \frac{\Delta x}{4\pi}\frac{\zeta''(x_n)}{\phi_n^2} \quad m = n,$$

$$L_{mn}^{(3)} = \Delta x(i/4)H_0^{(1)}\left(n_3\frac{\omega}{c}\big[(x_m - x_n)^2 + \big(\zeta(x_m) - \zeta(x_n)\big)^2\big]^{\frac{1}{2}}\right) \quad m \neq n$$

$$= \Delta x(i/4)H_0^{(1)}\left(n_3\frac{\omega}{c}\frac{\phi_n \Delta x}{2e}\right) \quad m = n,$$

where $\phi(x_n)$ is given by Eq. (2.1.113b).

APPENDIX C. THE SINGULARITY IN $G_\ell(r|r')$

In this appendix we obtain the form of the singularity of $G_\ell(r|r')$ in the limit as $r' \to r$. Our starting point is the definition of $G_\ell(r|r')$ given by Eqs. (3.1.118) and (3.1.130),

$$g_0(\mathbf{x}|\mathbf{x}')\Big|_{\substack{x_3'=H(r')\\ x_3=H(r)}} = \sum_{\ell=-\infty}^{\infty} \exp\bigl[-i\ell(\phi_x - \phi_{x'})\bigr] G_\ell(r|r'). \tag{C.1}$$

If we replace the difference $\phi_x - \phi_{x'}$ by θ, this equation yields $G_\ell(r|r')$ in the form

$$G_\ell(r|r') = \frac{1}{2\pi}\int_{-\pi}^{\pi} d\theta \, \exp(i\ell\theta) \frac{\exp[ik|\mathbf{x}-\mathbf{x}'|]}{|\mathbf{x}-\mathbf{x}'|}\Bigg|_{\substack{x_3'=H(r')\\ x_3=H(r)}}, \tag{C.2}$$

where we have set $k = (\omega/c)$. Now, $|\mathbf{x}-\mathbf{x}'|$ evaluated at $x_3 = H(r)$ and $x_3' = H(r')$ is given by

$$\begin{aligned}
|\mathbf{x}-\mathbf{x}'|\Big|_{\substack{x_3'=H(r')\\ x_3=H(r)}} &= \Bigl[\bigl(\mathbf{x}_\parallel + \hat{\mathbf{x}}_3 H(r)\bigr)^2 - 2\bigl(\mathbf{x}_\parallel + \hat{\mathbf{x}}_3 H(r)\bigr)\cdot\bigl(\mathbf{x}_\parallel' + \hat{\mathbf{x}}_3 H(r')\bigr) \\
&\quad + \bigl(\mathbf{x}_\parallel' + \hat{\mathbf{x}}_3 H(r')\bigr)^2\Bigr]^{\frac{1}{2}} \\
&= \bigl[r^2 + H^2(r) - 2\bigl(rr'\cos\theta + H(r)H(r')\bigr) + r'^2 + H^2(r')\bigr]^{\frac{1}{2}} \\
&= \bigl[R^2 + 4rr'\sin^2(\theta/2)\bigr]^{\frac{1}{2}},
\end{aligned} \tag{C.3}$$

where

$$R^2 = (r-r')^2 + \bigl[H(r) - H(r')\bigr]^2. \tag{C.4}$$

Therefore, we can write Eq. (C.2) as

$$G_\ell(r|r') = \frac{1}{\pi}\int_0^{\pi} d\theta\, \cos \ell\theta \frac{\exp\{ik[R^2+4rr'\sin^2(\theta/2)]^{\frac{1}{2}}\}}{[R^2+4rr'\sin^2(\theta/2)]^{\frac{1}{2}}}$$

$$= \frac{2}{\pi}\int_0^{\frac{\pi}{2}} d\phi \cos 2\ell\phi \frac{\exp\{ik[R^2+4rr'\sin^2\phi]^{\frac{1}{2}}\}}{[R^2+4rr'\sin^2\phi]^{\frac{1}{2}}}. \tag{C.5}$$

If we rewrite Eq. (C.5) in the following form,

$$G_\ell(r|r') = \frac{2}{\pi}\int_0^{\frac{\pi}{2}} d\phi \frac{\cos 2\ell\phi}{[R^2+4r'\sin^2\phi]^{\frac{1}{2}}}$$

$$+ \frac{2}{\pi}\int_0^{\frac{\pi}{2}} d\phi \cos 2\ell\phi \frac{\exp\{ik[R^2+4rr'\sin^2\phi]^{\frac{1}{2}}\}-1}{[R^2+4rr'\sin^2\phi]^{\frac{1}{2}}}, \tag{C.6}$$

we see that the first term is singular as $r' \to r$ $(R \to 0)$, while the second term is not. Therefore, if we neglect terms that vanish as $r - r'$ tends to zero in the latter term, we can rewrite Eq. (C.6) as

$$G_\ell(r|r') = \frac{2}{\pi}\int_0^{\frac{\pi}{2}} d\phi \frac{\cos 2\ell\phi}{[R^2+4rr'\sin^2\phi]^{\frac{1}{2}}}$$

$$+ \frac{1}{\pi r}\int_0^{\frac{\pi}{2}} d\phi \cos 2\ell\phi \frac{\exp(i2kr\sin\phi)-1}{\sin\phi}. \tag{C.7}$$

We consider each of these terms in turn.

We rewrite the first term on the right-hand side of Eq. (C.7) in the form

$$\frac{2}{\pi}\int_0^{\frac{\pi}{2}} \frac{\cos 2\ell\phi}{[R^2+4rr'\sin^2\phi]^{\frac{1}{2}}} = \frac{2}{\pi}\int_0^{\frac{\pi}{2}} d\phi \frac{1}{[R^2+4rr'\sin^2\phi]^{\frac{1}{2}}}$$

$$- \frac{4}{\pi}\int_0^{\frac{\pi}{2}} d\phi \frac{\sin^2\ell\phi}{[R^2+4rr'\sin^2\phi]^{\frac{1}{2}}}. \tag{C.8}$$

The first term on the right-hand side of Eq. (C.8) is singular as $r \to r'$ $(R \to 0)$; the second term is not. Therefore, if we again neglect terms that vanish as $r - r'$ tends to zero in the latter term, Eq. (C.8) becomes

$$\frac{2}{\pi}\int_0^{\frac{\pi}{2}} \frac{\cos 2\ell\phi}{[R^2+4rr'\sin^2\phi]^{\frac{1}{2}}} = \frac{2}{\pi}\int_0^1 dx \frac{1}{(1-x^2)^{\frac{1}{2}}(R^2+4rr'x^2)^{\frac{1}{2}}} - \frac{2}{\pi r}\int_0^{\frac{\pi}{2}} d\phi \frac{\sin^2\ell\phi}{\sin\phi}. \tag{C.9}$$

The first term on the right-hand side of Eq. (C.9) can be rewritten as

$$\frac{2}{\pi}\int_0^1 dx \frac{1}{(1-x^2)^{\frac{1}{2}}(R^2+4rr'x^2)^{\frac{1}{2}}} = \frac{2}{\pi}\int_0^1 dx \frac{1}{(R^2+4rr'x^2)^{\frac{1}{2}}} + \frac{2}{\pi}\int_0^1 dx \frac{1}{(R^2+4rr'x^2)^{\frac{1}{2}}}\left(\frac{1}{(1-x^2)^{\frac{1}{2}}}-1\right). \tag{C.10}$$

The first term on the right-hand side of Eq. (C.10) is singular as $r \to r'$ $(R \to 0)$; the second is not. Therefore, if we neglect terms that vanish as $r - r'$ tends to zero in the latter term, we find that

$$\begin{aligned}\frac{2}{\pi}\int_0^1 dx &\frac{1}{(1-x^2)^{\frac{1}{2}}(R^2+4rr'x^2)^{\frac{1}{2}}} \\ &= \frac{2}{\pi}\int_0^1 dx \frac{1}{(R^2+4rr'x^2)^{\frac{1}{2}}} + \frac{1}{\pi r}\int_0^1 dx \frac{1}{x}\left(\frac{1}{(1-x^2)^{\frac{1}{2}}}-1\right) \\ &= -\frac{1}{\pi\sqrt{rr'}}\ell n R + \frac{1}{\pi\sqrt{rr'}}\ell n\left[2\sqrt{rr'}+\sqrt{R^2+4rr'}\right] + \frac{1}{\pi r}\ell n 2.\end{aligned} \tag{C.11}$$

In the limit as $r - r'$ tends to zero, this result becomes

$$\frac{2}{\pi}\int_0^1 dx \frac{1}{(1-x^2)^{\frac{1}{2}}(R^2+4rr'x^2)^{\frac{1}{2}}} = -\frac{1}{\pi r}\ell n \frac{\{1+[H'(r)]^2\}^{\frac{1}{2}}|r-r'|}{8r}, \tag{C.12}$$

apart from terms that vanish as $r - r'$ tends to zero.

When we combine the results given by Eqs. (C.7), (C.9), and (C.12), we find that in the limit as $r' \to r$, $G_\ell(r|r')$ has the form

$$G_\ell(r|r') \sim -\frac{1}{\pi r}\ell n \frac{\{1+[H'(r)]^2\}^{\frac{1}{2}}|r-r'|}{8r} + \frac{1}{\pi r} C_\ell(kr), \tag{C.13}$$

where we have neglected terms that vanish as $r' \to r$, and

$$\begin{aligned} C_\ell(kr) &= \int_0^{\frac{\pi}{2}} d\phi \cos 2\ell\phi \frac{\exp(i2kr\sin\phi) - 1}{\sin\phi} - 2\int_0^{\frac{\pi}{2}} d\phi \frac{\sin^2 \ell\phi}{\sin\phi} \\ &\equiv C_\ell^{(1)}(kr) + C_\ell^{(2)}. \end{aligned} \tag{C.14}$$

The function $C_\ell(kr)$ can be written in a more explicit form. Thus, we have

$$C_\ell^{(1)}(kr) = \sum_{n=1}^{\infty} \frac{(i2kr)^n}{n!} \int_0^{\frac{\pi}{2}} d\phi \cos 2\ell\phi \sin^{n-1}\phi. \tag{C.15}$$

The integral in Eq. (C.15) has different forms depending on whether n is even or odd. Thus, if $n = 2k+2$, with $k = 0, 1, 2, \ldots$, [C.1]

$$\int_0^{\frac{\pi}{2}} d\phi \cos 2\ell\phi \sin^{2k+1}\phi = \frac{1}{2}\frac{(-1)^\ell 2^{k+1} k!(2k+1)!!}{(2k-2\ell+1)!!(2\ell+2k+1)!!} \quad k \geqslant \ell - 1 \tag{C.16a}$$

$$= \frac{1}{2}\frac{(-1)^{k+1}2^{k+1}k!(2\ell-2k+1)!!(2k+1)!!}{(2\ell+2k+1)!!} \quad k < \ell - 1. \tag{C.16b}$$

On the other hand, if $n = 2k+1$, with $k = 0, 1, 2, \ldots$, [C.2]

$$\int_0^{\frac{\pi}{2}} d\phi \cos 2\ell\phi \sin^{2k}\phi = \frac{\pi}{2}\frac{(-1)^\ell}{2^{2k}}\binom{2k}{k-\ell} \quad k \geqslant \ell \tag{C.17a}$$

$$= 0 \quad k < \ell. \tag{C.17b}$$

The function $C_\ell^{(2)}$ is [C.3]

$$
\begin{aligned}
C_\ell^{(2)} &= -2\int_0^{\frac{\pi}{2}} d\phi \frac{\sin^2 \ell\phi}{\sin\phi} = -\int_0^{\frac{\pi}{2}} d\phi \frac{1-\cos 2\ell\phi}{\sin\phi} \\
&= -\left\{ \ell n\left(\tan\frac{\phi}{2}\right) - \left[2\sum_{n=1}^{\ell} \frac{\cos(2n-1)\phi}{2n-1} + \ell n\left(\tan\frac{\phi}{2}\right)\right]\right\}\Bigg|_0^{\frac{\pi}{2}} \\
&= -2\sum_{n=1}^{\ell}\frac{1}{2n-1}.
\end{aligned}
\tag{C.18}
$$

In the case that $\ell = 0$, we have that

$$
C_0(kr) = \frac{\sqrt{\pi}}{2}\sum_{n=1}^{\infty}\frac{\Gamma(\frac{n}{2})}{n!\Gamma(\frac{n+1}{2})}(i2kr)^n. \tag{C.19}
$$

Finally, we note that $C_\ell^{(1)}(kr)$ and $C_\ell^{(2)}$ are even functions of ℓ.

References

[C.1] I.S. Gradshteyn, I.M. Ryzhik, Tables of Integrals, Series, and Products, Academic Press, New York, 1980 (Section 3.63, No. 13).
[C.2] Ref. [C.1], Section 3.63, No. 12.
[C.3] Ref. [C.1], Section 2.526, No. 1, and Section 2.539, No. 4.

SUBJECT INDEX

T

www.ingramcontent.com/pod-product-compliance
Lightning Source LLC
LaVergne TN
LVHW020056220926
848484LV00002BB/593